Humanity and the Sea

Series Editor
Andrew I.L. Payne
A&B Word Ltd
Halesworth, Suffolk, U.K.

For further volumes:
http://www.springer.com/series/11481

Mark A. Shields • Andrew I.L. Payne
Editors

Marine Renewable Energy Technology and Environmental Interactions

Springer

Editors
Mark A. Shields
Oceanlab
University of Aberdeen
Newburgh
Aberdeenshire
United Kingdom

Andrew I.L. Payne
A&B Word Ltd
Halesworth
Suffolk
United Kingdom

ISSN 2213-607X ISSN 2213-6088 (electronic)
ISBN 978-94-024-0775-4 ISBN 978-94-017-8002-5 (eBook)
DOI 10.1007/978-94-017-8002-5
Springer Dordrecht Heidelberg New York London

Printed on acid-free paper

Springer is part of Springer Science+Business Media (www.springer.com)

Contents

Contributors

David Ainsworth Marine Current Turbines, Bristol and Bath Science Park, Bristol, UK

Mathias H. Andersson Department of Underwater Research, Swedish Defence Research Agency, Stockholm, Sweden

Stuart Bearhop Centre for Ecology and Conservation, College of Life and Environmental Sciences, University of Exeter, Penryn, Cornwall, UK

Robert A. Beharie International Centre for Island Technology (ICIT), Heriot-Watt University, Scotland, UK

Michael C. Bell International Centre for Island Technology (ICIT), Heriot-Watt University, Scotland, UK

Peter A. Bowyer Environmental Research Institute, Centre for Energy and the Environment, North Highland College UHI, University of the Highlands and Islands, Thurso, Scotland, UK

Caroline Carter Scottish Association for Marine Science (SAMS), Oban, Argyll, UK

Nadja Christen Centre for Ecology and Conservation, College of Life and Environmental Sciences, University of Exeter, Penryn, Cornwall, UK

Thomas G. Dahlgren Uni Research, Bergen, Norway

Ian M. Davies Marine Laboratory, Marine Scotland, Aberdeen, UK

Matthew C. Easton Environmental Research Institute, Centre for Energy and the Environment, North Highland College UHI, University of the Highlands and Islands, Thurso, Scotland, UK

Bjoern Elsaesser School of Planning, Architecture and Civil Engineering, Queen's University Belfast, Belfast, UK

Ilker Fer Geophysical Institute, University of Bergen, Bergen, Norway

Frank Fortune Royal Haskoning DHV, Leith, Edinburgh, UK

Andrew B. Gill Environmental Science and Technology Department, School of Applied Sciences, Cranfield University, Cranfield, Bedfordshire, UK

Douglas M. Gillespie Sea Mammal Research Unit, Scottish Oceans Institute, School of Biology, University of St Andrews, St Andrews, Fife, UK

Ian Gloyne-Philips Centre for Marine and Coastal Studies Ltd (CMACS Ltd), East Ham, Wirral, UK

Jonathan C. D. Gordon Sea Mammal Research Unit, Scottish Oceans Institute, School of Biology, University of St Andrews, St Andrews, Fife, UK

Gordon D. Hastie Sea Mammal Research Unit, Scottish Oceans Institute, School of Biology, University of St Andrews, St Andrews, Fife, UK

SMRU Marine Ltd, New Technology Centre, North Haugh, University of St Andrews, St Andrews, Fife, UK

Sarah K. Henkel Hatfield Marine Science Center, Oregon State University, Newport, OR, USA

Rich Inger Environment and Sustainability Institute, University of Exeter, Penryn, Cornwall, UK

Robert Kennedy Marine Ecosystem Research Laboratory, Ryan Institute, School of Natural Sciences, National University of Ireland Galway, Galway, Ireland

Joel Kimber Centre for Marine and Coastal Studies Ltd (CMACS Ltd), East Ham, Wirral, UK

Barbara A. Lagerquist Hatfield Marine Science Center, Oregon State University, Newport, OR, USA

Olivia Langhamer Department of Biology, Norwegian University of Science and Technology, Trondheim, Norway

Rebecca Langton School of Biological Sciences, Institute of Biological and Environmental Sciences, University of Aberdeen, Aberdeen, UK

Paul A. Lepper School of Electronic, Electrical and Systems Engineering, Loughborough University, Leicestershire, UK

Jamie D. J. Macaulay Sea Mammal Research Unit, Scottish Oceans Institute, School of Biology, University of St Andrews, St Andrews, Fife, UK

Bernie J. McConnell Sea Mammal Research Unit, Scottish Oceans Institute, School of Biology, University of St Andrews, St Andrews, Fife, UK

Jason McIlvenny Environmental Research Institute, Centre for Energy and the Environment, North Highland College UHI, University of the Highlands and Islands, Thurso, Scotland, UK

Angus McRobert Water Management Unit, Northern Ireland Environment Agency, Lisburn, Co. Antrim, UK

Evelyn Philpott School of Biological Sciences, Institute of Biological and Environmental Sciences, University of Aberdeen, Aberdeen, UK

Kate E. Plummer Centre for Ecology and Conservation, College of Life and Environmental Sciences, University of Exeter, Penryn, Cornwall, UK

British Trust for Ornithology, The Nunnery, Thetford, Norfolk, UK

David Pratt Marine Scotland, Edinburgh, UK

Daniel W. Pritchard School of Planning, Architecture and Civil Engineering, Queen's University Belfast, Belfast, UK

Stephen P. Robinson National Physical Laboratory (NPL), Teddington, Middlesex, UK

Yuri Rzhanov Chase Ocean Engineering Laboratory, Center for Coastal and Ocean Mapping Joint Hydrographic Center, Durham, NH, USA

Aleksej Šaškov Coastal Research and Planning Institute, Klaipeda University, Klaipeda, Lithuania

Graham Savidge Marine Laboratory, Queen's University Belfast, Portaferry, Co. Down, UK

Marie-Lise Schläppy EPFL, Lausanne, Switzerland

Uni Research, Bergen, Norway

Beth E. Scott School of Biological Sciences, Institute of Biological and Environmental Sciences, University of Aberdeen, Aberdeen, UK

Mark A. Shields Oceanlab, University of Aberdeen, Aberdeenshire, UK

Department of Energy and Climate Change, Aberdeen, UK

Jon C. Side International Centre for Island Technology (ICIT), Heriot-Watt University, Scotland, UK

Peter Sigray Swedish Defence Research Agency—FOI, Stockholm, Sweden

Carol E. Sparling SMRU Marine Ltd, New Technology Centre, North Haugh, University of St Andrews, St Andrews, Fife, UK

Robert M. Suryan Hatfield Marine Science Center, Oregon State University, Newport, OR, USA

James J. Waggitt School of Biological Sciences, Institute of Biological and Environmental Sciences, University of Aberdeen, Aberdeen, UK

Andrew Want International Centre for Island Technology (ICIT), Heriot-Watt University, Scotland, UK

Trevor J. T. Whittaker School of Planning, Architecture and Civil Engineering, Queen's University Belfast, Belfast, UK

Dan Wilhelmsson Swedish Secretariat for Environmental Earth System Sciences, Royal Swedish Academy of Sciences, Stockholm, Sweden

Ben Wilson Scottish Association for Marine Science (SAMS), Oban, Argyll, UK

David K. Woolf Environmental Research Institute, Centre for Energy and the Environment, North Highland College UHI, University of the Highlands and Islands, Thurso, Scotland, UK

International Centre for Island Technology, Institute of Petroleum Engineering, Heriot-Watt University, Orkney, Scotland, UK

An Introduction to Marine Renewable Energy

1

Mark A. Shields

Abstract

It is now widely recognized that there is a need for long-term secure and suitable sustainable forms of energy. Renewable energy from the marine environment, in particular renewable energy from tidal currents, wave and wind, can help achieve a sustainable energy future. Our understanding of environmental impacts and suitable mitigation methods associated with extracting renewable energy from the marine environment is improving all the time and it is essential that we distinguish between natural and anthropocentric drivers and impacts.

Keywords

Environmental impact · Marine renewable energy · Sustainable energy · Tidal · Wave · Wind

Fossil fuel sources are finite and establishing long-term global energy security requires suitable alternative and sustainable forms of energy (Mackay 2008). Moreover, concerns are growing that our continued reliance on the burning of fossil fuels and the associated release of greenhouse gases may be contributing to the rapidly changing climate of our planet. Total and *per capita* energy usage has grown and energy demand is anticipated to burgeon along with the human population (Mackay 2008). Society is currently very reliant on fossil fuels as a source of energy, but this reliance will lead eventually to the depletion of the reserves of fossil fuels (Mackay 2008). Securing a sustainable energy future will help reduce reliance on fossil fuels, however, and renewable energy offers a secure energy pathway. Many countries now recognize the need for sustainable energy and are adopting policies that encourage investment in and the development of technology for the extraction of energy from renewable sources (Edenhofer et al. 2011).

In 2008 it was estimated that 13 % of the global energy supply, some 490 Exajoules, was met by renewable energy (Edenhofer et al. 2011). By 2050, however, renewable energy could contribute close to 80 % of the world's energy demands if supported by policies that support the development of renewable energy (Edenhofer et al. 2011). Even without the right supporting policies, though, the use of renewable energy worldwide is predicted to increase. Six renewable energy sources with global potential have been identified by the Intergovernmental Panel on Climate Change (IPCC), including bioenergy, direct solar energy, geothermal energy, hydropower, ocean energy and wind energy (Edenhofer et al. 2011). The marine environment alone has extensive renewable energy sources and there is therefore burgeoning interest in developing means for extracting it.

Wind energy and ocean energy are two of the IPCC renewable energy sources with a marine component (Lewis et al. 2011; Wiser et al. 2011). Wind energy includes both onshore and offshore wind technology and is currently at a more advanced stage of development than ocean energy. Offshore wind utilization has been expanding rapidly and the growing interest in expansion offshore is because of the large resource of wind energy available there. Further, the

M. A. Shields (✉)
Department of Energy and Climate Change, Atholl House, Aberdeen, UK
e-mail: mark.shields@decc.gsi.gov.uk

M. A. Shields
Oceanlab, University of Aberdeen, Newburgh, Aberdeenshire, UK

M. A. Shields, A. I. L. Payne (eds), *Marine Renewable Energy Technology and Environmental Interactions,* Humanity and the Sea,
DOI 10.1007/978-94-017-8002-5_1,

development of wind energy on land has had to address issues such as limits on suitable space and social concerns about the visual impact of windfarms (Wiser et al. 2011). The capacity of wind turbines being installed and operating offshore currently are typically 5 MW each, and their blade diameter exceeds 100 m. Research is being undertaken, though, to increase the capacity of individual wind turbines and to target new locations not currently suitable for the technology now available. Globally, the technical potential of wind energy exceeds current global electricity production (Wiser et al. 2011). Within the EU alone, plans for the development of offshore wind include nearly 40 GW of installed capacity by 2020 and the installation of another 100 GW between 2020 and 2030 (EWEA 2011).

Ocean energy, as defined by the IPCC, includes a diverse range of technology for the extraction of energy from sources including waves, tidal ranges, tidal currents, ocean currents, ocean thermal energy conversion and salinity gradients (Lewis et al. 2011). Most ocean energy technology, with the exception of that relating to tidal barrages, is either at the conceptual phase, undergoing research and development or in the pre-commercial and demonstration phase. Hence, it is believed that ocean energy will not start to contribute significantly to global energy supply until 2020 (Lewis et al. 2011). Notwithstanding, several ocean energy technologies are at the demonstration phase and there is potential for the number of installations to increase rapidly. Unlike wind energy, however, there is no single characteristic design for ocean energy technology, nor is it likely that there will be because of the technical differences in extracting the various ocean energy sources. The theoretical potential of ocean energy nevertheless easily exceeds present human energy requirements (Lewis et al. 2011) and several countries have already undertaken assessments identifying resource potential and suitable locations for the deployment of ocean energy technology.

Over the past decade the development of ocean energy technology has accelerated, largely because of the recognition of the potential of ocean energy and government support provided by policies implemented at both national and regional levels (Lewis et al. 2011). Government policies supporting ocean energy extraction have included the development of marine infrastructure to support the industry, such as the establishment of test sites. Other policies have included research and development grants, implementation of industry standards and protocols, the streamlining of the consenting process for ocean energy and the inclusion of ocean energy in marine spatial plans (Lewis et al. 2011). As the technology continues to develop there is potential for ocean energy not only to contribute towards a sustainable energy future but also to create jobs and economic growth in countries with suitable resources (European Commission 2012)

With the development of any industry there will be social and environmental interactions and impacts associated directly and indirectly with it. The focus of this book is on the environmental interactions and the impacts of the technology suitable for extracting energy from waves, tidal currents and offshore wind. Here, tidal currents, wave and offshore wind energy are referred to as marine renewable energy. Many topics discussed within the book relate to the environmental issues of other ocean energy technology, particularly relating to tidal barrages and ocean currents. The reader should note, though, that the terminology describing the renewable energy source and technology does vary between chapters. Standardization of terminology between chapters has intentionally not been attempted because terminology naturally evolves and becomes standard through time in the peer-reviewed literature. The terminology for offshore wind energy is consistent between chapters and offshore wind energy is the most advanced technology discussed. In contrast, the terminology describing the technology for extracting energy from both tidal current and wave energy resources differs between chapters. Therefore readers should consider chapter by chapter rather than throughout the book the description of the marine renewable energy source and technology provided, for example tidal current energy may be referred to as tidal flow, tidal stream, hydrokinetic or simply tidal energy.

The use of the term "impact" here may well create a negative image, but it is essentially a term employed to describe a change to the marine environment as a direct or indirect result of human activity. A "change" can be in either positive or negative direction in terms of the environment, and that is how it is interpreted in this book. Further, the temporal and spatial scales of an environmental impact depend on a number of factors including the environmental sensitivity of the location, timing and type of activity proposed. Impacts can be short term, i. e. during the construction phase only, and limited to the local environment, i. e. within 10–100 m of the activity. However, there is also the potential for impacts to be permanent and cover a large spatial area, perhaps > 10 km from the activity. Operators planning a new marine renewable energy development therefore have to consider and review all potential environmental impacts and satisfy the consenting authority that the development will not result in any detrimental environmental impacts and that all impacts are indeed minimized. During the same consenting process, the operator would be expected to consider and implement mitigation methods that adopt best practice for addressing the impacts identified.

Environmental impacts associated with marine renewable energy vary during the life cycle of a development, so for example, the impacts associated with the installation phase will differ from those in the operation and decommissioning phases (Boehlert and Gill 2010, Wilhelmsson et al. 2011). A detailed description of potential impacts is not provided in this Introduction, but other chapters of the book provide detailed overviews. Research into the environmental impacts associated with marine renewable energy is still in its infancy (Boehlert and Gill 2010; Wilhelmsson et al. 2011). At present there is greater understanding of the environmental impacts of the installation and operational phases of offshore wind energy than for tidal current and wave energy (Wilhelmsson et al. 2011). This difference in understanding is largely attributable to the offshore wind energy industry being at a more advanced stage of development than wave and tidal current energy. However, understanding of the environmental impacts associated with marine renewable energy and potential mitigation methods for addressing those impacts is improving with the deployment of demonstration projects. Most offshore wind energy developments and associated environmental impact studies have taken place to date in coastal locations characterized by shallow water (<50 m) with a soft sedimentary seabed. However, offshore wind energy operators are researching and developing new technology suitable for installation in deeper water and/or highly energetic marine environments.

Both tidal current and wave energy technologies are designed specifically for deployment in highly energetic marine environments, our understanding of which is often very limited, traditionally both scientists and marine industries having avoided such locations because of the difficulties associated with working in such harsh conditions. However, research undertaken in other offshore sectors can help inform on the potential impacts associated with marine renewable energy activities, such as direct disturbance of the seabed or the discharge of chemicals to the sea. It is worth noting that if marine renewable energy is to make an important contribution to a sustainable energy future, then space will need to be allocated in the marine environment to allow it. Then, of course, there will be potential for conflict with other stakeholders to be created, particularly if other marine stakeholders are excluded from sites for marine renewable energy developments. Clearly, marine renewable energy will continue to grow, but it is as yet unknown how the expansion of the industry might impact on the marine environment and resource users.

Worldwide, several research programmes are now being undertaken to improve the environmental understanding of highly energetic marine environments, with a view to supporting the development of marine renewable energy. Understanding the resource is fundamental to the successful development and extraction of marine renewable energy; the industry will be installing technology in highly energetic locations and ultimately the technology will need to survive the extremes of that environment. Devices requiring little maintenance or that are easily maintained at sea are likely to be preferred by operators because the health and safety risks and financial costs of working in highly energetic marine environments will be great. It is inevitable that as the numbers of marine renewable energy devices increase, then the likelihood of cumulative environmental impacts will increase. The environmental impacts from a single device or a small number of devices are likely to be of little concern, but it is unclear at the moment what will be the impacts associated with the establishment of arrays (>50 devices) or multiple arrays, if any. The development of monitoring programmes will assist in identifying any long-term environmental impacts associated with the expansion of the marine energy industry. Generating good baseline environmental data will be fundamental for any monitoring programme, ensuring that observed changes to the marine environment caused by other pressures, such as climate change, are not misinterpreted as impacts associated with marine renewable energy, distinguishing them from natural and anthropocentric drivers and impacts. What is clear is that marine renewable energy has the potential to help secure a sustainable energy future, but it is important that associated environmental impacts are understood, minimized and suitable mitigation methods employed.

Acknowledgements I thank Alexandrine Cheronet for initiating and supporting the idea of producing an edited book on marine renewable energy, series editor Andrew Payne for his editorial guidance and the reviewers of each chapter for their constructive feedback and recommendations, all of which ultimately benefitted the finished product.

References

Boehlert GW, Gill AB (2010) Environmental and ecological effects of ocean renewable energy developments. Oceanog 23:68–81

Edenhofer O, Pichs-Madruga R, Sokona Y, Seyboth K, Matschoss P, Kadner S, Zwickel T et al (2011) IPCC special report on renewable energy sources and climate change mitigation. Cambridge University Press, Cambridge, 1076 pp

European Commission (2012) Blue growth—scenarios and drivers for sustainable growth from the oceans, seas and coasts. European Commission, Brussels, 202 pp

EWEA (2011) Pure power—wind energy targets for 2020 and 2030. A report by the European wind energy association, 97 pp. http://www.ewea.org/fileadmin/ewea_documents/documents/publications/reports/Pure_Power_Full_Report.pdf. Accessed 20 May 2013

Lewis A, Estefen S, Huckerby J, Musial W, Pontes T, Torres-Martinez J (2011) Ocean energy. In: Edenhofer O, Pichs-Madruga R, Sokona Y, Seyboth K, Matschoss P, Kadner S, Zwickel T et al (eds) IPCC special report on renewable energy sources and climate change mitigation, Cambridge University Press, Cambridge, pp 497–534

Mackay DJC (2008) Sustainable energy—without the hot air. UIT Cambridge, Cambridge, 384 pp. ISBN 978-0-9544529-3-3. www.withouthotair.com

Wilhelmsson D, Malm T, Thompson R, Tchou J, Sarantakos G, McCormick N, Luitjens S et al (2010) Greening blue energy: identifying and managing the biodiversity risks and opportunities of offshore renewable energy. IUCN, Gland, Switzerland. 102 pp

Wiser R, Yang Z, Hand M, Hohmeyer O, Infield D, Jensen PH, Nikolaev V et al (2011) Wind energy. In: Edenhofer O, Pichs-Madruga R, Sokona Y, Seyboth K, Matschoss P, Kadner S, Zwickel T et al (eds) IPCC special report on renewable energy sources and climate change mitigation. Cambridge University Press, Cambridge, pp 535–608

The Physics and Hydrodynamic Setting of Marine Renewable Energy

2

David K. Woolf, Matthew C. Easton, Peter A. Bowyer and Jason McIlvenny

Abstract

Increasing interest is apparent in marine energy resources, particularly tidal and wave. Some TeraWatts of energy propagate from the world's oceans to its marginal seas in the form of surface waves ($\approx$2 TW) and tides ($\cong$2.6 TW) where that energy is naturally dissipated. The seas and coastlines around the UK and its neighbours are notable for dissipating a significant fraction of the global energy of waves ($\approx$50 MW km^{-1} on the Atlantic coast) and especially tides (>250 GW north of Brittany). Displacing a significant fraction of the natural dissipation by energy capture is a tempting and reasonable proposition, but it does raise technical and environmental issues. Sustainable exploitation of the energy needs to consider diverse effects on the environment, waves and tides having a role in maintaining the shelf sea, coastal, estuarine and shoreline environment through associated advection, stirring and other processes. Tides are particularly significant in controlling the stratification of shelf seas and their flow characteristics. Surface waves are more important in determining conditions nearshore and in the intertidal zone. Also, the exploitation of wave and tidal resources is only practical economically and technologically at a limited number of energetic and accessible sites, and societal and ecological considerations inevitably narrow the choice.

Keywords

Hydrodynamics · Marine physics · Marine renewable energy · Shelf seas · Tides · Waves

Introduction and the Global Energy Context

The modern human's thirst for energy together with concerns over the influence of fossil fuel burning on atmospheric carbon dioxide, the "greenhouse effect" and climate has led to a wide search for alternative forms of energy. At the turn of the century, total global power use was estimated at about 12 TW (Royal Commission on Environmental Pollution 2000), with 310 GW in the UK alone, equating to ~2 kW per person globally or 5 kW per person in the UK (which is fairly typical for an industrialized nation). These values are typically a factor of seven greater than electricity consumption alone and cover a diversity of needs, including transport and heating (Mackay 2008). As populations and appetites grow, both total and *per capita* energy use have risen and are expected to continue to rise, with consequences for fossil fuel reserves and the climate. A sustainable future requires that energy must be provided ultimately from naturally renewing supplies without severe environmental damage, and sustaining 12 TW from renewable energy resources is challenging. The total radiation absorbed by the Earth from the sun is ~10,000 times greater than this value, but it is distributed over the entire surface of the Earth and the direct conversion of solar energy (for example by photovoltaics to electricity) is not the

D. K. Woolf (✉) · M. C. Easton · P. A. Bowyer · J. McIlvenny
Environmental Research Institute, Centre for Energy and the Environment, North Highland College, University of the Highlands and Islands, Ormlie Road, Thurso KW14 7EE, Scotland, UK

D. K. Woolf
International Centre for Island Technology, Institute of Petroleum Engineering, Heriot-Watt University, The Old Academy, Back Road, Stromness, Orkney KW16 3AW, Scotland, UK
e-mail: d.k.woolf@hw.ac.uk

M. A. Shields, A. I. L. Payne (eds), *Marine Renewable Energy Technology and Environmental Interactions,* Humanity and the Sea,
DOI 10.1007/978-94-017-8002-5_2, © Springer Science+Business Media Dordrecht 2014

only option and may not be the most practical either. Other forms of energy are more concentrated than the sun locally and are promising, but individually their practical exploitation cannot supply 12 TW.

A progressive electrification of most energy use including transport and heating is foreseen with sustainable energy sources directed mainly to generating electricity. Most studies of future energy supply envisage a major or complete supply of energy by a broad portfolio of sustainable energy conversion methods. The Special Report on Renewable Energy Sources and Climate Change Mitigation, SRREN (IPCC 2011) identifies six classes of renewable energy source and capture technology, of which ocean energy is one. Ocean energy is the internationally accepted term for a broad range of energy sources including surface and internal waves, tides, other ocean currents, ocean thermal energy and salinity gradient. Offshore wind is included in wind energy, another of the six classes, so for the purposes of this chapter, marine renewable energy (MRE) is here defined to consist just of ocean surface waves (hereafter, waves) and tidal energy.

In what follows, we focus first on the physics of tides and waves and the gross magnitude of the MRE resource, then look at the environmental function of tides and waves before considering the factors that lead to favourable circumstances for energy capture and the magnitude of the practical MRE resource. No consideration is made here of the detailed interactions with individual species or with ecosystems because they are discussed in other chapters and more broadly in the published literature (e.g. Shields et al. 2009, 2011; Scott et al. 2010; Burrows 2012), but we do discuss the role of waves and tides in processes such as transport, morphodynamics, stratification and dispersion that have ecological consequences.

Physics and Energy

Tides

Tides have been the subject of long and extensive study although, paradoxically, there have been relatively few studies in the energetic tidal channels where currents are strongest. Here the basics of tides are reviewed with emphasis on understanding energy balance and flux and on the conditions leading to favourable circumstances for energy exploitation. A broader and more complete description of tides is given by Pugh (1987), and Simpson and Sharples (2012) explain the major role of tides in the physical and biological oceanography of shelf seas.

Tides are forced globally by gravitational forcing (primarily associated with the Earth–Moon–Sun system). For example, the generation associated with the Moon results from an imbalance in two forces acting on elements of the Earth (including parcels of water): the gravitational attraction of the moon and the centrifugal force associated with rotation about the combined centre of gravity of Earth and Moon. The combined equilibrium effect of these forces is towards the moon on the side of the earth nearest the moon, and away from the moon on the other side. This pattern underpins the "equilibrium theory of tides", where an ellipsoid of fluid would rotate such that the major axis is always aligned towards the Moon. This equilibrium theory ignores the rotation of the earth beneath the tidal forcing, so is limited in predictive power because any fluid (including that in the oceans) on a rotating Earth cannot respond fully to the astronomical forcing that can be represented as a twice daily rocking of an ocean basin. However, it does provide a useful explanation for the dominance of periodicities in tidal flow that coincide with certain periodicities in the Earth–Moon–Sun system. For example, it can be inferred that two "bulges" in the equilibrium tide described above will pass a given point on the Earth's surface in the period between consecutive instances of that point facing the moon. Those instances will be separated by slightly more than one day as the Earth needs to rotate slightly more than once to "catch" the orbiting moon. The frequency at which that alignment arises is the principal lunar semi-diurnal frequency, designated as the "M2 tide". Hence, the M2 tide is semi-diurnal with a period slightly exceeding half a day. Similarly, the same alignment between a face of the Earth and the sun occurs precisely twice every day, and this semi-diurnal tide, the principal solar tide, is designated as S2.

Tides are forced at a number of frequencies characteristic of the Earth–Moon–Sun system. At most locations and for most purposes, the M2 tide is greatest in importance followed by the S2 tide, but there are locations (near M2 and S2 amphidromic points) where other tides are of similar or greater importance, typically the major diurnal tides. The S2 and M2 tides will be in phase twice per lunar month (29.6 days), so there is about a 2-week cycle between stronger spring tides (when M2 and S2 are in phase) and weaker neap tides (when M2 and S2 are in anti-phase). Ideally, all calculations relating to tides would be over very long periods (and based on measurements over those periods; Pugh 1987), but being more pragmatic, the absolute minimum is to resolve accurately the M2 and S2 tides, and to base calculations on the two-week spring–neap cycle (or more precisely, a semi-lunar month). Typically, the amplitude of both elevation and current for S2 will be one-third of the M2 equivalents, so there will be variation of typically a factor of two in both tidal ranges and current speeds between springs and neaps. Kinetic energy flux across a plane perpendicular to the predominant flow direction is the theoretical resource for very limited extraction of tidal stream energy and is proportional to the cube of current speed. Consequently, kinetic energy flux can vary by a factor of eight between springs and neaps. The statistics of kinetic energy flux have to be calculated across a minimum of a semi-lunar month and use at a minimum information on M2 and S2 currents (conveniently, that is the maximum information that can be inferred from tidal diamonds on a chart). For major exploitation of tides, the kinetic energy flux can be very misleading (we will discuss this further in the Effects section below), and more useful though simplistic measures are the total energy and power in the tides. Energy and power are each proportional to the square of the amplitude of the elevation or current speed and will typically vary by a factor of 4 across a semi-diurnal month.

Useful indications of the energy (and resource) associated in total and with specific tidal components can be extracted from various studies of global and regional tides. Only two examples are summarized here. It is known that the work done by the moon on waters in the oceans is at the principal lunar semi-diurnal frequency, M2, whereas the sun mostly releases energy at S2. Generally, most dissipation will be in the same tidal components, although some energy can be transferred to different components (e.g. shallow-water tidal components) through various interactions. The tides are primarily generated in the deep ocean, but dissipate mainly after they cross the shelf edge. A total of 3.7 TW of work is done on the tides (2.5 TW at M2), of which 2.6 TW (1.8 TW at M2) is dissipated in the marginal seas (Munk and Wunsch 1998). These global values suggest that some 70 % of the energy is contained in the M2 component. Apart from the M2 tide, most of the energy is at other semi-diurnal frequencies and at diurnal frequencies (periodicities close to one day). There can be local variations in the partition of energy between tidal components. For example, Robinson (1979) reports on calculations for part of a section in the Celtic Sea (on a line between the south coast of Ireland and the north coast of Cornwall, near the southwest tip of England) based on suitable observations over a 709-h period. Those calculations reveal that M2 accounts for $>80\,\%$ of the flux and that M2 and S2 together account for $>96\,\%$. Below, we describe how the tidal resource can be understood in terms of regional tidal energy fluxes, taking the highly studied seas surrounding the UK as an exemplar.

In the modern era, radar satellite-borne altimeters have provided a means to map globally the dissipation of tidal energy (Egbert and Ray 2001). A small number of shelf sea areas is responsible for most of the dissipation of global tidal energy (Green 2010; Simpson and Sharples 2012). Tides and tidal dissipation are relatively high in the Atlantic and the seas bordering Europe, especially around the British Isles. The geographic resolution possible using satellite data is limited, however, and greater geographic resolution of fluxes and dissipation requires *in situ* instrumentation (or a numerical model adequately validated by data). Estimates of tidal energy flux and dissipation have a history dating back to calculations made for the Irish Sea by Taylor (1920), but they are limited to a few sea areas. The study of the seas around the British Isles is unusually thorough, and Cartwright et al. (1980) provide an excellent summary (see Fig. 2.1). They estimated M2 fluxes across a large number of sections surrounding the British Isles and additionally across several key sections within the bounded area, e.g. across Dover Strait. Many of the outer sections were defined close to the edge of the continental shelf, but necessarily some sections cross the shelf to convenient headlands. A southern boundary was defined by several sections from the tip of Brittany (Ouessant) to southwest Ireland (near Valentia). A flux of 190 GW is estimated across that boundary, with 45 GW penetrating to the northern part of the Celtic Sea, Bristol Channel and Irish Sea. Significant flux (16 GW) is estimated through the Straits of Dover to the southern North Sea, but most of the remainder (approaching 130 GW) has to be dissipated in the southern part of the Celtic Sea and English Channel. A northern boundary is defined between Malin Head (Ireland) and Florø (Norway), and an estimated flux of 60 GW across this boundary has to be dissipated in Scottish shelf and coastal waters and in the North Sea (note also the additional 16 GW into the North Sea through the Straits of Dover, which must also be dissipated).

The methods pioneered by Taylor (1920) and developed by Cartwright et al. (1980) and others greatly elucidate the flows of tidal energy, but they are relatively scarce. At a more local scale, an excellent example is given by Robinson (1979) who, within a study of the tidal dynamics of the Irish and Celtic seas, includes a fairly detailed study of energy flux vectors. More-detailed studies (following Robinson 1979) need to include minor energy interaction terms such as the work done locally by the currents on astronomical bodies, although the transports of tidal energy originating from the open ocean and the frictional dissipation terms are usually dominant. Such studies can be enlightening as one contemplates exploiting the energy.

As energy and power is the focus here, it is worth calculating the intensity of energy flows implied by the fluxes estimated by earlier workers. Hence, Fig. 2.1 depicts the average tidal power per unit length of each section (a convenient parameter to compare with estimated offshore wave-energy fluxes that are readily available and discussed below). At the southern boundary the fluxes are relatively large, between 200 and 300 kW m^{-1}. Fluxes across the northern boundary are weaker, though, consistently <70 kW m^{-1}, but even these lesser fluxes are generally similar to or greater than the wave fluxes at like locations.

A satisfactory comparison with the intensity of solar fluxes (typically several hundred W m^{-2} across a horizontal plane during direct sunlight) is difficult to achieve, but on the face of it there is significant concentration of energy within tides, although the total global dissipation of tidal energy is

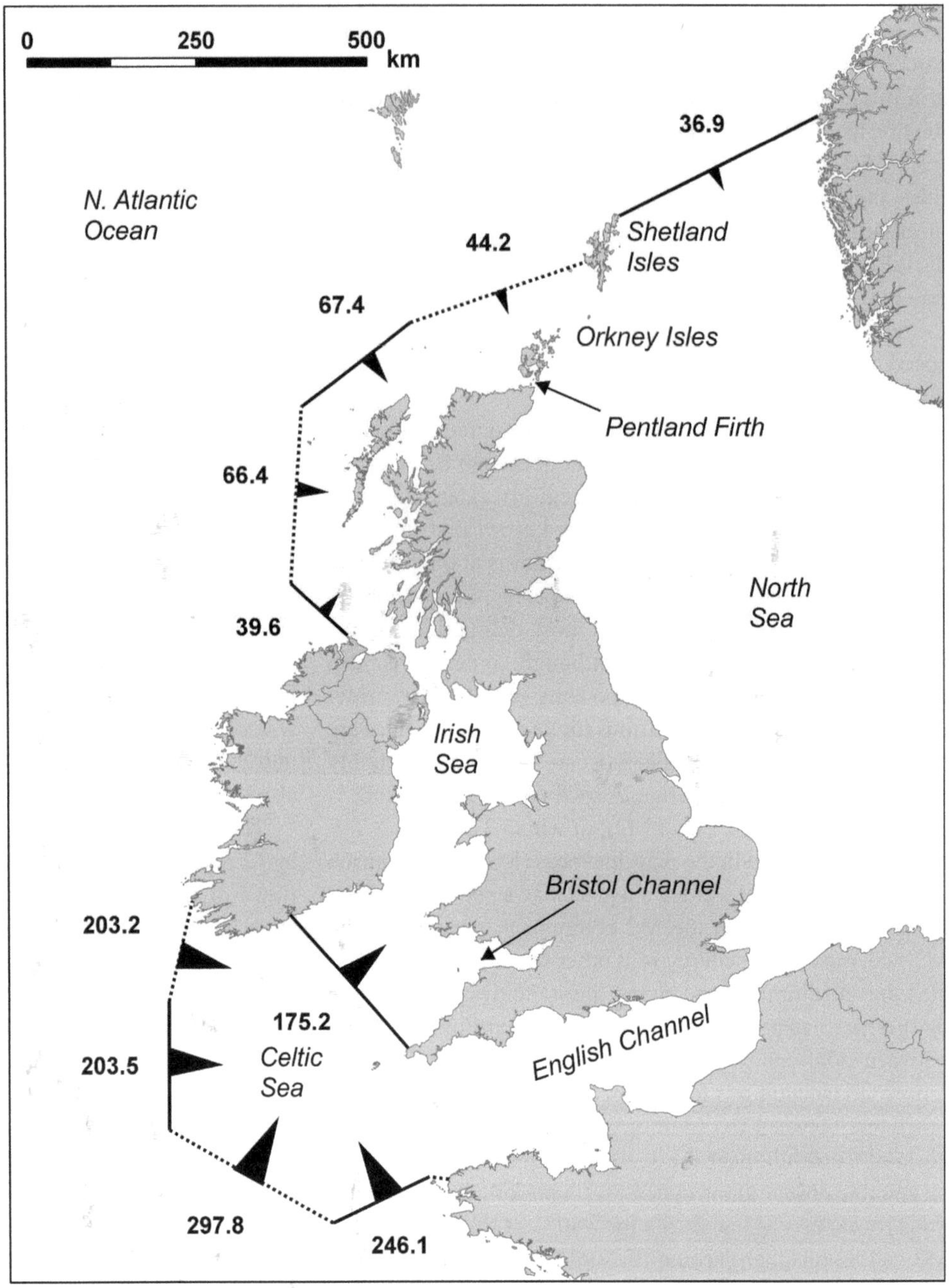

Fig. 2.1 Estimates of M2 tidal energy flux across the boundaries enclosing the British Isles, in kW m^{-1} (adaptation of a schematic developed by Cartright et al., 1980, who estimated values across key sections in GW, with values recalculated to power per unit boundary length to aid comparison with similar estimates for wave power; see Fig. 2.2)

tiny relative to solar energy. The global tidal energy dissipation (3.7 TW) is less than global energy use (12 TW in 2000, but rising thereafter), but considerable. Adding additional tidal harmonics to the 250 GW known for M2, the tidal energy flux from the deep ocean dissipated in the waters around the UK is coincidentally close to total energy use in the UK. Therefore, there is certainly sufficient tidal energy to be of interest, but given practical considerations, one should not expect it to solve the world's energy problems. We return to the "practical resource" later. Also, it can be inferred that the energy of the tides needs to be dissipated naturally within bounded areas, but tidal dissipation will be far from uniform and will be particularly great in some coastal areas, where strong currents interact with the seabed. This issue is addressed further below after completing a more general summary of tides.

The tidal energy fluxes across the shelf edge west of Europe can be viewed as the limb of an Atlantic tidal system. As described above, those tides and the tides in all the world's oceans are a response to astronomical forces, but that response is modified from the equilibrium tide by inertia and the rotation of the Earth. Simpson and Sharples (2012) provide a cogent description of tides in shallow seas, so the following description is merely a brief summary. The tides are generally described as Kelvin waves, which are long or shallow-water waves (i.e. wavelength » water depth) affected by the Earth's rotation. If rotation were ignored, then ocean tides might be regarded as long waves reflecting at continental boundaries to form patterns of standing waves, with nodal lines of low tidal range interspersed with anti-nodes of large tidal range (Simpson and Sharples 2012). Note that a different set of patterns is formed by each tidal harmonic, although harmonics close in frequency (and hence wavelength) tend to have similar patterns. The effect of rotation (and the resulting Coriolis force) can be thought of as replacing the nodal lines with amphidromic points, or simply "amphidromes", with tides rotating around them and zero tidal amplitude at the amphidrome. An amphidromic system can be described by two sets of lines, "co-tidal" and "co-range". Co-tidal lines radiate from each amphidrome and join points of identical phase in tidal elevation (e.g. high water will coincide at each point on a co-tidal line). The radiating lines are spaced such that they describe the full rotation of the tide around the amphidrome in a single tidal cycle. Lines of co-range connect points of equal tidal range, and the innermost of these lines are concentric to the amphidromes so describe the increasing tidal range outwards from the amphidrome. Farther from the amphidromes, maps of range are complicated by interactions of neighbouring amphidromes and interactions with the continental margins. The tides on the continental shelf west of Europe are primarily a function of an M2 amphidromic system centred far away in the western part of the North Atlantic. The anticlockwise rotation of the M2 tide around that amphidrome implies tides propagating northwards up the European continental shelf. Tides north of the UK and in the North Sea are also strongly influenced by a M2 amphidrome east of Iceland. Basin-scale amphidromic systems in the seas around the British Isles further complicate the tides around the UK.

The propagation of long-water waves depends on water depth. The phase and group velocity of such waves (and, for a given wave period, the wavelength) depend on the square root of the water depth. The tides, which are Kelvin waves with typical wavelengths of 8,000 km, are altered greatly where they cross the continental shelf edge from waters thousands of metres deep to shelf waters typically 100 m or so deep. Long waves propagate more slowly in shallow water and the motion is necessarily restricted by the water depth. As a result, tidal elevation (and tidal range) is typically elevated by a factor of 2.5 on the shelf, whereas tidal currents are typically elevated by a factor of 16 (Simpson and Sharples 2012).

Another significant effect of rotation apparent for tides (Kelvin waves) on the shelf and in coastal waters is an interaction with a coastline that amplifies the amplitude (and tidal range) near the coast. Consider a Kelvin wave propagating anticlockwise (the direction of rotation applicable to the northern hemisphere) around an amphidrome within an ocean basin. At the margins of that basin, the tide will propagate with the continental margin on the right and the interaction will result in an amplification of tidal range adjacent to that margin (again, an effect attributable to rotation). The amplification will only fall off gradually with distance from the coast—the appropriate length scale is the Rossby radius (equal to the speed of the shallow wave divided by f, the Coriolis parameter), which is typically 250 km for the depth of shelf seas and > 1,000 km for deep-ocean water—and the amplification is rather general for coastal and shelf water where the tide propagates with the coast on the right (i.e. in the northern hemisphere).

As noted above, Kelvin waves will reflect when they meet a coast at right angles to their direction of propagation, and the tides on the shelf will generally be a superposition of incident and reflected waves. At some sites (e.g. on the north coast of Scotland, where the tides propagate eastwards towards Pentland Firth and Orkney), the incident wave is dominant and the net effect resembles a simple progressive wave. In most cases, the reflected waves are highly significant and in some locations the superposition can resemble a standing wave with equal and opposed incident and reflected waves. An interesting case is the resonant system, where a natural periodicity dictated by the coastline and bathymetry coincidentally matches the periodicity of one of the major ocean tides. The resonant condition is that the distance between the shelf edge and the closed end of a gulf (or any bay or closed channel) should be an odd multiple of the quarter wavelength of the particular tide. In principle, the tidal range within a resonant system can grow indefinitely as it is fed by energy from the ocean tide, but in practice the tidal range and currents will be large but finite such that the energy input is balanced by frictional dissipation. The most famous example of this is in the Bay of Fundy, where the quarter-wavelength of the M2 tide almost precisely matches the length of the bay and tidal ranges up to 16 m are experienced. The Irish Sea is an example of a system that is not far from resonance and the tides are also very large, but which could in principle be nudged closer to resonance by civil engineering works (Pugh 1987).

Waves

Ocean wind waves and swell are a result of the action of the wind on the sea surface. A wind sea is the response to contemporary winds, and swell is the relic of earlier winds. The distinction between a wind sea and a swell is not always clear, but it is not essential because the same physical laws and formulae apply to both. Both types of waves arise at a range of wavelengths, but although a wind sea will include short ripples, swell is restricted to longer waves of tens or hundreds of metres wavelength. Waves propagate across the sea surface, but the motion of fluid elements within a wave is primarily orbital, with the frequency of the wave and a surface diameter equal to the wave height. Water far below the sea surface will also follow an orbit but with a monotonically decreasing diameter with increasing depth. In shallow water the orbits are compressed in the vertical, first to an ellipse and finally to a line near the seafloor. Some of the energy of a wave is present and can be extracted from far below the sea surface, but a diminishing fraction.

The generation of the wind is a complex process resulting from solar heating and the Earth's rotation, and a fraction of the energy of the wind over the ocean surface is cascaded into wave energy. Very little of the Sun's power is diverted into waves (only exceeding 1 W m^{-2} in very strong winds). The potential utility of wind waves for sustainable energy arises first from the property of waves of accumulating energy over hours or days, and second, in the case of ocean swell, from energy being stored efficiently for several days (in the absence of strong wind forcing) and being propagated thousands of kilometres over many days to the ocean margins and an accessible location for exploitation. The physics of wave generation and propagation are largely neglected below, although we describe some formulae and results that help explain the practical value of wave energy. A specialist text on ocean waves (e.g. Holthuijsen 2007) provides a more-complete description of wave physics and statistics.

Where the wind blows steadily on the sea surface, waves may grow from nothing to a maximum, or fully developed wave height, over many hours, and the energy within the waves grows with the square of wave height. In addition to the increase in wave height, waves generally grow longer as non-linear interaction between waves of different wavelength directs most of the energy to longer waves. All waves move, i.e. they propagate, and longer waves move faster. A dispersion relationship relates the period and wavelength of a deep-water wave, with the wavelength proportional to the square of the period. A phase velocity can be defined as the speed of individual waves, and group velocity describes the speed of a packet of waves and of the transport of the energy within the wave field. The phase velocity is twice the group velocity for surface waves in deep water, and both are proportional to the period of the waves or to the square root of their wavelength. Therefore, in the open ocean, a wave field can be expected to grow and propagate downwind in response to a storm. If the wave field travels a similar path to the storm (ideally, if the group velocity of the dominant waves coincides with the propagation velocity of the storm centre), then the wave field accumulates energy from the storm winds. The flux of energy within a propagating wave front is given by the product of the energy density and the group velocity, so is proportional to the square of the wave height multiplied by the period. The wave field will lose energy by deep-water wave-breaking, or whitecapping, and at some point a terminal or fully-developed wave height will be reached, though only if the winds are sufficiently sustained and there is enough sea room for the propagating wave group. Hence, both the duration of the winds and the fetch are important.

It is useful to complement the qualitative description above with some practical values, based here on sustained wind speeds of 10 m s^{-1} (fairly strong, but common) and 20 m s^{-1} (very strong winds, typical of an Atlantic storm). Some empirical formulae are used for the height and period of wind waves, as proposed by Carter (1982) on the basis of observations. The fully developed significant wave height (the average height of the highest one-third of waves) increases with the square of wind speed, reaching 2.5 m for 10 m s^{-1} winds and 10 m for 20 m s^{-1} winds, and the period of the dominant waves increases to >7 s and 14–15 s, respectively. These waves will travel with group velocities of ~5 m s^{-1} or ~10 m s^{-1}, respectively, when fully developed and will require the wind speed to be sustained for >20 h or >40 h and over a fetch of 225 km or 900 km, respectively. The energy density of 2.5 m waves is ~4 kJ m^{-2} of sea surface and if that energy is accumulated in >20 h, an average energy accumulation rate of only ~0.05 W m^{-2} is implied (note that this is a net rate and energy lost through wave-breaking is subtracted from the energy picked up from the wind, and that any spreading of the wave packet is also neglected). For 10 m waves, the energy density is ~60 kJ m^{-2} and the average accumulation rate is 0.4 W m^{-2} over 40 h. As noted already, waves accumulate energy slowly, but given persistent strong winds and enough sea room they eventually store a huge amount of energy that is rapidly propagated across the oceans.

The property of the wave field most directly applicable to sustainable energy is the flux of energy crossing a line parallel to the coast, or perpendicular to the main direction of propagation, in kW/m length of that line (see Fig. 2.2). From the values given above it can be calculated that a swell of 2.5 m height and 7 s period will contribute an energy flux of ~20 kW m^{-1}, whereas a swell of 10 m and 14–15 s will contribute ~600 kW m^{-1}. It is apparent from these values that even quite rare instances of very large waves may contribute considerably to the average energy flux, and that more-common waves of 1 or 2 m play a less important role in determining energy flux. That simple insight has a few implications. First, one would normally expect wave energy capture to be sited where a large ocean swell is relatively common (e.g. on the Atlantic margin of Europe where swell from Atlantic storms may propagate). Conversely, a coastal site in a windy area may be of little interest, if there are no or few instances where a wave field can propagate over several hundreds of kilometres to that site. Also, one should beware of simple estimates of average flux, because these may be affected by a few instances of very high flux (which may be entirely useless if they exceed the operating tolerances of the capture devices).

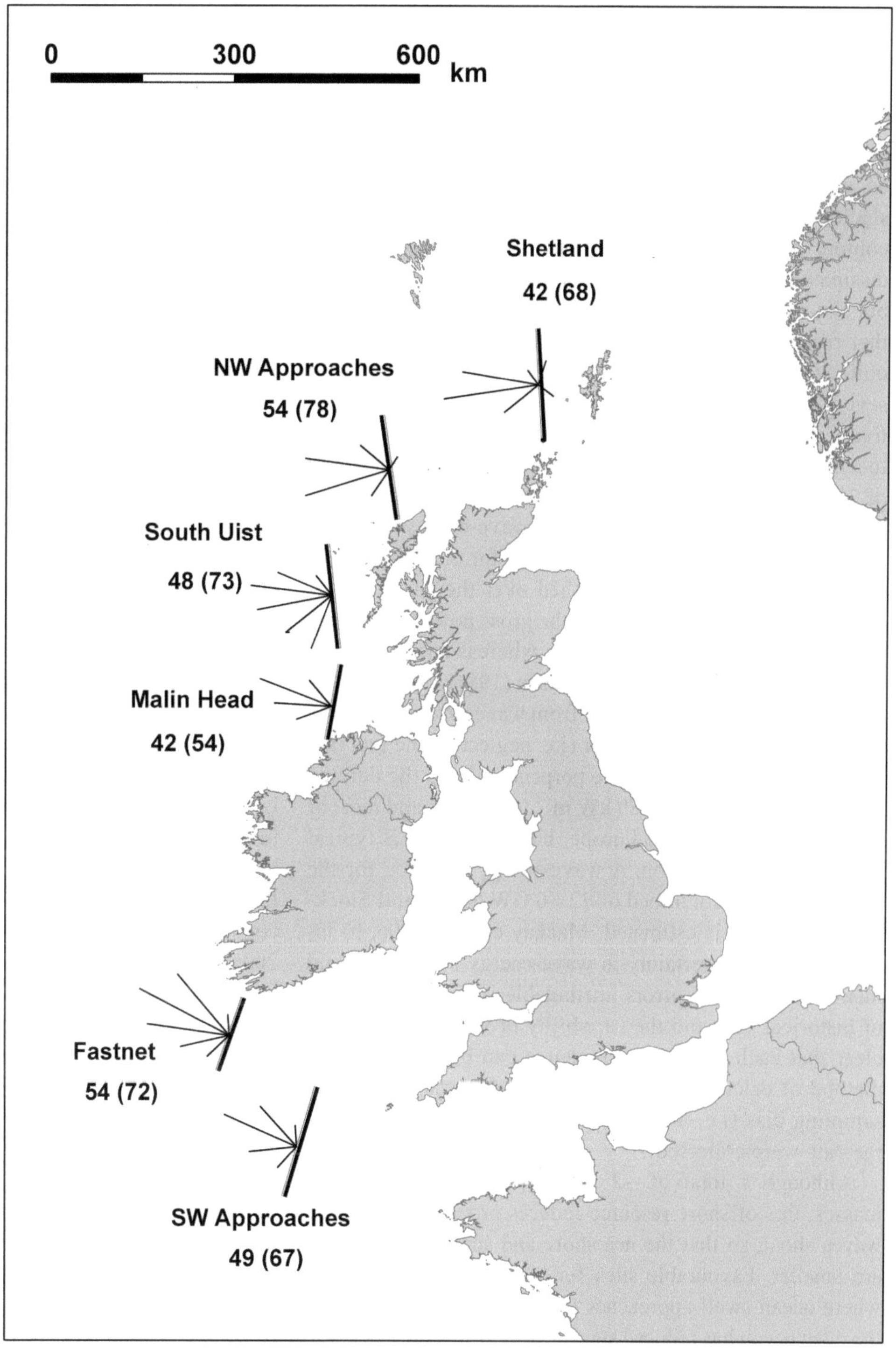

Fig. 2.2 Wave energy fluxes offshore of the UK (redrawn from Mollison 1991). The thick lines define arbitrary sections demarcating the UK offshore resource and the numbers denote the calculated wave flux across each section per unit length of that section in kW m^{-1}. The numbers in parenthesis neglect the angle of approach of the waves (effectively they are assumed to cross perpendicular to the section) and the more accurate values precede the uncorrected values. The thin lines schematically show the wave energy from each 30° sector contributing to the total

Many available wave-power statistics are presented as average flux. Note also that although we have presented some indicative values above based on simple unidirectional waves with a dominant frequency, real wave fields consist of a confusion of directions and wavelengths. In other words, a wave field should at a minimum be described by a directional wave spectrum that describes the distribution of energy across wavelengths and directions. A thorough treatment of the theoretical potential of wave power should integrate across this wave spectrum in calculating the total wave power across a defined line. Calculation of practical resource should also consider thresholds for capture. For example, the calculation should account for saturation of capture devices at a design limit and should exclude instances where the wind speed or wave height is impractically high for safe operation. Historical estimates were often limited to approximate calculations that additionally relied on sufficient data. Some of the more-recent calculations are more complete, but are also based on limited data.

Wave-energy resources are typically quoted as the wave power reaching an ocean margin at or within the shelf edge (the offshore resource) or for technology suitable for shallow water or the shoreline, reaching a particular depth contour or the shoreline. As noted above, there is a natural dissipation of wave energy associated with whitecapping, but the useful resource is the energy contained in waves that reach the ocean margins. Some global estimates are quite vague. Gunn and Stock-Williams (2012) reviewed several historical estimates and completed their own relatively thorough calculation, estimating a global wave flux to a line 30 nautical miles offshore from a defined coastline of 2.11 ± 0.05 TW, consistent with earlier estimates on the order of 2 TW. Regional estimates of wave-energy resource, for example for the UK, for Ireland, or for the UK and Ireland together have been published over the past few decades. Early estimates calculate the gross power, which ignores the direction of propagation, whereas later estimates used directional spectra, and Mollison (1986, 1991), for example, used directional spectra from wave-model hindcasts (see Fig. 2.2). The gross power (i.e. neglecting the fact that many waves do not propagate perpendicular to the defined line) yields values of 50–80 kW m^{-1} for the UK and most of the Atlantic seaboard of Europe, but 40 kW m^{-1} is typical accounting for direction. A wave-energy resource for the UK and Ireland combined of 72 ± 6 GW (Gunn and Stock-Williams 2012) is estimated. Mackay et al. (2010a, b) investigated the uncertainty in wave-energy assessment and identified potential errors attributable to both limitations of historical data and the variability of wave climate. It is clear that each estimate of resource can be biased by the method of calculation, by errors in the source data, or by sampling bias (i.e. the values will reflect the period when the data were collected).

Although a total of ~2 TW approaches global landmasses, this offshore resource reduces progressively as the waves shoal, so that the nearshore and shoreline resources are smaller. Favourable sites for wave energy tend to be where ocean swell approaches the coast. Once the swell in the open ocean has reduced significantly below the steepness of newly generated wind waves, it propagates with almost no wave-breaking or loss of power. The swell steepens on the continental shelf, because the group velocity of waves is less in shallower water, and may eventually steepen to a stage where shallow-water wave-breaking is a substantial sink of energy. Also, a turbulent boundary layer will be set up by the interaction of the seabed with the deep motion associated with the waves (an orbital motion that decays with depth), resulting in some frictional dissipation. Therefore, a slow loss of wave power may be noticed on the continental shelf. Mollison (1986) reviewed a few estimates for water of intermediate depth (20–100 m) that suggest that a loss of 1 % per km may be typical.

As noted previously, wave power can vary dramatically. At a favourable site frequented by swell, there is likely to be <1 kW m^{-1} of wave power at least 10 % of the time. At the same time, peak wave powers of the order of 1 MW m^{-1} are likely, which will at best be superfluous and at worst catastrophically destructive. Wave energy will vary on the same time-scales as the weather (i.e. weather systems passing through every few days), but will also exhibit seasonal and interannual variations reflecting the variability of ocean winds. For example, the UK wave climate features higher, more-energetic waves in winter and shows strong variation between winters paralleling the variation between wet and windy winters and drier, calmer ones (Woolf et al. 2003). The interannual variation is associated with the North Atlantic Oscillation (NAO) and other large-scale modes of variation in regional climate (Woolf et al. 2002; Mackay et al. 2010b). The behaviour of these modes under climate warming and hence the statistical characteristics of Atlantic cyclones and other factors affecting the waves reaching the Atlantic seaboard of Europe (Wolf and Woolf 2006) is difficult to predict. The greater wave power in winter may be helpful because it coincides with greater energy demand in the UK. Unfortunately, however, interannual variations are unhelpful because wind, wave and hydro-energy generation will all be lower in relatively cold, dry winters. Clearly, therefore, future climate is a significant consideration in evaluating the commercial prospects of wave energy (Harrison and Wallace 2005), although changes in wave climate related to anthropogenic climate change are anyway likely to be dwarfed by natural variation within the lifetime of a wave farm (Mackay et al. 2010b).

Tides, Waves and the Environment

Shelf seas and coastal waters are an important environment that supports a rich ecosystem that is economically significant and enriches the human experience. Tides and waves are an important part of the physics of this environment and below we explain how the physics partly determines the environment and the nature of the ecosystems.

Tides

Some of the simplest effects of tides on both the water column and benthic habitats are related simply to tidal currents. It has already been said that tidal currents vary from one site to another. Fast tidal currents present problems both for benthic organisms that need to attach to the seabed or for organ-

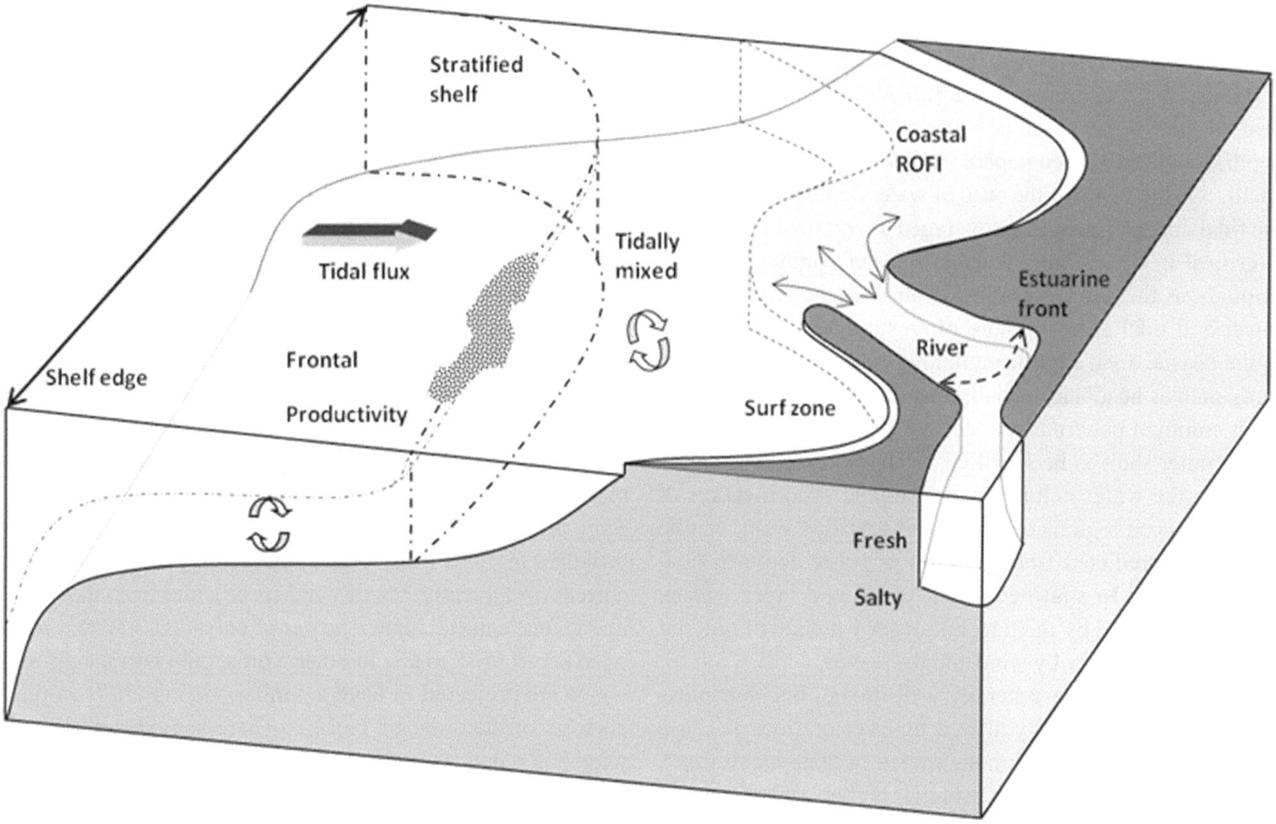

Fig. 2.3 A schematic of shelf and coastal seas, depicting a number of different environments. The effects of tides and waves on the environments are discussed in the text. ROFI refers to a region of freshwater influence, a region of coastal water where river water influences the stratification

isms within the water column that need to resist transport or dispersal. They may also be closely related to the seabed sediment, because in fast currents bare rock, or episodes of disturbance, removal and smothering, might be expected. Hence, speciation of fast-current regions should reflect adaptation to these specific conditions (Shields et al. 2011). Fast currents may also assist or hinder migration, with colonization depending on net transport over many tidal cycles (residual currents).

Apart from fairly direct responses to currents, most of the important effects are mediated through the influence of the tides (and to a lesser extent, the influence of wind and waves) on the zonation of the shelf seas into some distinct environments. The diversity of the physical environment from the shelf edge to estuaries is depicted in Fig. 2.3, and this illustration needs to be borne in mind when reading the paragraphs below.

Tides are an important part of the shelf-sea environment, are responsible for significant transport and most importantly stir the shelf seas. As illustrated in Fig. 2.3, there tend to be two distinct regions within the shelf seas, with fronts between them. Where the tidal current is strong and the water shallow, the sea will be mixed from top to bottom throughout the year. In other areas (deeper, or slower current regimes), summer heating can partly isolate a warmer upper layer from a lower layer stirred by tides (seasonal stratification). Tidal currents certainly play a key role in determining the location of seasonal shelf sea fronts (Simpson and Hunter 1974; Simpson and Sharples 2012), and this can be understood first from the insight that vertical mixing depends on the stirring phenomenon. When the sea surface cools, the water column is overturned by convective instability, but when the surface warms, near-surface density reduces and the resulting buoyancy flux tends to stratify the water column, and that can only be overcome by vigorous stirring. The most effective stirring process is usually the stirring by turbulence induced by tidal flow over the seabed (we return briefly to the secondary roles of wind- and wave-induced stirring later). The power going into stirring by a tidal current cannot exceed the rate of dissipation of tidal energy and is proportional to the cube of the tidal current and a drag coefficient. For a typical drag coefficient, this dissipation rate will be ~2.5 W m^{-2} at a current speed of 1 m s^{-1}. For tidal stream sites, the dissipation rate is high, averaging over a tidal cycle up to the order of 100 W m^{-2} (Fig. 8 of Easton et al. 2012). These gross figures are misleading, however, because it appears that the power is not used efficiently in the mixing process. Simpson and Sharples (2012), for instance, estimate a mixing efficiency of just 0.5 % for one case, and it is assumed that most of the remainder generates heat in the turbulent boundary layer. Note also that the power (per unit depth) available to overturn the stratification will be inversely proportional to water depth, so where the geographic variation in buoyancy flux is small, it is the value of the ratio of water depth to the cube of the tidal current (most often averaged over a tidal cycle) that is critical in determining whether a water column is mixed throughout the year. As water depths are invariably shallower and tidal currents more often stronger inshore (also, in the coastal area, significant mixing is associated with features such as headlands and dissipation of wave energy), the most common pattern, depicted in Fig. 2.3, is for the water of the outer shelf to be stratified in summer, whereas closer to shore the water column is well mixed. The location of fronts between regions of mixed and stratified water can be effectively predicted from the ratio of water depth to tidal current cubed. In stratified water, the lowest layer will be effectively stirred by the tide, but upper waters will only by stirred intermittently by wind and the waves.

Coastal waters are generally well mixed, but this situation may be dramatically altered by river outflow, because freshwater can be an enormous source of "buoyancy flux". Stratified zones exist within estuaries and coastal regions of freshwater influence, or ROFIs. Again, there is effectively competition between stratification by the buoyancy flux and stirring by tidal currents, and numerical models are quite effective in predicting the resulting zonation, given adequate bathymetry, tidal physics and estimates of river flow.

Various states of mixing and stratification affect phytoplankton and hence water-column ecology, because they affect the distribution of nutrients and the amount of light received. Frontal regions between fully mixed and stratified waters tend to be regions of convergence and vertical movement, so are conducive to high levels of primary productivity. Stirring may also influence encounter rates, feeding and reproduction (Shields et al. 2011).

Currents will also resuspend and transport sediment. In some areas dominated by fine sediments (e.g. within the Bristol Channel), resuspension of sediment can be so great as to make the water column opaque, lowering the light available for photosynthesis. Sediment accumulation will vary greatly according to sediment resuspension and transport, with consequences for benthic habitat.

A simple but important result of tides is a variation in sea level (tidal range), implying a variety of water depths and environmental conditions for benthic organisms below the low-water line and creating intertidal habitats. Both range and currents have a role in ecological zonation in benthic and intertidal habitats. For example, Burrows (2012) noted a statistical association between a shift from macroalgae to filter-feeders and tidal current speed, but only in areas where chlorophyll concentrations were high. As described below, waves may be a more substantial cause of varying zonation nearshore and onshore than tides.

Waves

The influence of wind and waves on stirring the shelf seas is generally considered to be much weaker than the effect of tides. The relative strength of stirring can be understood from a calculation of the power dissipated per unit area of the sea surface associated with each process. Above, it was noted that tidal dissipation amounts to a power of about 2.5 W m^{-2} at a current speed of 1 m s^{-1} but can reach an order of 100 W m^{-2} at tidal stream energy sites. Simpson and Sharples (2012) only discuss the direct effect of wind-stirring. They provide formulae for calculating the power going into turbulent motions from wind stress acting on a wind-driven current (of typically 2 % of wind speed), and from these formulae, one can calculate a power of only ~0.03 W m^{-2} for a wind speed of 10 m s^{-1}. In other words, only hurricane-force winds are projected to have a similar stirring effect as tidal currents of the order of 1 m s^{-1}. There is evidence that the wind has a noticeable if secondary role in governing stratification (Simpson et al. 1978). The significant contribution of wind is partly attributable to a greater proportion of the energy from wind going into turbulent motion than is the case for tide (2.3 vs. 0.4 %; Simpson et al. 1978).

Simpson and Sharples (2012) did not consider the effect of waves on stirring and stratification, and this subject seems rather to have been neglected by physical oceanographers, although it does have the attention of coastal engineers (Nielsen 1992). Above, we noted that although the power going into waves is small, waves accumulate energy over large distances and durations. The accumulation and propagation of large amounts of energy (a flux of the order of 100 kW m^{-1} of wave front is common for ocean waves) implies that when and where this energy is lost to turbulent motion in the upper ocean, the strength of stirring must be considerable. Energy is lost from a wave field as ocean waves approach the coast through either direct frictional interaction with the seabed, or because the steepening of waves in shallower water results in wave-breaking (note that the longest ocean waves respond to the seabed even in depths of 100 m or more). Note again that Mollison (1986) suggested a loss of energy of 1 % per km of wave may be typical in water of intermediate depth (20–100 m). Such a rate of loss implies a loss of 1 W m^{-2} from a wave field of 100 kW m^{-1}, suggesting that in wave-exposed shelf areas, the contribution of wave-induced stirring may be substantial. If an energetic wave field comes nearshore, then the rates of energy loss may be massive; for example, if a 100 kW m^{-1} wave field is wholly dissipated in a surf zone 100 m wide, then an average rate of dissipation of 1 kW m^{-2} is implied.

It follows from the calculations above that although waves are rarely considered in the context of shelf-sea stratification and stirring, their influence is unlikely to be negligible in shelf areas exposed to ocean waves. The prevailing paradigm that competition between surface heating and tidal stirring generally dictates shelf-sea stratification should remain largely unchallenged, but wave-induced stirring is relevant. Closer to shore, the dissipation of large amounts of wave energy should become progressively more important. The influence may occasionally include breaking down stratification (for example, if a ROFI is impinged upon). More frequently, though, the interaction of the wave field with the seabed or turbulence from breaking waves is likely to disturb the seabed and result in the suspension, dispersion and resettlement of sediment (Nielsen 1992). Organisms would then also need to contend with a fairly aggressive environment especially in a surf zone or very close to shore. Hence, although the depiction of shelf seas in Fig. 2.3 emphasizes the role of tidal mixing, the influence of waves should be considered too, especially nearshore.

Extraction, Production and Technical and Practical Resources

Tides

Typical tidal currents of <1 m s^{-1} in most shelf seas are generally insufficient for commercial tidal energy generation (Couch and Bryden 2006). That limit may be understood partly by considering the intensity of the available energy. Tidal energy fluxes of up to 300 kW m^{-1} at the shelf edges may sound impressive, but this is the total (potential and kinetic) energy contained within the wave over the full water depth (typically 200 m near the edge of the shelf). A more appropriate value would be the amount of kinetic energy passing per unit cross-sectional area across a vertical plane perpendicular to the flow. That kinetic energy flux is proportional to the cube of current and will be just 500 W m^{-2} at a current speed of 1 m s^{-1}. It is this hydrokinetic energy flux that is directly exploited by most energy-capture devices (e.g. tidal turbines), and a current of 1 m s^{-1} is an optimistic minimum for a business case even at a fairly accessible site. Relatively few areas are therefore likely to be suitable for tidal energy extraction (Couch and Bryden 2006); strong tides are certainly essential and, at least in the near future, sites are likely to be limited to water depths of 25–45 m.

The few special conditions that naturally yield currents exceeding 1 m s^{-1} in reasonably accessible locations have been discussed by Couch and Bryden (2006). They identify three classes: tidal streaming; resonant system; hydraulic current. Very fast currents are encountered in resonant systems such as the Bay of Fundy, although it may be even more attractive to consider the potential energy associated with the large tidal range there. Tidal streaming refers to a local acceleration of flow where there is some constriction of that flow (somewhat similar to squeezing the end of a hosepipe to speed the flow) and arises naturally where a headland or an island steers the flow. An unambiguous example of a hydraulic current can be imagined by considering two large, quiescent reservoirs; if the surfaces of the connected reservoirs are at different heights, then water will flow from the higher to the lower one, accelerated by the potential energy of the height difference. In the natural world, the common situation is two basins (e.g. North Sea and North Atlantic connected by the Pentland Firth) with different amplitudes and phases. At most phases of the tide, there will be height difference to accelerate the flow in the Pentland Firth in one direction or the other. Many sites such as the Pentland Firth should be regarded therefore primarily as hydraulic in nature, but often there is a further enhancement by tidal streaming and the external properties of the tidal system set up the height differences (Easton et al. 2012).

The practical obstacles to tidal energy development are considerable and may limit the potential for growth, at least in the immediate future. The simple criterion of directly exploiting hydrokinetic energy only at sites where currents are fast greatly reduces the potential relative to the total energy available within a tidal energy system. The power in the M2 tide alone around the UK is estimated at 250 GW, but much lower figures are calculated for technical or practical resource. In the case of the UK, recent obstacles to the development of the Severn Barrage concentrated efforts on in-flow technologies, where power is extracted directly from hydrokinetic energy in natural tidal currents. Many technologies have been suggested for this energy extraction, and the practicality, density and efficiency of a tidal array will depend on the selected technology, although generally all technologies will have a similar effect on the large-scale flow that scales with the amount of energy extracted. We refer hereafter to all projects involving the in-flow capture of hydrokinetic energy as tidal stream projects and the resulting total energy production as the tidal stream resource.

Technical resource can be defined as "the energy … that can be harvested from tidal currents using envisaged technology options and restrictions (including project economics) without undue impact on the underlying tidal hydrodynamic environment" (Black and Veatch Ltd 2011b). A necessary step is to infer the changes of hydrodynamic parameters associated with energy extraction at each site. The first step in the method used by Black and Veatch Ltd (2011b) is to assign each site to one of the three classes listed above (after Couch and Bryden 2006); it is postulated that the effects of energy extraction will differ according to the class of site. Garrett and Cummins (2005) established an analytical solution for the hydraulic current case, which then yielded a strict theoretical limit on the amount of energy that can be harvested (the theoretical resource). The maximum yield is proportional to both the maximum flow and the amplitude of the height difference driving the hydraulic current. The approach has been broadened by Black and Veatch Ltd (2011b) to include practical formulae for calculating the technical resource, with a separate formula for each generic tidal current regime. The technical resource is based on restricting the harvesting to keep changes in the flow characteristics within reasonable (but arbitrary) limits. Black and Veatch Ltd (2011b) propose limiting the reduction in peak current speed to <10% and the reduction in tidal range to <0.2 m (or 5% of the spring tidal range, whichever is less). As these limits are quite low, the technical resource will always be much less than the theoretical resource. Each formula describes a proportionality to maximum flow and to amplitude, but with varying definitions and proportionality constants for each regime. The formulae have been shown to be quite robust with respect to parameters such as channel length. Black and Veatch Ltd (2011a, b) identified each recognized potential tidal stream energy site with one of the three regimes and calculated a resource for each using available data on amplitudes and flow rates. Calculation of the technical resource requires a few additional steps including consideration of the cost of extracting the energy (see pp. 26–27 of Black and Veatch Ltd 2011a, for a worked example for the Pentland Firth).

Black and Veatch Ltd (2011a) estimates a base UK tidal stream technical resource of 29 TW-h year^{-1}, with alternative pessimistic and optimistic values of 16.4 TW-h year^{-1} and 38.4 TW-h year^{-1}, respectively. However, a practical resource includes external constraints such as the impact of shipping. Each proposed site has to be considered individually, and in total the practical resource is reduced to some 70% of the technical resource. The resulting practical resource of 20.6 TW-h year^{-1} is from a range of tidal stream projects, including examples of all three tidal current regimes deemed to be economically and environmentally feasible (Black and Veatch Ltd 2011a). Almost half that figure (10 TW-h year^{-1}) is calculated for the appropriate exploitation of the deep Pentland Firth. Note also that although each 1 GW of capacity could in principle supply nearly 9 TW-h year^{-1} from steady, strong currents, the reality that currents vary requires much greater capacity to achieve the required yield. For example, the calculation of Black and Veatch Ltd (2011a) implies that 10 TW-h year^{-1} will be generated from a farm rated 4.2 GW. Also, in each case the energy lost from the flow has to exceed the energy produced, because some energy will be lost though drag around the generating structure.

There is in principle the possibility of extracting more energy using civil engineering to manipulate flows (using barrages or lagoons), usually where the tidal range is already large. Such tidal capture technologies could add greatly to the total resource (theoretical, technical and practical).

The extraction of tidal energy by either tidal stream or tidal capture projects (or tidal range projects; e.g. involving barrages or lagoons) interests many around the world, for example on both coasts of North America and in New Zealand. Associated with that interest is great interest in calculating technical and practical resource, and new studies are progressively illuminating the several factors that affect the size of the resource. For example, Vennell (2012) demonstrated that a number of factors will determine how many turbines can be inserted in the flow within a channel before the return diminishes notably. Factors among natural characteristics important to the theoretical and technical resource include the relative importance of inertia and friction and constrictions within the channel. Practical factors include the proportion of the cross-section that can be filled. Although most attention has shifted to tidal stream projects, there still remains interest in tidal capture schemes. For example, Yates et al. (2013) considered a large range of options for extracting energy from the western waters of the UK, including tidal stream arrays and tidal lagoons, but with emphasis on estuary barrages. Along with specific local modelling, simple models have been developed to calculate the technical limits of energy extraction by tidal capture.

The available power from tides is not constant, and this needs to be considered as part of the integration of tidal power within an energy grid. As noted in the Tides section of "Physics and energy" above, semi-diurnal tides are usually dominant, so High Water, ebb flow, Low Water and flood flow each recur over slightly more than half a day. The phase relationship between, for example, High Water and peak flood varies, but peak and ebb flows will usually be separated by slightly more than 6 h. Typically, at a fast-current site, there will be 4-h periods (strong flood or ebb flow) when currents are strong enough for energy extraction, punctuated by 2-h periods when currents are relatively weak and therefore uneconomic. Fortunately, the timing of flows will vary between sites, although it is possible that flows at sites attractive to development may coincide (Iyer et al. 2013). Tidal capture will also be subject to such variation, although there is some scope for manipulating the timing of flow in capture projects. Also as explained in the same tides section above, tidal range and currents will vary over a two-week period between springs, when the lunar and solar forcing reinforce each other, and neaps, when they conflict. This astronomical forcing is universal, so one site cannot compensate for another. Tidal range typically varies by a factor of two between springs and neaps, implying a factor of four in the theoretical resource (again see above) for tidal stream or tidal capture.

Waves

As noted earlier, Gunn and Stock-Williams (2012) estimate a global wave flux to a line 30 nautical miles offshore from a defined coastline of 2.11 ± 0.05 TW. They continue to calculate an extractable resource of 4.6% of that value (i.e. 96.6 ± 1.3 GW) for a chosen wave-energy conversion configuration, and that configuration, as with most practical configurations, is spread widely along the ocean margin, so the reduction in wave energy in inshore waters behind it is never intense.

Leading estimates of the total exploitable wave energy resource for UK include 30.8 GW (Winter 1980) and 43 ± 4 GW (Gunn and Stock-Williams 2012). The former suggested that 7 GW of electricity could be supplied from the 30 GW calculated, but that this would represent rather intensive exploitation. The siting of wave energy devices will be widespread, so there has been fairly limited interest in determining the maximum safe fraction of wave energy that can be extracted. If intense localized exploitation is proposed, however, then finding that safe limit will be important.

Far-greater wave resources are available internationally. Gunn and Stock-Williams subdivided the global coastline by region and demonstrated a large resource in many cases, e.g. for both coasts of Canada and the USA, Chile, Australia and the rest of Europe. EPRI (2011) revisited the US wave resource, also finding large available and recoverable resources, especially for Alaska and the west coast states. Interest in exploiting these resources is now growing, notably in Oregon.

A diverse range of wave-energy extraction technologies is proposed for deployment (according to type) between the shelf edge and the shore. Generally, however, there needs to be a trade-off between maximizing the resource (which will typically be greatest at the shelf edge) and the economic advantage of accessibility (which will generally be better at the coast). Some studies envisage a specific wave-energy conversion device (see Gunn and Stock-Williams 2012), but other studies simply presume that it will be practical to capture the energy. The enormous variability of wave power is not always considered, and accurate calculations require the efficiency of the devices in various sea states to be considered, including evaluating the reliability of the systems and allowing for periods in survival mode for the worst weather or for maintenance or repair.

As described above, there will be a general loss of power from an ocean wave field as the coast is approached. There may be specific locations nearshore with a relatively high wave resource because of the refraction of the waves towards shallower water (e.g. on shoals or neighbouring headlands). Inshore too, most of the remainder of the energy may be lost within surf zones where wave-breaking is intense, so a shoreline energy-capture device can only be effective in specific locations where a reasonable proportion of the power is preserved to the shore. Some estimates of wave-energy resource suggest that the resource is reduced to such an extent nearshore to be of only niche interest. For example, Carbon Trust (2006) estimates UK practical nearshore wave resource at 18 TW-h year^{-1} and shoreline resource at 0.2 TW-h year^{-1}. This perspective has been disputed in other studies (see Folley et al. 2010), however.

The theoretical resource for waves, although substantial, is much lower than the total energy in the tidal system around the British Isles. Also, the global value for wave energy reaching the continental margin (2.1 TW) is notably less than the tidal energy dissipated in marginal seas (2.6 TW). This comparison is somewhat misleading, however, because as discussed above, the tidal energy only becomes readily available in certain cases. Therefore, comparable resource estimates (theoretical, technical or practical) for wave energy are sometimes higher than for tidal stream energy alone, or even for tidal stream and tidal range together. One of the most influential reports (Carbon Trust 2006) estimates the UK practical offshore wave-energy resource at 50 TW-h year^{-1}, far in excess of the technical UK tidal stream resource. Note, however, that as described earlier, Black and Veatch (2011a) revised estimates of the tidal stream technical and practical resources; see also p. 107 of Mackay (2008) for a depiction of several estimates of each for the UK. Therefore, at least in the more-optimistic scenarios, offshore wave energy could make a significant contribution to energy supply in the UK.

Effects of Energy Extraction

There are many potential effects on the environment and ecosystems of all offshore engineering developments. For example, one may displace one habitat but supply a new one, or one may create a collision risk between organisms and the new structures. Here, we limit discussion to considering how extracting energy, and therefore changing the flow, could alter the environment and hence change the pressures on ecosystems. Further consideration of these pressures and the ecosystem response should eventually lead to calculations of a safe bearing capacity and detailed maritime planning, to displace simple calculations of technical and practical resources.

Tides

An important feature of the effect of engineering on tides is that the effect can be felt at a great distance from the engineering work. Introducing an obstacle in a channel will affect the entire current flow through the channel, not just in the wake zone immediately downstream of the obstacle.

Similarly, a tidal barrage in an embayment will generally affect tidal range and currents, both seaward and landward. On a large scale, construction of large tidal barrages can alter basin geometry, causing a shift in the tidal regime in the basin. Hence, for example, construction of barrages in the Dee, Mersey, Morecambe Bay and Solway Firth would significantly perturb the tides in the Irish Sea (Wolf et al. 2009). Therefore, in principle it is possible for a major engineering project to influence the ecology from deeper shelf waters (primarily through influence on stratification and the location of fronts) to intertidal habitats. As tidal capture projects explicitly require notable manipulation of the environment, they tend to attract more environmental concerns. However, alteration of the basic biogeography of shelf seas seems unlikely, whereas effects close to (primarily coastal) actual developments are far more likely.

In the act of extracting energy from a source, one needs to reduce the energy in that source. However, there are a few surprises when one considers taking energy from tides. One notable example is the case of building a barrage across an embayment that is near resonance. For example, Pugh (1987, pp. 174 and 299) predicts that tidal power stations in the Bay of Fundy would be likely to bring the tidal system closer to resonance (for the primary, M2, tidal frequency) raising the tidal range even more. In general, any engineered modification of water depths or boundaries needs to alter the propagation of tides and hence tidal ranges and currents. Pugh (1987, pp. 298/299) states that "Since the most favourable sites are those where large tidal amplitudes are generated by local dynamic resonances, they are particularly vulnerable to imposed changes". The precise level of change (e.g. for the case of a Severn Barrage) requires numerical modelling. In most cases, this is readily achievable, but Pugh (1987, p. 174) notes that in the case of tuning close to resonance, the response may be difficult to predict because of the uncertainties in estimating dissipation, which will be critical close to resonance.

A contrasting example to resonant systems and tidal barrages is the case where it is proposed that energy be taken from the stream in a channel. For that case, calculation of the technical resource, which includes consideration of the effects on current speed and tidal range, has been described above. Here, however, we note a few specific effects of placing devices in the flow not mentioned previously.

The flow in a channel as a whole must be generally reduced by energy extraction, although local effects may be more complicated. In the case of hydraulic flow between two large, multiconnected basins, the sea level and tidal range at either end of the channel will be unaltered, but because there will be a level drop across the devices (typically turbines), sea level needs to be increased upstream of the turbines and decreased downstream. Applying a principle of continuity to the channel, the flow then needs to be stronger where the water is shallower (for a given channel width) and the current speed will be most-strongly reduced upstream of the devices.

Rather more local effects on flow can be modelled numerically or in physical laboratory simulations (Harrison et al. 2010; Myers and Bahaj 2012). For a single turbine in a steady flow, there will be slower flow and a turbulent wake behind the turbine and an enhanced flow outside. Where an array of turbines is placed in a steady stream, the enhanced flow from between each pair of turbines in the front row of an array can be used by staggering the distribution of the turbines in the second row, and so forth. Some device designs also use a ducted turbine, where the flow is first funnelled before entering the turbine to accelerate the flow. In general, any objects associated with a development including buoys, mooring lines and foundations are likely to alter the flow. Modifications of time-averaged current and the intensity of turbulence may be expected, and these will feed through to the resuspension, transport and accumulation of sediment (so altering habitat). The reality of energetic tidal channels is that they will be turbulent before disturbance and that the nature of flow will be far more complicated than encountered in the laboratory or in simple numerical simulations (Lu and Lueck 1999a, b). There is extraordinarily little published on the flow at proposed sites (particularly energetic tidal stream sites), and the nature of such sites requires proper investigation before development. Lu and Lueck (1999a, b) reported on complex features in a fairly energetic tidal channel (Cordova Channel, peak currents exceeding $1\ \mathrm{m\,s^{-1}}$), which included secondary circulation and intense up- and downwelling events, but similar information is rarely available for even more-energetic tidal channels beyond the casual observations of local mariners. Recently, Goddijn-Murphy et al. (2013) showed that the complex flow within a tidal channel can be mapped effectively by a combination of underway surveying and numerical modelling, but the task to map all tidal sites at sufficient resolution is obviously immense.

Sediment-loading in the water column and the sites of sediment accumulation are sensitive to currents, and this may be a significant effect of tidal energy extraction. Some schemes may also have an effect on mean sea level and/or tidal range. The estuarine environment is also greatly influenced by tides, with mixing and stratification usually a competition between density-driven stratification and stirring from the tides. Barrages or lagoons that neighbour estuaries may significantly change such environments. Neill et al. (2009) identified a particular sensitivity of large-scale tidal dynamics to tidal energy extraction in regions of strong tidal asymmetry; greater changes in seabed level are predicted in such cases. Tidal arrays are also expected to affect the dynamics of headland sandbanks (Neill et al. 2012).

Waves

Wave-energy conversion devices can be efficient, so it is possible that an array of devices will greatly reduce wave heights. Large waves will usually propagate from seawards and the effects of wave-energy devices on hydrodynamics are almost entirely restricted to shorewards of the devices. Most wave-energy devices are tuned to extract energy from swell or longer wind waves, which generally represent a greater source of power than shorter waves. A suitable wave model can be tuned to represent the frequency-dependent energy absorption (Smith et al. 2012). Shorewards of the device, the energy (and therefore the height) of the long waves is inevitably reduced. Orbital motions are associated with waves, and these will reduce in proportion to wave height. A reduction in the height of long waves will also reduce the associated stress on the seabed and where sediment suspension was being caused by wave action, this will be diminished. A reduction in wave energy will also generally reduce the amount of wave-breaking and associated turbulence. The exposure to wave breaking and turbulence can be directly important to organisms. Perhaps the most important effect will be on sediment suspension and sediment transport. In particular, longshore transport of material (and therefore the sites where sediment accumulates or disappears) ultimately depends on the size and direction of incoming waves. Therefore, by reducing waves in general and particularly those from a specific direction (i.e. in the wake of the device), the longshore drift of material and ultimately beach morphology and shallow-water bathymetry and substrata will be altered. Note that the effects on all of the above are already caused by natural interannual variation of wave energy and direction (Woolf et al. 2002, 2006) and may be caused by climate change (Harrison and Wallace 2005; Tsimplis et al. 2005; Wolf and Woolf 2006) on a regional basis, so the local effects of a wave-energy development need to be seen in that context.

In addition to the intentional extraction of energy by a device, there will be other effects, broadly common to putting any large solid body in the water. As the designs of wave energy conversion devices are diverse, it is difficult to be more specific, but certainly where there are currents, devices will generate a wake, as for the tidal energy devices discussed above.

In considering the possible influence of wave-energy devices further, it is worth remembering that natural variation and climate change may also have an effect on the waves reaching the shores. The loss of energy by waves from offshore to nearshore results from interaction with the seabed. There is a stress and often a turbulent boundary on the seabed associated with waves and sediment resuspension, and transport can result, but generally the effect of tides is greater. Waves are more important nearer the coast where the associated stress and turbulence will be greater. Where the waves shoal and break the turbulence will be particularly intense. These physical considerations seem to be consistent with empirical experience showing that the impact of wave physics is most manifest near the shore (Burrows et al. 2011).

The Broader Context

It is clear that extraction of (wave or tidal) energy from the seas changes the environment and that there is potential for ecological impact. The changes do need to be seen, however, in the context of alternatives for energy supply (Mackay 2008) as well as other pressures on marine ecosystems, especially in a changing climate (Burrows et al. 2011).

Acknowledgements We acknowledge the support of Highlands and Islands Enterprise, the Scottish Funding Council and the European Regional Development Fund through the Supergen Plus and Marine Renewable Energy and the Environment (MaREE) projects.

References

Black and Veatch Ltd (2011a) UK tidal current resource and economics. Report CTC799, The Carbon Trust, London
Black and Veatch Ltd (2011b) UK tidal current resource and economics—Appendix C. Report CTC802, The Carbon Trust, London
Burrows MT (2012) Influences of wave fetch, tidal flow and ocean colour on subtidal rock communities. Mar Ecol Prog Ser 445:193–207
Burrows MT, Schoeman DS, Buckley LB, Moore P, Poloczanska ES, Brander KM, Brown C et al (2011) The pace of shifting climate in marine and terrestrial ecosystems. Science 334:652–655
Carbon Trust (2006) Future marine energy. Results of the marine energy challenge: cost competitiveness and growth of wave and tidal stream energy. Report to the Carbon Trust, London
Carter DJT (1982) Prediction of wave height and period for a constant wind velocity using the JONSWAP formulae. Ocean Eng 9:17–33
Cartwright DE, Edden AC, Spencer R, Vassie JM (1980) The tides of the north-east Atlantic Ocean. Philos Trans Roy Soc A 298:87–139
Couch SJ, Bryden I (2006) Tidal current energy extraction: hydrodynamic resource characteristics. Proceedings of the Institution of Mechanical Engineers. Part M. J Eng Maritime Environ 220:185–194
Easton MC, Woolf, DK, Bowyer PA (2012) The dynamics of an energetic tidal channel, the Pentland Firth, Scotland. Cont Shelf Res 48:50–60
Egbert GD, Ray RD (2001) Estimates of M-2 tidal energy dissipation from TOPEX/Poseidon altimeter data. J Geophys Res 106:22475–22502
EPRI (2011) Mapping and assessment of the United States ocean wave energy resource. EPRI, Palo Alto, USA. 176 pp
Folley M, Elsaesser B, Whittaker T (2010) Analysis of the wave energy resource at the European Marine Energy Centre. http://www.aquamarinepower.com/resource-library/
Garrett C, Cummins P (2005) The power potential of tidal currents in channels. Proc Roy Soc A 461:2563–2572
Goddijn-Murphy L, Woolf DK, Easton MC (2013) Current patterns in the Inner Sound (Pentland Firth) from underway ADCP data. J Atmos Ocean Technol 30:96–111

Green JAM (2010) Tides and ocean resonance. Ocean Dyn 60:1243–1253
Gunn K, Stock-Williams C (2012) Quantifying the global wave power source. Renew Energy 44:296–304
Harrison ME, Batten WMJ, Myers LE, Bahaj AS (2010) Comparison between CFD simulations and experiments for predicting the far wake of horizontal axis tidal turbines. IET Renew Power Gener 4:613–627
Harrison GP, Wallace AR (2005) Climate sensitivity of wave energy. Renew Energy 30:1801–1817
Holthuijsen LH (2007) Waves in oceanic and coastal waters. Cambridge University Press, Cambridge, 387 pp
IPCC (2011) IPCC special report on renewable energy sources and climate change mitigation. Prepared by Working Group III of the Intergovernmental Panel on Climate Change. In: Edenhofer O, Pichs-Madruga R, Sokona Y, Seyboth K, Matschoss P, Kadner S, Zwickel T et al. Cambridge University Press, Cambridge, 1075 pp
Iyer AS, Couch SJ, Harrison GP, Wallace AR (2013) Variability and phasing of tidal current around the United Kingdom. Renew Energy 51:343–357
Lu Y, Lueck RG (1999a) Using a broadband ADCP in a tidal channel. 1. Mean flow and shear. J Atmospheric Ocean Technol 16:1556–1567
Lu Y, Lueck RG (1999b) Using a broadband ADCP in a tidal channel. 2. Turbulence. J Atmospheric Ocean Technol 16:1568–1579
Mackay DJC (2008) Sustainable energy—without the hot air. UIT Cambridge, Cambridge. ISBN 978-0-9544529-3-3, 384 pp. www.withouthotair.com
Mackay EBL, Bahaj AS, Challenor PG (2010a) Uncertainty in wave energy resource assessment. 1. Historica data. Renew Energy 35:1792–1808
Mackay EBL, Bahaj AS, Challenor PG (2010b) Uncertainty in wave energy resource assessment. 2. Variability and predictability. Renew Energy 35:1809–1819
Mollison D (1986) Wave climate and the wave power resource. In: Evans D, de Falcao AFO (eds) Hydrodynamics of ocean wave-energy utilization, IUTAM Symposium, Lisbon, 1985. Springer, Dordrecht, pp 133–156
Mollison D (1991) The UK wave power resource. In: Reilly JW (ed) Wave energy. Institution of Mechanical Engineering, London, pp 1–6
Munk W, Wunsch C (1998) Abyssal recipes. 2. Energetics of tidal and wind mixing. Deep Sea Res I 45:1977–2010
Myers LE, Bahaj AS (2012) An experimental study of flow effects within 1st-generation marine current energy converter arrays. Renew Energy 37:28–36
Neill SP, Jordan JR, Couch SJ (2012) Impact of tidal energy converter (TEC) arrays on the dynamics of headland sand banks. Renew Energy 37:387–397
Neill SP, Litt EJ, Couch SJ, Davies AG (2009) The impact of tidal stream turbines on large-scale sediment dynamics. Renew Energy 34 2803–2812
Nielsen P (1992) Coastal bottom boundary layers and sediment transport. Advanced Series on Ocean Engineering, 4. World Scientific, Singapore, 324 pp
Pugh DT (1987) Tides, surges and mean sea level. Wiley, New York. ISBN 0 471 91505 X
Robinson IS (1979) The tidal dynamics of the Irish and Celtic Seas. Geophys J Roy Astron Soc 56:159–197
Royal Commission on Environmental Pollution (2000) Energy—the changing climate. Report to Parliament, CM4749, HM Stationery Office, London
Scott BE, Sharples J, Ross O, Wang J, Pierce GJ, Camphuysen CJ (2010) Sub-surface hotspots in shallow seas: fine scale limited locations of marine top predator foraging habitat indicated by tidal mixing and sub-surface chlorophyll. Mar Ecol Prog Ser 408:207–226
Shields MA, Dillon LJ, Woolf DK, Ford AT (2009) Strategic priorities for assessing the ecological impact of marine renewable devices in the Pentland Firth (Scotland, UK). Mar Policy 33:635–642
Shields MA, Woolf DK, Grist EPM, Kerr SA, Jackson AC, Harris RE, Bell MC et al (2011) Marine renewable energy: the ecological implications of altering the hydrodynamics of the marine environment. Ocean Coast Manag 54:2–9
Simpson JH, Allen CM, Norris NCG (1978) Fronts on continental-shelf. J Geophys Res 83:4607–4614
Simpson JH, Hunter JR (1974) Fronts in Irish sea. Nature 250:404–406
Simpson JH, Sharples J (2012) Introduction to the physical and biological oceanography of shelf seas. Cambridge University Press, Cambridge. ISBN 978-0-521-70148-8
Smith HCM, Pearce C, Millar DL (2012) Further analysis of change in nearshore wave climate due to an offshore wave farm: an enhanced case study for the Wave Hub site. Renew Energy 40:51–64
Taylor GI (1920) Tidal friction in the Irish Sea. Proc Roy Soc London A 220:1–33
Tsimplis MN, Woolf DK, Osborn TJ, Wakelin S, Wolf J, Flather R, Shaw AGP et al (2005) Towards a vulnerability assessment of the UK and northern European coasts: the role of regional climate variability. Philos Trans A Math Phys Eng Sci 363:1329–1358
Vennell R (2012) The energetics of large tidal turbine arrays. Renew Energy 48:210–219
Winter AJB (1980) The UK wave energy resource. Nature 287:826–828
Wolf J, Walkington IA, Holt J, Burrows R (2009) Environmental impacts of tidal power schemes. Proc Inst Civil Eng Marit Eng 162:165–177
Wolf J, Woolf DK (2006) Waves and climate change in the north-east Atlantic. Geophys Res Lett 33:L06604. doi:10.1029/2005GL025113
Woolf DK, Challenor PG, Cotton PD (2002) Variability and predictability of the North Atlantic wave climate. J Geophys Res 107:3145. doi:10.1029/2001JC001124
Woolf DK, Cotton PD, Challenor PG (2003) Measurements of the offshore wave climate around the British Isles by satellite altimeter. Philos Trans A Math Phys Eng Sci 361:27–31
Woolf DK, Gommenginger C, Sykes N, Srokosz MA, Challenor PG (2006) Satellite and other long-term data sets on wave climate; application to the North Atlantic region. Full Proceedings of World Renewable Energy Congress IX, MT11. Elsevier, Amsterdam. ISBN 008 44671 X
Yates N, Walkington I, Burrows R, Wolf J (2013) Appraising the extractable tidal energy resource of the UK's western coastal waters. Philos Trans A Math Phys Eng Sci 371. doi:10.1098/rsta.2012.0181

Baselines and Monitoring Methods for Detecting Impacts of Hydrodynamic Energy Extraction on Intertidal Communities of Rocky Shores

3

Andrew Want, Robert A. Beharie, Michael C. Bell and Jon C. Side

Abstract

As part of the UK government's objective to deliver an increasing proportion of electricity from renewable sources, West Mainland, Orkney, is at the forefront of the development of wave-energy extraction devices. Exposure to wave energy plays a dominant role in shaping the Orkney landscape and determining the ecological community, but little is known of the consequences of commercial scale removal of energy from the environment. An extensive long-term monitoring programme to assess the impacts of altering wave-energy exposure on these rocky shores alongside responses to other systemic forcing agents such as climate change is continuing. Within the programme are photographic surveys, including quadrat and fixed viewpoint techniques, littoral studies of sentinel species, and the development of cost-effective wave-energy quantifying devices. Software has been developed to analyse images efficiently, to produce quantitative data on species and biotope coverage. Additionally, extensive surveys along the shoreline provide detailed image records, including areas without prior scientific description, and have helped identify locations of environmental sensitivity. Collectively, the data provide a comprehensive pre-development baseline along this important coast.

Keywords

Climate change · Ecological monitoring · Environmental variables · Marine renewables · Rocky shoreline · Wave energy

Introduction

Despite growing interest in extracting energy from waves and tides, installations of marine-energy-converting devices are limited, especially on a commercial scale, and there remains a paucity of studies addressing the potential environmental and ecological consequences of wave and tidal energy developments (Frid et al. 2012). A potentially important locus for responses to wave-energy extraction is the shoreline, so focus here is on monitoring methods for describing baseline conditions and detecting biological responses against a background of other potential changes.

A. Want (✉) · R. A. Beharie · M. C. Bell · J. C. Side
International Centre for Island Technology (ICIT),
Heriot-Watt University, Stromness,
Orkney, KW16 3AW, Scotland, UK
e-mail: a.want@hw.ac.uk

The potential ecological consequences of deploying wave-energy converting devices (WECs) can be divided broadly into impacts and interactions which might arise at the location of a device and resulting from the action or presence of the device and effects downstream of the device as a consequence of energy extraction (Lohse et al. 2008). The former are described elsewhere in this volume and might include physical impacts on the seabed during construction, maintenance and decommissioning, risks of collision and disturbance for marine vertebrates, changes in biodiversity attributable to the presence of new structures, and changes in sediment suspension and deposition. Downstream effects from the deployment of WECs, especially in large "farms", would be expected to stem from changes in wave climate, principally reduced energy levels reaching the shore. It is relevant to ask whether or not this would be sufficient to produce observable changes in the littoral or sublittoral communities.

M. A. Shields, A. I. L. Payne (eds.), *Marine Renewable Energy Technology and Environmental Interactions,* Humanity and the Sea,
DOI 10.1007/978-94-017-8002-5_3,

The consequences of reduced hydrokinetic energy reaching shorelines and nearshore environments are not well understood, but it is generally considered that observable changes following reduced exposure to wave energy can be expected in the intertidal zone (Shields et al. 2011). This is not to suggest that wave energy dissipates only in the surf zone or that only intertidal organisms are likely to be affected. In fact, prior to breaking on the shore, waves begin inducing flows along the seabed, with possible consequences for sublittoral organisms at depths of up to 75 m (Denny 1987); other possibilities for interaction between marine-energy extraction and the sublittoral benthos at the boundary layer are described by Shields et al. (2011). However, focus in this chapter is on detecting responses in the intertidal zone that may be indicative of wider ecological changes.

In terms of monitoring studies, intertidal areas have several advantages over other marine zones, including ease of access from land and the use of relatively inexpensive methodologies for direct observations. Intertidal communities of rocky shores also have the precedent of extensive studies without the major stressor of commercial fisheries exploitation seen in many other marine environments (Broitman et al. 2008). Further, the areas are characterized by the presence of eurythermic species (that have a wide range of temperature tolerance) for which one might expect population responses more closely related to changes in hydrokinetic energy than to direct temperature fluctuations, otherwise an important ecological factor. This is a key consideration in relation to the role of other environmental forcing agents such as concurrent global climatic change.

The biota of intertidal areas already experience considerable seasonal fluctuations (owing to cycles in temperature, light, nutrients, etc.), and longer term trends linked to the local effects of large-scale climatic variables such as the North Atlantic Oscillation (NAO) may play some role in community changes (Hiscock et al. 2001; Broitman et al. 2008). Of particular concern for designing monitoring strategies are changes in the global marine climate, which are manifested most obviously in increased sea temperature, but include such other important considerations as sea level increases and increased winter storminess. Recent predictions of global sea temperature changes suggest a 2 °C increase by 2100 (Solomon et al. 2007). Several studies have provided evidence of trends in seabed populations being influenced by climate change (e.g. Barry et al. 1995; Mieszkowska et al. 2006). Long-term monitoring is essential to understanding community dynamics, notably the relative roles of localized wave-energy removal and systemic climatic changes.

The rocky shoreline of West Mainland, Orkney, includes part of a National Scenic Area and is characterized by dramatic sandstone cliffs, complex geomorphological features including sea stacks and caves, plus a few embayments. With a westerly fetch of > 3000 km, wave energy plays a dominant role in shaping the landscape and determining the biological community, in this case creating an important assemblage of diverse organisms adapted to an extreme energy environment. The exposure to wave action striking the shore is a major determinant of species composition, growth rates, reproductive success and other aspects of life histories (Lewis 1964). Monitoring indices of these variables can provide valuable data in assessing changes at an individual as well as a community level. It is along this rocky shoreline that the testing and early commercial development of wave-energy extraction is taking place.

This chapter describes the approach taken by the team at the International Centre for Island Technology (ICIT), the Stromness, Orkney-based campus of Heriot–Watt University, in developing a monitoring programme for detecting biological responses to changes in wave energy reaching rocky shores in an area where large-scale deployment of WECs is anticipated in the near future. Efforts are also made to distinguish between the relative roles played by different environmental variables such as energy extraction and climate change. The aim is to provide comprehensive pre-impact baseline data for the monitored area and at the same time to address common issues allowing the methodologies to be transferable to different shores with the same or different species.

The use of multiple indicators to establish pre-development background conditions and to monitor the impacts of wave-energy extraction on the rocky shores of West Mainland, Orkney, includes:

- quantitative assessment of localized wave-energy exposure at an ecologically meaningful scale;
- characterization of biotopes, including difficult to access cliff bases and skerries that can be used to identify potentially environmentally sensitive areas;
- image analysis to quantify species cover and spatial zonation;
- selection of candidate sentinel species and the use of a paired species protocol for long-term monitoring of biological responses to environmental change;
- choice of multiple species pairs to differentiate the effects of wave-energy extraction from other forcing agents such as global climate change.

Rocky Shore Biotopes Characteristic of Exposed Conditions

Wave energy is one of the key factors determining marine species distributions and community structure along shorelines. Typical species within the intertidal zone of high energy shores on the Atlantic coast of Europe include barnacles (*Chthamalus* spp. and *Semibalanus* spp.), limpets (*Patella* spp.), small red algae (*Mastocarpus* spp. and *Palmaria* spp.), encrusting coralline algae and high-energy variant fucoids. The lower intertidal fringe often features encrusting red algae and *Alaria esculenta* in the most energetic sites and *Himanthalia elongata* at moderate energy sites. The infralittoral (immediately below the low tide line) is dominated by the kelps *Laminaria digitata* and *L. hyperborea*, with epiphytic algae such as *Palmaria palmata* growing on their stipes. This vertical distribution of organisms is correlated with energy exposure (Lewis 1964) upper littoral zones being more closely linked to wave energy than lower zones (Thomas 1986). Although the organisms mentioned above are generally characteristic of northeastern Atlantic rocky shores of similar, high-exposure levels, community differences are increasingly apparent along a latitudinal axis. Rocky shores generally have good water clarity because of the low levels of suspended particulate matter in the water column or fine sediment close to shore. This allows light to reach kelps and red algae, permitting them to survive at greater depths than near shores dominated by sediment outflow. In certain areas, such as Scotland's Solway Firth and Clyde Sea, higher levels of suspended sediment or phytoplankton can reduce clarity. The break-up of organic matter and the subsequent undercurrents created by wave action can favour benthic filter-feeders such as sponges and molluscs.

The term habitat is generally used to describe an environment in which a particular species or community lives, and biotope a particular combination of biological assemblage and physical variables, although in practice the two words are often used interchangeably. Marine habitat and biotope classification systems allow the categorization of areas based on physical characteristics (e.g. sediment type), environmental conditions (e.g. exposure to waves or currents) and species characteristics. Important early studies included the essential contributions of zonation characterization (Stephenson and Stephenson 1949) and the development of abundance scales for indicator species (Crisp and Southward 1958). Subsequent developments by Ballantine (1961) led to the assignment of key assemblages to an arbitrarily defined series of subjective energy levels. These concepts have since been refined further and recently the Marine Habitat Classification for Britain and Ireland (MHCBI) produced as part of the Marine Nature Conservation Review (MNCR) details of biotopes found within the British Isles (Connor et al. 2004), including those associated with approximated energy levels, and provide the basis for the European Nature Information System (EUNIS), part of the biodiversity data centre. The latter contains additional biotopes found within European Union countries (EEA 2008) and underpins the European Commission's NATURA2000 protected habitats directive (EC 2012).

The marine sections of the MNCR and the EUNIS habitat system broadly group habitats within three qualitative categories (levels) of high, medium and low energy. These levels form a hierarchical base for eight subdivisions of exposure level, decreasing in energy level from extremely exposed to ultra-sheltered; these in turn are based upon the three variables of fetch, bathymetry and aspect. Fetch is the linear unbroken distance of water from a particular coastal location, bathymetry the profile of water depth over the fetch, and aspect the directional orientation of the shoreline to prevailing winds. The MNCR exposure levels of extremely exposed, very exposed and exposed fall under the high energy hierarchy, moderately exposed is under medium energy, and sheltered, very sheltered, extremely sheltered and ultra-sheltered are placed in the low energy category. Within each of these arbitrary energy levels, individual biotopes may be expected to be spatially interchangeable, albeit limited by abiotic physical factors such as aspect, but it has not yet been investigated whether these communities can be established as comparable through using measured wave-energy data.

There are six levels of water stream/current classes included within the three main energy classifications in EUNIS, which identify biotopes associated with impacts from moving sediment. These sublittoral biotopes would generally be near sandy coves on rocky coastlines or within tidal channels between islands, and they feature more opportunistic algae such as *Saccorhiza polyschides* and *Chorda filum*. Further details on species within high-energy biotopes can be found online on the MNCR and the EUNIS databases (http://jncc.defra.gov.uk/page-1596; http://eunis.eea.europa.eu/index.jsp).

Methods of Measuring Exposure

The earliest approaches to deriving biologically meaningful indices of shore exposure to wave action were based simply on measuring fetch from charts and maps (Baardseth 1970). Later improvements involved using wind data to weight the fetch in different directions from a site (Sjøtun et al. 1998; Burrows et al. 2008). Such methods remain the basis for habitat classification systems and are useful in characterizing exposure and predicting habitat for some species in fetch-limited locations (Bekkby et al. 2009), where local wind conditions and aspect are important determinants of wave action. However, they do not account for ocean swell dominating the wave regime at more open locations (Westerbom and Jattu 2006), the typical location where commercial scale WEC arrays are designed to operate. Coastal areas bordering large bodies of deep water are subject to prevailing long-frequency swell waves that propagate high energy over great distances. Although they have greater stability in average energy levels, their direction usually bears little correlation with local and intermittent wind climate (EMEC 2006), in some locations not even displaying any correlation with prevailing winds on eastern UK shores (Angus Council 2010).

Direct measurements can be made of offshore wave climate using modern electronic technology such as satellite altimeter data together with weather hindcasts. This has been used to produce a detailed worldwide average wave-climate atlas, but lacks the resolution needed for specific shoreline site assessments. More accurately, spherical wave buoys and acoustic Doppler current profilers (ADCPs) have been used to measure offshore and nearshore wave fields, allowing direct measurement at renewable energy sites and providing data for wave-propagation models. Both these technologies have serious problems when used in the littoral to sublittoral fringe, mainly because breaking waves create excessive buoy mooring line forces or entrained bubbles, which result in acoustic opacity for ADCPs. These problems can lead to erroneous or missing data during high-energy events. Moreover, both these *in situ* devices can create limitations to projects because they are expensive to buy and install. The high probability of total loss or severe damage by debris impacts in the littoral zone can quickly become uneconomical for continuous data acquisition over prolonged periods.

Inexpensive devices for the measurement of littoral hydrodynamics were created for intertidal studies in the 1960s, providing a method for measuring maximum wave forces (Jones and Demetropoulos 1968). These are based on the spring extensional force (Denny 1983; Fuji 1988; Bell and Denny 1994; Castilla et al. 1998) designed to record maximum water velocity (the intensity of wave force), to determine the link between hydrodynamics and survivorship, mechanical strength and distribution of particular species. One study using these maximum flowmeters showed that it was possible to correlate wave-action measurements with offshore significant wave height (Denny 1995). Some have used the dissolution of plaster blocks to estimate average water flow and were used as recently as 2005 (Lindegarth and Gamfeldt 2005), although they do suffer from limitations of time-consuming methodologies and water temperature/flow regimes affecting the rates of dissolution. The key drawback to all these devices is their susceptibility to impact damage and erroneous data induced by coarse sediment and flotsam. The problems described above led directly to the development of the Terobuoy device described below.

Measuring Exposure at Ecologically Meaningful Scales

The descriptive term wave exposure is historically and currently widely used to describe the forcing hydrodynamic stress mechanism that can have a modifying effect on the form and abundance of species within the littoral to sublittoral zones. If we presume that this term is a direct alternative for wave action, then exposure varies both spatially and temporally and is not characterized simply by a site's quantifiable openness to a fixed level of wave action. Measurements of fine-scale *in situ* wave forces and directions are important for reliably determining how wave energy can influence community structure and are essential for predicting how they may be changed by the future installation of WEC arrays. When trying to predict any effect, it is important to take WEC technology into consideration; alteration to a wave regime downstream of buoyant arrays will be produced by the operational characteristics of all the devices that make up the array. It is not correct to assume that wave energy is permanently reduced to a limited or constrained level (analogous to fetch reduction) for the duration of a development's presence. In the example of WECs designed to minimize interactions during high-energy storms, allowing them to survive the high stresses imposed, this will allow the usual winter storm waves to reach the shore with minimal anthropogenic attenuation. This will effectively increase the difference in shoreline wave action between lowered energy in summer and unaffected high energy in winter. To monitor these seasonal changes in wave action effectively requires long-term measurements that can be difficult to make using maximum flow devices.

At this stage it is not computationally feasible to use hydrodynamic models to assess the wave-induced forces acting at the spatial scale of individual rocky shore biotopes, even if bathymetric data were able to populate these models at sufficient resolution or accuracy. A quantitative value of wave action associated with each classified biotope, capable of being measured in the field, would allow prediction of how biotic assemblages at certain locations might respond to changes in wave energy. We describe here the development of a new cost-effective new device that can support such measurements.

The Terobuoy wave-action gauge (Fig. 3.1) is designed to quantify both the level of wave action and its directional component and can survive in the harsh high-energy rocky shore environment where entrained sediment will damage and interfere with other types of measurement equipment. The gauge consists of a bracket (fixed to the substratum) retaining a black, high-density polyethylene block combined with an abrading ring (placed around the polymer block) attached via a length of rope to an expanded ethylene vinyl acetate float.

The length of the ring-float assembly allows for maximum interaction of water motion at various states of tide while preventing possible rope entanglement. Measurement by the device is during the interaction of water within the surf zone and when the buoyant float is providing a lift force of the metal ring against the polymer block. The float moves in direct response to water motion removing material from the block in a controlled manner, so the quantity of eroded material lost from the sacrificial polymer block becomes a function of the total hydrodynamic energy of the water acting upon it over a given period. Most material is lost from

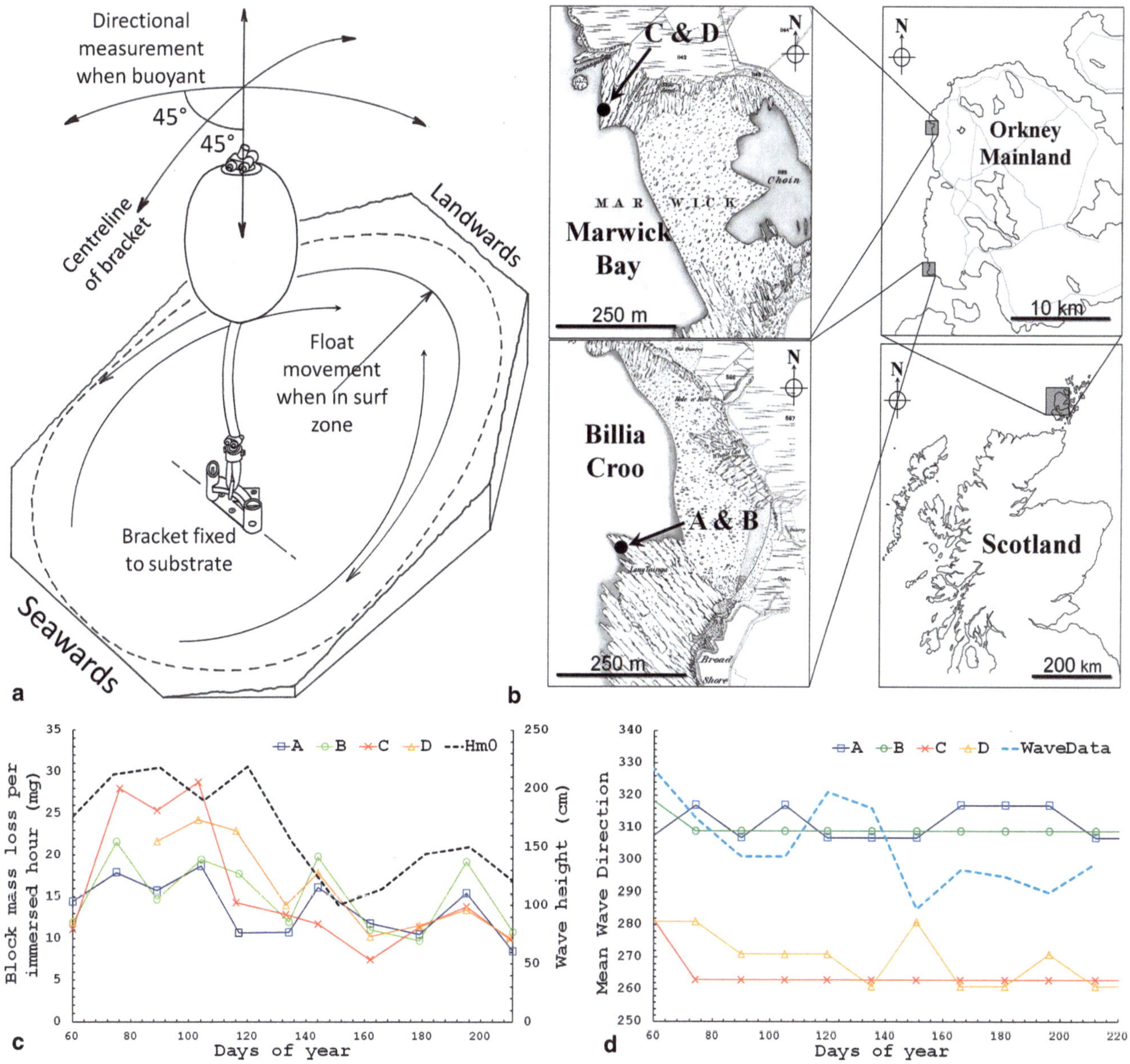

Fig. 3.1 **a** Diagram of the movement of a Terobuoy unit when in the surf zone and buoyant attitude when submerged. **b** Wave-action measurements were made at several sites including Marwick Head and Billia Croo on West Mainland, Orkney. **c** Wave-action measurements, 1 March to 12 August 2010, given in mass loss per immersed hour showing three discrete energetic events over this period. At higher levels of offshore *Hm*0, a significant difference is found between the shoreline wave action at the two sites. **d** The mean direction of waves impacting the test sites is stable over a range of offshore wave directions

the block on its curved lower forward edge; this forms a 90° baseline arc of measurement to establish the directional component of the wave regime. Measurements are taken using digital callipers of the linear thickness between front lower edge and back upper edge at 10° directional increments on each block prior to deployment and after retrieval. In comparison with other measurement devices, the Terobuoy unit measures a cumulative level of wave energy over the entire sacrificial block installation period as opposed to flowmeters measuring a single maximal value between site visits. Depending on the wave energy to which the unit is subjected, or the monitoring strategy, data intervals can be from one week to several months, with the removal and replacement of each block requiring ~2 min. Subsequent examination of the block can be performed in the safe environment of an office or laboratory. The reduced time needed for each field visit is potentially a benefit, especially to long-term monitoring studies and those over a wide geographic range. Terobuoy units have been deployed at several sites in Orkney, encompassing a range of exposure levels.

Data from initial deployments at two, high-exposure survey sites 14 km apart on the west coast of Orkney have produced measurements correlated with both significant wave height (*Hm*0) and direction from concurrent wave buoy data (Fig. 3.1). Results at one of the sites indicate that wave directionality is enhanced by bathymetric features such as deep water close to the shore and in line with mean wave direction, demonstrating that both bathymetric and topographic features are an important consideration when modelling shoreline wave-energy propagation. Replicate units at these two monitored sites show comparable energy levels during summer as indicated by similar mass losses; during the stormier conditions experienced through winter, total mass lost is much greater. Interestingly, there is an approximately twofold difference in mass loss between these two sites during winter. That these two sites, despite their seasonal difference in received energy levels and dissimilar species assemblages, are currently classified as equivalent according to both MNCR and EUNIS reflects the lack of discriminatory power within the classification systems rather than that the difference lacks ecological significance.

Biotope Mapping of Exposed Intertidal Areas

The littoral environment of West Mainland, Orkney, has been the subject of a number of key studies, some examining individual species (Powell 1963; Baxter 1983); and others looking at broader groups of organisms (Wilkinson 1975; Baxter et al. 1985; Wells et al. 2003). In the mid- to late 1990s, thorough characterization of marine biotopes in Orkney was carried out as part of the Marine Nature Conservation Review (MNCR) programme (Murray et al. 1999). The review included a large portion of southwest Mainland, specifically high-energy coastline from south of Billia Croo (prior to its development as the European Marine Energy Centre's wave-test site) north to the Bay of Skaill. In all, seven littoral sites were surveyed north of Billia Croo, along the coast where commercial scale WEC deployment is expected. Communities are defined in these surveys using the MNCR marine biotope classification system described above.

Ongoing work by ICIT is continuing to characterize this shoreline in greater detail, extending the surveying to include the entire region within the West Mainland leasing sites defined by the Crown Estate, i. e. to continue north from Skaill Bay to Costa Head. Where possible, detailed examination is made at low spring tides from sites accessible by land, but most of this coastline is accessible only by boat. Boat-based observations and photographic records have been complemented with frequent landings from a rigid-hulled inflatable boat to allow more complete, direct sampling. In including these areas, we have been able to provide the first comprehensive description of biotopes along a large portion of the coast. Continued studies of selected sites along the shore will allow monitoring of the areas expected to be developed for commercial scale removal of wave energy, as well as of similar areas at a considerable distance from the developments. In producing a comprehensive survey along an extensive coast, important background data have been gleaned for both control and impact study sites.

There is general homogeneity of biological communities along large areas of the shores of West Mainland, characterized by organisms adapted to extreme wave-energy exposure. The most common littoral biotopes described feature *Alaria esculenta*, *Chthamalus stellatus*, *Corallina officinalis*, *Mastocarpus stellatus* and *Mytilus edulis*, with the shallowest edge of an extensive *Laminaria hyperborea* forest exposed at low tide. There is an important geomorphological difference in the highest energy areas between essentially vertical cliff faces and the wave-swept platforms at the bases of some cliffs. On average, these platforms dip to the west, towards the sea, at ~ 12°. Importantly for the monitoring programme, wave-swept platforms, which extend sufficiently to allow emersion at mid-tide, are frequently home to stands of the rare boreal seaweed *Fucus distichus anceps*. Spatial assessment and evaluation of that species' environment complements sentinel species monitoring of it described below. Presumably, differences observed between certain littoral organisms on vertical vs. more-horizontal substrata are in part attributable to contrasts in draining patterns, local sediment transport, predation and exposure to differing directionality of hydrodynamic forces. Burrows et al. (2008) showed horizontal shore extension to be a crucial factor in models of fetch-based community prediction.

Within these areas of extreme wave energy along the coast are less extensive habitats created by caves and small rocky inlets (known locally as geos), as well as a few larger embayments, which allow significant sediment deposition. In most cases, the reduction in wave energy created by these geological features appears to be the major factor in determining community composition.

A library now exists of digital images, species identification records and geomorphological measurements, all georeferenced to a precision of 3 m. This will be merged into a geographic information system (GIS) along with future sublittoral and bathymetric data records. Collectively, this record of pre-development conditions can be used to identify sites of greater environmental sensitivities, which may inform marine spatial planning and consenting processes. Fixed view-point photography is another important strategy for seasonal and long-term monitoring (Moore 2001). Once suitable sites have been identified, bedrock is incised with an identifiable mark, and camera height and lens length is recorded to provide a reliable means of capturing images from an identical perspective on subsequent visits. Fixed view-point photography is being used to monitor seasonal and long-term changes in mid-littoral macroalgae, with particular interest in *F. distichus anceps*.

Image Analysis for Quantifying Biotopes

Essential to the biotope monitoring programme is the capability to make quantitative measures of baseline biotope coverage. The approach followed in Orkney has been based on semi-automated analysis of photographic images (Fig. 3.2). In brief, the procedure

i. identifies an area of the image within which the biotopes are to be quantified;
ii. transforms the image to a bird's-eye view, whereby the location of each pixel in the image is mapped and scaled to actual shore coordinates using a transformation matrix estimated from known reference points within the image;
iii. uses principal components analysis (Digby and Kempton 1987) to pre-process the red, green and blue components of pixel colours in the transformed image to up-weight the more subtle aspects of colour variation that may be more important than overall light intensity in discriminating biotopes;
iv. allocates pixels to squares of a grid imposed on the image, each of which is to be assigned to a biotope type according to the (pre-processed) colour composition of its component pixels;
v. manually selects a sample of grid squares representative of each biotope known to be present, the "supervised" element of the classification;
vi. applies canonical variates analysis (Digby and Kempton 1987) to the (pre-processed) colour composition data of the manually sampled grid squares, deriving weighting factors for allocating other grid squares to the most closely matching biotope in terms of colour composition;
vii. back-transforms the classified grid squares on to the original image.

The final step here is used only for the purpose of displaying the classification outcome (bottom panel in Fig. 3.2). The cover of each biotope is estimated by counting the classified grid squares at step (vi).

Experience using software developed in-house by ICIT has shown that biotopes can be separated and quantified successfully in photographic images using this method. The approach is suitable for fast, cost-efficient photographic surveys of difficult-to-access areas such as cliff bases, intertidal reefs and skerries (provided preliminary observations are available to inform species and biotope identification). The method is being used to establish baseline data for long-term monitoring along West Mainland, Orkney, including the analysis of images recorded during boat-based comprehensive assessment of the littoral habitat along this extremely exposed coastline. Similar approaches to habitat and biotope mapping have been applied using aerial photographs, satellite imagery and other remote sensing data in comparable marine environments (e.g. Mumby et al. 1997). Work is ongoing at ICIT to determine the repeatability and comparability of biotope discrimination between and within locations and the extent to which the approach is able to detect short- and long-term changes.

Biological Responses to Environmental Change

Changes in the abundance of a species in relation to a gradient of environmental conditions are often envisaged as taking the form of a bell-shaped curve. According to this simple model, abundance peaks at an optimum on the environmental gradient that represents the conditions that are most suitable for the species. The whole ecological niche of a species is defined in terms of responses to many such environmental gradients. In reality, these multiple gradients can be complex and interactive in their effects, but the simple idea that environmental optima exist has important implications for the design of monitoring programmes. It means that, depending on starting conditions, responses may be either an increase or a decrease in abundance with any given direction of change in environmental conditions. If conditions are close to the optimum for a species at a given location, any change in environment will likely lead to an eventual decline in its abundance at that location. Unless the environmental preferences of a species are known precisely, however, and unless the baseline environmental conditions can be measured with equal precision, it is hazardous to base monitoring programmes on the responses of individual species.

Measuring changes in the composition of assemblages of multiple species is often used in benthic impact studies, but this type of multivariate response is inherently variable and often difficult to interpret in terms of causality. Our approach to measuring biological responses to changes in wave energy acting on rocky shores is instead to consider carefully chosen pairs of species, where each sentinel species (see below) is matched with a second similar species with known differences in environmental preferences. Knowing that one species differs from another one in terms of its occurrence in relation to wave exposure (or other environmental variable) is much less challenging than having to identify how conditions at a site match up with the precise preferences of a single species.

The use of species pairs, and of indicator species more generally, to index changes in marine environments has a long pedigree. Russell (1935), for example, demonstrated that changes in the proportions of two chaetognath species in plankton samples from the western English Channel were related to large-scale patterns of movement of oceanic and shelf water bodies. More recently, Southward et al. (1995) related long-term changes in the proportion of warm water *Chthamalus* species in intertidal barnacle populations at a site near Plymouth (a barnacle index) to changes in water temperature. Species pairs may be of greater utility

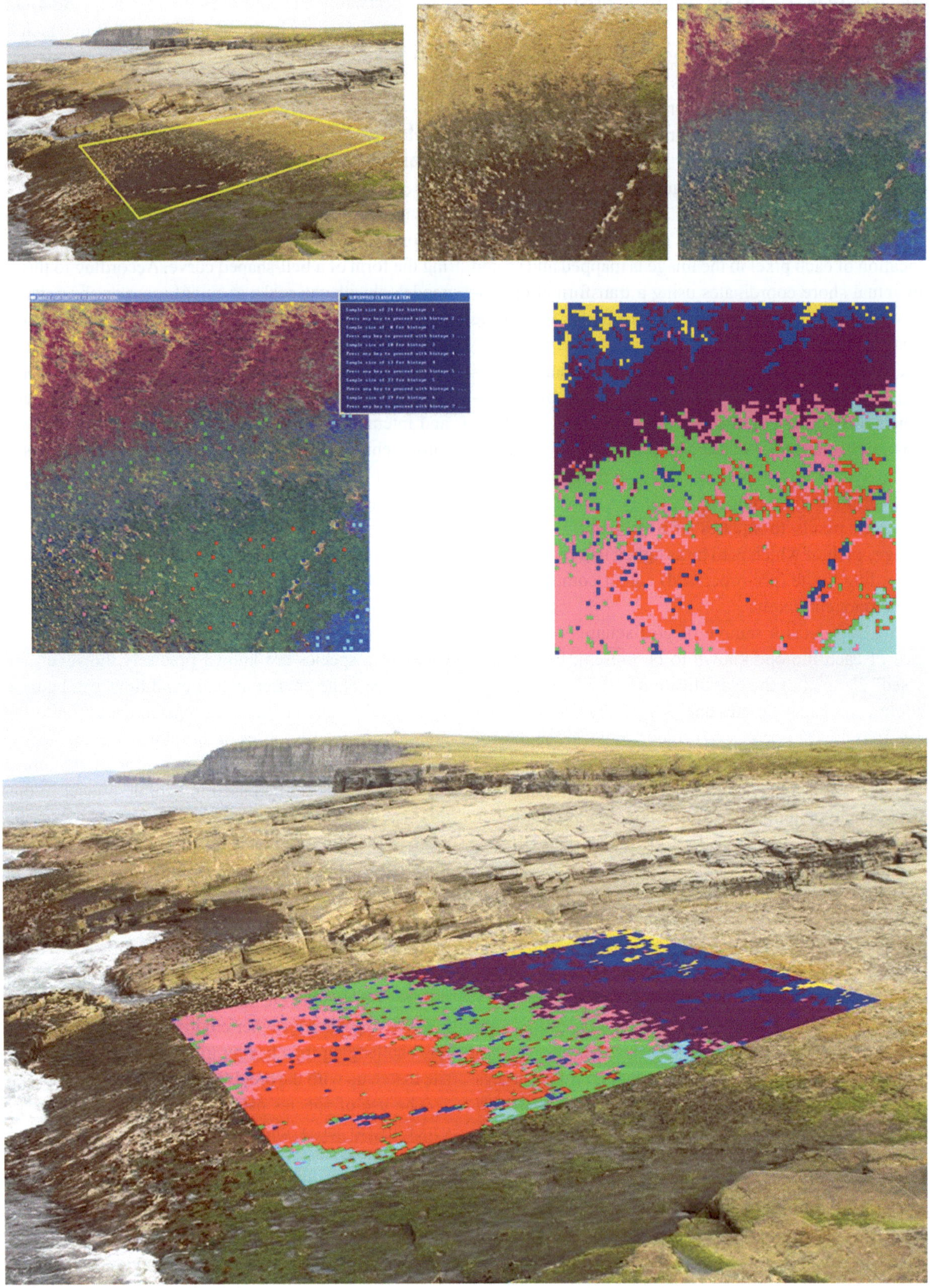

Fig. 3.2 Fixed viewpoint photographic monitoring of *Fucus distichus anceps* and other midlittoral macroalgae at Northside using semi-automated image-analysis software

than single species in designing effective monitoring programmes, and as shown in Fig. 3.3a, it is not hard to see why. Although the abundances of individual species are likely to show unimodal patterns of variation in relation to a given gradient of environmental change such as exposure to wave energy, i. e. over different ranges of the environmental variable they show increases as well as decreases for any given direction of change, the proportion of each species in a pair is much more likely to change in one direction only. All that is necessary to interpret a change in species proportion is to know that one species differs from another in its preference (or tolerance) for a given environmental variable.

The use of paired species also offers at least two more advantages over single-species monitoring protocols. First, measuring species proportions does not depend on determining absolute densities, i. e. numbers per unit of habitat area. Quantifying habitat areas is a notoriously difficult task in structurally complex rocky shore environments. All that is required for estimating meaningful proportions is that the sampling efficiencies are comparable between the two species, or are at least stable in relation to changes in the environmental variable of interest. This is a reasonable assumption for pairs of species that are similar in conspicuousness (e.g. limpet species). The second advantage is less intuitive, but it emerges from computer simulations of the statistical sampling properties of paired and single-species monitoring protocols. The advantage is that the statistical power to detect changes and correctly identify their direction is more consistently high across the range of environmental conditions for species proportions than for single-species abundance. If, as is likely for closely related species, there are features of the environment (other than the variable of interest) that favour both species rather than one at the expense of the other, the statistical power of the proportional variable to detect change will increase further.

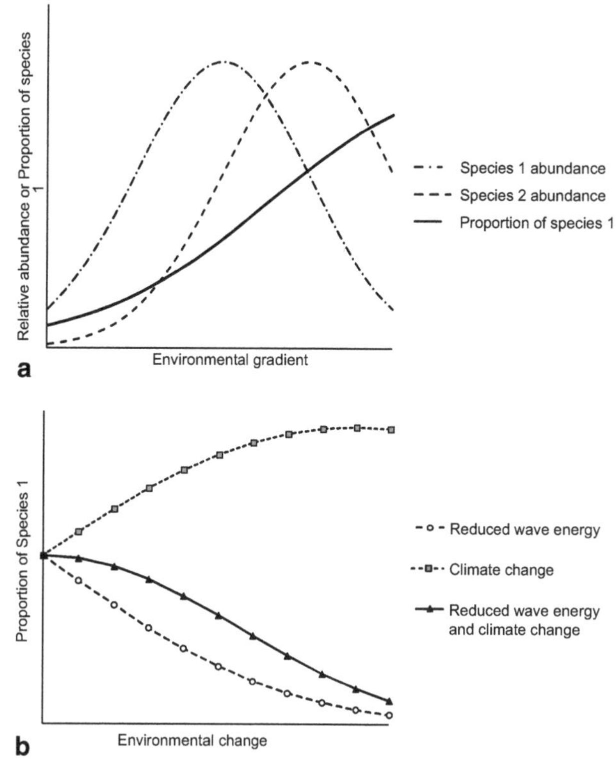

Fig. 3.3 Use of paired species monitoring to detect biological responses to environmental change: **a** abundance of individual species and proportion of one species in relation to an environmental gradient. **b** modelled response of species proportions to environmental changes, when species 1 is disadvantaged by reduced wave energy and favoured by climate change

Detecting Responses Against a Background of Other Environmental Changes

In designing a monitoring programme to detect biological responses to wave-energy extraction, it is important to take cognizance of the fact that reduced exposure is probably only one among many ongoing environmental changes that might influence life on rocky shores. Climate change has already been highlighted as an important driver of change for marine organisms; depending on the species, this has the potential either to augment the expected direction of response to energy extraction, making it more difficult to assign causality, or else to dampen, cancel out or even reverse the expected response. Without accounting for concurrent climate change, there is a real risk of drawing wrong conclusions from monitoring outcomes, either concluding that energy extraction has had no effects, when climate change has obscured or compensated the response, or wrongly ascribing changes to the effects of energy extraction when climate change has been the real forcing agent.

The potential for multiple and confounding influences on response variables does not in itself invalidate the use of sentinel species and paired species protocols to monitor the biological consequences of energy extraction. However, it does mean that care must be taken in the selection of species for monitoring to ensure that the contributions of different forcing agents are clearly separable. The approach taken here is to consider several pairs of species, differing in their expected magnitude and directions of response to climate change and reduced exposure to energy. Figure 3.3b illustrates notional responses for a species pair in which species 1 is expected to be disadvantaged by reduced wave energy and favoured by climate change. In this example, concurrent climate change would make it more difficult to detect responses to reduced wave energy. Other examples could have been constructed showing very different patterns of response, depending on the location of optimum environmental conditions for each species in relation to baseline conditions. By careful selection of a suite of sentinel species, the capability to detect biological responses to wave-energy extraction is provided by the differential patterns of response expected under any given scenario of environmental change.

The Selection of Sentinel Species

Important considerations for selecting sentinel species are listed below.

i. Species favouring more sheltered conditions will benefit from wave extraction, whereas species adapted to higher energy environments will be adversely affected.
ii. The rate of response to environmental change will vary between species, depending on factors such as mortality, mobility and reproductive strategy. Observable changes will be first apparent in species most capable of dispersal. Among benthic species, the greatest response will be in those with a planktonic stage in their life history (Hiscock et al. 2001).
iii. With increased sea temperature, southern species will move north providing there are no biogeographic or hydrographic barriers (Crisp and Southward 1953); presumably the strong tidal currents and width of the Pentland Firth may act as a barrier to some organisms extending their range from northern Scotland to Orkney. The importance of temperature in distribution is particularly demonstrated in prosobranch molluscs, decapod crustaceans and barnacles (Hiscock et al. 2001).

Several life cycle factors can be examined to determine whether or not an organism is of value as a sentinel of environmental change. The following criteria, adapted from Hiscock et al. (2001), have been used to select species for consideration in this long-term monitoring programme:

i. mobility of existing populations, i.e. do they depend on larval dispersal to increase distribution or can individuals propel themselves?
ii. contribution of viable larvae from individuals—non-gamete-producing outliers will not contribute;
iii. type of reproductive and/or dispersal mechanisms, which favour extending distribution, i.e. motility of larvae or length of larval stage;
iv. larval and adult temperature tolerance—less tolerant species should experience more rapid distribution changes following temperature changes;
v. suitable habitats for extension of range during dispersive stages
vi. geographic barriers preventing spread—near Orkney, the Pentland Firth may retard the movement of certain species;
vii. favourable currents;
viii. longevity of individual organisms: short-lived species will retreat from areas faster than those with greater longevity;
ix. population at or near distribution limits in local waters;
x. association with exploitable wave-energy profile: can they be studied in areas open to potential WEC deployment?

Ideally, candidate species will include a broad range of mobile, short-lived, temperature-sensitive, high-energy-adapted organisms living at the extremes of their distributions which produce motile larval and juvenile forms.

Based on the selection criteria described above and on detailed field observations (see the section on littoral site selection below), the following four species have been selected for establishing a monitoring programme: *Chthamalus stellatus*, *Fucus distichus anceps*, *Gibbula umbilicalis* and *Patella ulyssiponensis*. All have been identified as potentially valuable indicators of long-term global climate change and specifically increased sea temperature (Southward et al. 1995), so it is likely that they will be good candidate sentinels for changes following both energy extraction and climate change along the West Mainland of Orkney.

In all the above cases, the selected organisms are located at or near their distribution limit in local waters, which should increase their sensitivity to environmental stressors manifest in observable population changes. For each organism, a congener or similar species has been assigned that has overlapping habitat, including the high-energy rocky shoreline. The exact nature of the interaction between the closely related species is not fully known, and to what extent there is direct competition is also uncertain. We are currently monitoring the selected organisms and their pairs as follows: *Chthamalus stellatus* and *Semibalanus balanoides*; *Fucus distichus anceps* and *Fucus vesiculosus* f. *linearis*; *Gibbula umbilicalis* and *Gibbula cineraria*; and *Patella ulyssiponensis* and *Patella vulgata*. In all pairings, one of the species is better adapted to high-exposure coastlines (see below under species description for further details) where removal of wave energy may be expected to have a detrimental effect.

In terms of selecting littoral sites, several along the wave-exposed west coast of Orkney Mainland were examined and evaluated at low spring tide for the following criteria: high-energy environment; presence of appropriate organisms; stable substratum; accessibility for observation (including health and safety issues); potential for deployment of WECs. The sites included Billia Croo, Marwick and Northside, as well as areas of more moderate exposure.

Using previous studies as a guide (e.g. Kendall and Lewis 1986; Southward 1991; Firth and Crowe 2010), monitoring of potential sentinel species is based on established methodologies; as necessity requires, novel approaches particular to local conditions have been explored. Reproductive or recruitment failure is important in determining the distribution limits of species (Kendall et al. 1987). The inclusion of studies of reproductive and recruitment processes in these organisms is of particular importance because early periods in the life cycle may be the most sensitive to environmental stressors and changes may be observable over a shorter period of time (Hiscock et al. 2001).

Chthamalus stellatus

The dominant littoral barnacles on the rocky shores of the British Isles belong to the genera *Semibalanus* and *Chthamalus* (Southward 1991). The former is represented by the wide-ranging *S. balanoides*; the latter by *C. stellatus* and *C. montagui*, two organisms sufficiently similar to have been raised to separate species status relatively recently. All three species are present in Orkney, with *C. stellatus* more closely associated with high-energy shores and exposed headlands and *C. montagui* better adapted to embayments (Crisp et al. 1981) and "drier" habitats (Power et al. 2011). In northern Scotland, the two barnacle genera are typically found in close proximity to chthamaloids in the upper zone, relative to the balanoids, which dominate closer to the sublittoral zone (Lewis 1986). The interface between the two species may be a valuable indicator of environmental changes.

Examining key reproductive and recruitment factors, Hiscock et al. (2001) recommended chthamaloid barnacles as potential indicators of climate change owing to their sensitivity to changes in sea temperature. The major determinant of *Chthamalus* distribution in the British Isles appears to be temperature (O'Riordan et al. 2010). Further, following rocky shore community data analysis from Shetland, Burrows et al. (2002) described *S. balanoides* as a particularly valuable indicator of large-scale environmental change. Reductions in wave exposure are likely to affect the pattern of vertical distribution of organisms most specialized for high-energy environments, such as *C. stellatus*. Therefore, some hypothesize that extraction of wave energy will reduce abundance and cover of *C. stellatus*, indicated by observable changes in settlement success, growth rates and zonation relative to *S. balanoides*. Overall abundance and cover percentage of barnacles would be expected to remain constant, with the *S. balanoides* population replacing *C. stellatus* (Power et al. 2011).

Suitable study localities for barnacle monitoring were chosen based on the following criteria: vertical distribution of both major barnacle species (*Chthamalus stellatus* and *Semibalanus balanoides*); inclusion of the zone where the two species overlap; the relative absence of other encrusting organisms; few complex rock features. Marwick Head was selected as the main site for barnacle studies. This shoreline features a lower zone dominated by *S. balanoides*, with lower to mid-level mixing with *C. stellatus*, and the latter species dominating from mid- to higher portions of the shore. Within the barnacle-dominated zone are less-abundant patellid limpets, small patches of *Mytilus edulis*, especially in small clefts and depressions, and small tufts of *Corallina officinalis*. The site is in close proximity to the proposed deployment of a farm of WECs.

To survey a site, detailed photographs are taken along vertical transects, perpendicular to the shore, at different sites from the lowest accessible areas, bordering on sublittoral, to the top of the barnacles. Quadrats measuring 10×10 cm are imaged at intervals of 1 m on a rock surface, in this example, dipping at $\sim 13°$, resulting in vertical intervals of ~ 22 cm. The surveys prioritize the identification of the vertical zone where the two species overlap. An example of an analysis of relative abundance (based on individual counts) from Marwick is presented in Fig. 3.4, which shows quadrats photographed for analysis at eight levels, descending down the slope at intervals of 1 m. There is a distinct reversal of dominant barnacle species between levels 3 and 4, where the potential for paired species monitoring is greatest, because predicted population shifts between the organisms will be expected to be most observable there. Such areas are the focus for clearance studies.

Barnacle studies conducted already have focused on photographic monitoring of cleared and intact columns on the rocky shore, including: clearance of a column of the rock surface using paint scraper and wire brush following final propagule settlement (in Orkney, balanoid settlement is typically in April and chthamaloid settlement in August); monthly observation of subsequent larval settlement and juvenile growth; a photographic study (Canon EOS 50 with 50 mm lens) of population changes for the cleared areas, and comparative assessment of the adjacent, intact population, using 10×10 cm quadrats; individual study using photograph-based biometrics, in particular the lengths of the operculum (Burrows et al. 2010) and, where applicable, the rostro-carinal axis (i.e. the maximum length; see Fig. 3.4) used by Barnes (1956); spatial comparisons between sites of varying energy.

Species identification is facilitated by close observation of such features as the shape of the operculum and the ontogeny of the rostrum. Many such features become less distinct in older individuals, as the shell plates tend to fuse, and also in more crowded populations where competition for space leads to modification in shell shape. In certain conditions of light intensity and "wetness", the optical qualities displayed by the two species are sufficiently distinctive to allow quantification using supervised classification (see above). However, in most cases, quantification of percentage cover needs to be performed manually, which is extremely labour-intensive. Image-analysis software is currently being developed specifically for the current research study in collaboration with the School of Mathematics and Computer Sciences at Heriot-Watt University to investigate the feasibility of using species-specific differences in shell shape to inform image-recognition software. Similar to face-recognition software, the team at ICIT has recommended using several features, in particular the shape of the operculum and the intersection of the tergal and scutal plates along the longitudinal axis of the operculum (see Fig. 3.4). These structures are more conserved, i. e. their features less affected by the individual's age or density of settlement, and may prove to be reliable markers for the software to differentiate and allow for

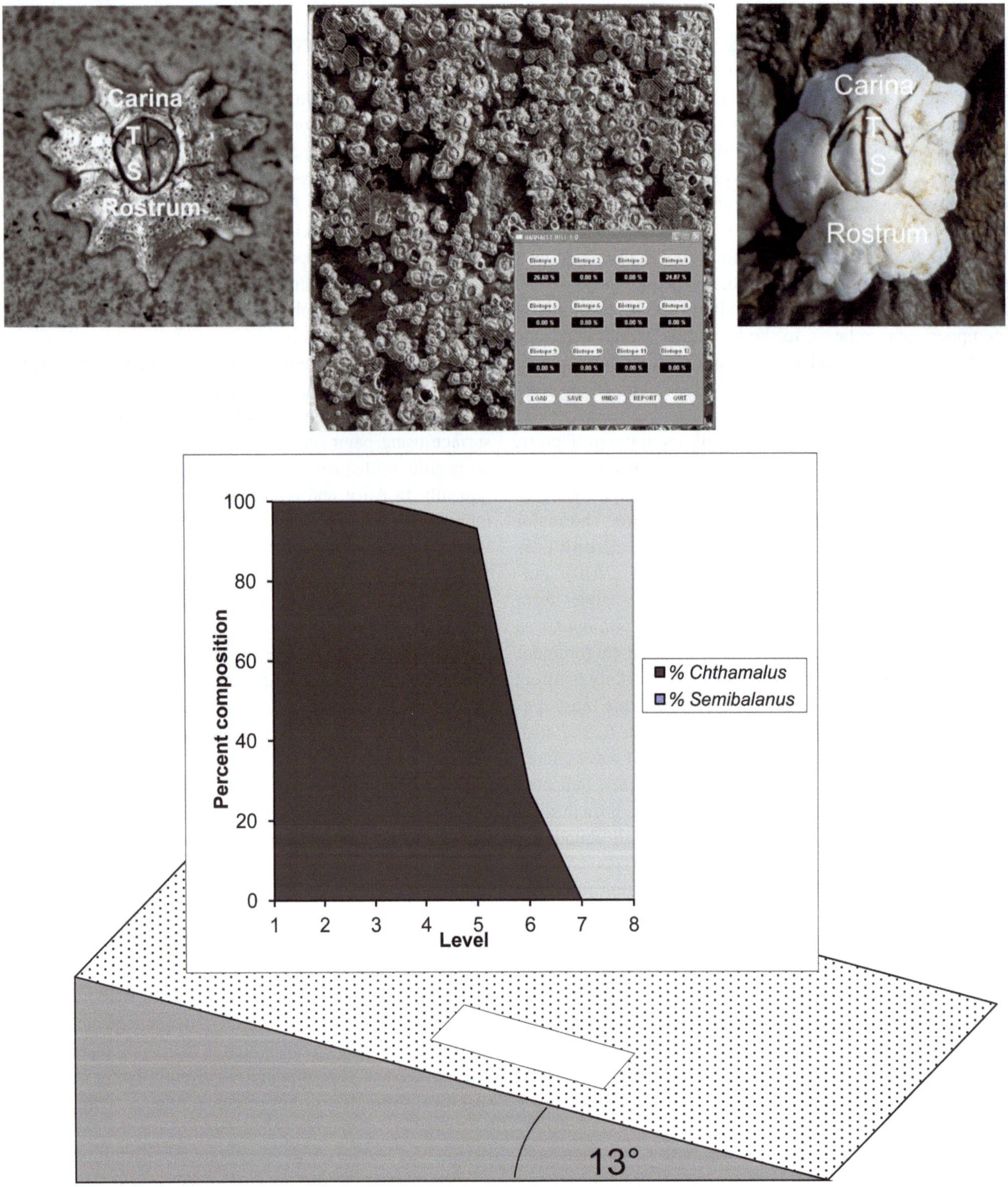

Fig. 3.4 Assessment of barnacle composition at Marwick Head: left, *Chthamalus stellatus* (*T* tergal plates; *S* scutal plates); right, *Semibalanus balanoides*; centre, 10 × 10 cm quadrat quantified by hand-selection software used in graphic representation of population shift between the two species between upper and lower littoral zone; bottom, a schematic demonstrating clearance of the species-transition zone on a typical West Mainland, Orkney, rocky shore

accurate analysis delivered in a fraction of the time needed to carry out the analysis by eye.

Fucus distichus anceps

Fucus distichus is a littoral brown alga found in a variety of forms in the northern hemisphere in a wide-range of exposure levels (Powell 1957). Only one morphological form has been identified in Orkney and described by Powell (1957) as *F. distichus anceps*. For the remainder of this article, this organism will be referred to as *F. distichus anceps* (references to *F. distichus* reflect studies at a broader species level). In the British Isles, *F. distichus anceps* is limited to a few rocky shores exposed to extreme wave conditions on the northern and western coasts of Scotland and Ireland. The geographic distribution of *F. distichus* does not extend below the 13 °C summer isotherm, except for the St Kilda population (Hiscock et al. 2001), but this may be due to photoperiod or desiccation rather than to temperature *per se*. With restricted frond width, stiffened stipe and lower overall height compared with other fucoids, *F. distichus anceps* possesses many of the morphological adaptations associated with survival in extreme exposed conditions (Denny 1987). Both the removal of wave energy and warming seas should have unfavourable consequences for this seaweed, especially at the southern limit of its distribution, although increased storm events predicted with global temperature increase may have a mitigating effect, depending upon how much average wave energy vs. extreme storm events dictates the presence of *F. distichus anceps*; the precise relationship remains unknown.

In Orkney, this fucoid is confined to the extremely exposed west coast (Wells et al. 2003), in particular on platforms below west-orientated cliffs dipping ~12° seawards with sufficient emersion at low tide (AW, pers. obs.). Whereas stands of other fucoid species may experience localized fluctuations over time (Lewis 1964), *F. distichus anceps* at several of these locations appears to remain stable over many decades (Powell 1957, 1963; Wilkinson 1975). A monitoring approach has been adopted using the monthly growth rate of individual plants, the monthly density of cover using quadrat photography, and quarterly zonation measurements using fixed point photography of extensive stands.

Individual plants have been identified and measured in triplicate at Northside and the high-energy variant *F. vesiculosus* f. *linearis* at Billia Croo using maximum frond height to monitor growth cycling (Edelstein and McLachlan 1975). Subsequent relocation of selected individuals is made using detailed laminated photographs including adjacent plants and stable, easily recognizable rock features to aid as accurate visual cues. Particular care is taken in identifying plants, which are clumped together and may appear as a single discrete unit. Like *F. distichus anceps*, the morphology of *F. vesiculosus* f. *linearis* is characterized by drag-reducing, strength-increasing adaptations associated with macroalgae on extreme exposed shorelines (Wernberg and Thomsen 2005).

Quadrat photography provides a relatively simple and repeatable method for long-term monitoring (Glanville 2001) and has been employed previously in population studies of *F. distichus* (Ang 1999). Density of cover is determined using photographic images of triplicate 1 m^2 quadrats for *F. distichus anceps* (at Marwick Head and Northside) and *F. vesiculosus* f. *linearis* (at Billia Croo). Site selection is recorded using GPS at the centre of each quadrat, and markers for two opposing corners are chiselled into the bedrock and painted bright yellow to aid repositioning during subsequent sampling. Quadrat photographs are then analysed to determine long-term variations in percentage cover of the plant. These complement larger scale fixed-point imaging of vertical zonation changes and percentage cover in *F. distichus anceps* relative to other mid-littoral macroalgae, using the image-analysis techniques described above.

Gibbula umbilicalis

Trochid gastropods are represented on the rocky shore of West Mainland, Orkney, by *Calliosotoma zizyphinum* and two species of *Gibbula*: *G. cineraria* and *G. umbilicalis*. The distribution of the latter species in the British Isles extends north to include Orkney, but the species is absent from the North Sea (Lewis 1999). The preferred habitats of *G. umbilicalis* are the rocky shores of sheltered to moderately exposed coasts (Hawkins and Jones 1992). Owing to Orkney's place at the extreme of this species' range and its association with different wave-energy activity from that of *G. cineraria*, this species has been selected as a potential indicator of ecological change resulting from both energy extraction and climate change. We believe that both reduced energy exposure and increased sea temperature should favour the abundance and individual growth rates of *G. umbilicalis* over the closely related *G. cineraria*.

The waters off Marwick Bay are being considered for deployment of a farm of WEC devices as part of the same leasing site that includes the adjacent headland used for studying barnacles and fucoids. The area has been surveyed extensively for the project: the embayment is dominated by an extensive boulder field characterized by abundant cover of *Fucus serratus* and red coralline algae. *Spirorbis* sp. features prominently on many of the rock surfaces, and the gastropods *Nucella lapillus* and *Littorina obtusata* are common above and below the waterline, respectively. Many boulders are quite large and, with the cementing effects of coralline algae and the presence of several other encrusting organisms, the substratum is considered immobile. Despite facing west to the open North Atlantic, these large boulders can be used reliably as reference points for return visits.

Small scale mark-recapture experiments to determine sampling efficiency between *Gibbula* species have informed a collection methodology in use at Marwick Bay. Topshells are sampled approximately quarterly at six permanent stations (GPS coordinates recorded and reference marks incised into a central boulder). A sampling area of 2 m radius is searched methodically and all the topshells within it collected. After 5 min, all the collected individuals are placed in a labelled jar for examination at the laboratory. Sampling is resumed with an exhaustive timed search of the remaining sample field; specimens collected in that search are similarly taken to the laboratory. An estimate of percentage suitable habitat is then made. Although monitoring studies are typically designed to avoid or minimize disturbance (unless this is part of the aim), completion of topshell data collection requires their removal from the field because of both the limited time available during low tides and practical constraints produced by inclement weather. Species identification can also be problematic, especially for smaller specimens where the distinguishing features of umbilicus shape and shell profile have not yet developed sufficiently. The mark-recapture experiments conducted earlier indicate that species mobility is likely to mitigate local depletion resulting from this methodology.

Ongoing monitoring of *Gibbula* at Marwick Bay includes species identification, absolute and relative abundance determination, and size comparisons using maximum whorl diameter.

Patella ulyssiponensis

This species is found along exposed coasts on the lower shore or in midshore pools (Neal and Skewes 2004). Study sites for patellid limpets have been selected in Orkney and feature shallow midshore pools (~15–30 cm deep) created by the dip, and subsequent erosion, of sandstone strata. The pools remain as the tide ebbs and provide a higher-shore habitat for *P. ulyssiponensis*, as well as red coralline algae, which dominate the immersed substratum within the pools. The species is also found fully emerged at the immediate periphery of the pools but quickly disappears beyond ~20 cm of the pool edges. By contrast, the area peripheral to the rock pools is home to plentiful *P. vulgata*, with typically smaller limpets fully immersed within the pools at low tide. A study by Delany et al. (1998) on the west coast of Ireland examining the same species in mid-shore rocky pools concluded that the *P. vulgata* use these pools as nursery grounds and that, upon reaching larger size, the adults migrate out of the pools. Those authors believe that the emigrating adults remain in the immediate vicinity; close examination of limpet distribution on the shore at Billia Croo supports this conclusion.

We have developed temporal and spatial monitoring studies of these two limpet species to test our hypothesis that reduction of wave energy impacting on midshore rock pools will favour the population of *P. vulgata* manifested by observable changes in abundance and growth rates between the two species. The following population variables are being considered: absolute and relative densities of both species; mortality rates of both species; individual growth rates of both species; shell morphometrics. The temporal studies require a robust and easily identifiable tagging procedure, for which logistically the greatest challenges are the reliability of tag adhesion and mortality following removal of the limpets for identification. Laboratory and field tests with a variety of adhesives and technical improvements during limpet removal have helped address both issues. Tags are produced from plastic discs etched with a soldering iron to produce deep, grooved numbers, subsequently marked with a red, indelible pen. In a procedure modified from Firth and Crowe (2010), small midshore rock pools are selected for study and drained by siphon to expose the limpets. Using a small paring knife, individual limpets are removed carefully for identification following examination of the mantle before being quickly returned to their "home scars". Shells are blotted dry with absorbent paper and acetone applied to further aid in drying the area for marker adhesion. Numbered plastic markers are attached using Cerebond cranioplastic adhesive (PlasticsOne). To account for and measure the rate of tag loss, limpets are tagged with two discs (Southwood and Henderson 2000).

In addition to using tagged individuals to study long-term population changes in patellid limpets, we have also conducted short-term, spatial monitoring. Surveying midshore rock pools for the two species of *Patella* at sites of varying energy exposure in Orkney may further understanding of the relationship between energy exposure, species dynamics and shell morphometrics. Various external shell features have been described as potential tools for identifying patellid limpets to species, but distinction can be problematic without the removal of the limpet from its rock (Fish and Fish 1989), a process traumatic to the animal. Whereas shell shape may be adapted for anti-predatory defence purposes (Lowell 1986) and to prevent desiccation (Baxter 1983; Denny 2000), it may be less of a contributing factor to species differences on high-energy shores. Both limpet species exhibit variation in several of these key diagnostic features resulting from shell modifications to external forces, i.e. wave and current exposure level (Baxter 1983; Denny 2000; Cabral and Jorge 2007). On high-energy sites, similar profiles of energy exposure may well result in similar external shell morphological adaptations in coexisting patellid limpet species. The study includes collection of the following data:

species identification; shell width and length; length of the anterior edge to the apex (i. e. with the measurement below, the adjacent and opposite lengths of a triangle to determine shell height); length of the posterior edge to the apex; position of the apex on the horizontal axis relative to the anterior and posterior edges (i.e. to determine its position along the anterior–posterior axis); a subjective assignment (scale of 1–5) of smoothness vs. "starriness" of the shell edge (this relates to the prevalence of ridges radiating from the apex); habitat position relative to tidal pool depth (scale of 1–5, 1 being deep in the rock pool, 2 fully immersed but in relatively shallow water, 3 straddling the surface, 4 emerged only slightly above, and 5 emerged beyond the immediate periphery); rock pool dimensions.

Midshore rock pools are cleared of all patellid limpets, individual species determined and shell measurements taken. Individual limpets are not returned to the study site, however. The procedure will be followed at rock pools at varying exposure from the high energy shore on West Mainland through moderately exposed sites to more sheltered locations. The procedure is repeated semi-annually, during which time the rock pools are allowed to repopulate. It is proposed that competition between the species to repopulate is influenced by their adaptation to energy level, as well as sea temperature, and monitoring the repopulation might provide an observable measure of ecological response produced by the environmental changes. Although this may provide a practical monitoring procedure applicable to studying the impacts of wave-energy extraction, any change observed would need to be understood in terms of interspecific population dynamics. There would still remain unanswered questions regarding the life histories and specific competitive responses of patellid limpets, so further research will be needed to determine the potential competitive nature of the relationship between the species.

Conclusions

The deployment of WECs is expected to contribute significantly to government projections of electricity generation from renewable sources (Scottish Government 2009). The ecological consequences of reducing exposure on rocky shores at the large scale associated with this development is, however, not yet well understood. The importance of energy level on rocky shore community dynamics has been known for some time (Lewis 1964), and the removal of wave energy by WECs will have the potential to alter community structure downstream, observable by changes in the littoral zone. The design and properties of WEC devices are often related to particular aspects of the wave climate, such as wave length and frequency, so devices might be expected to differ in their consequences for different species, depending upon species-specific adaptations to exposure. Complicating these potential impacts are normal seasonal fluctuations, less predictable community changes and long-term changes in global climate.

Climate change is a multifactorial alteration in the marine environment, involving changes in sea temperature and increased storminess, and it may alter shoreline communities to those favouring higher-energy, wave-exposed conditions. Exactly how increased storm events might affect the community is not well understood (Thomas 1986), but it will depend upon species adaptations to exposure: some organisms may be more responsive to changes in average, long-term wave exposure, others more to acute, extreme storm events. Collection of meteorological (including offshore wave) data will allow component factors underlying any observed changes to be addressed. In addition to sea temperature and storm condition considerations, long-term monitoring protocols may need to account for potential changes in sea level. This issue is particularly important in studies using fixed reference points in the vertical plane, such as barnacle clearance studies or intertidal seaweed imaging. Recently published analyses of the relationship between post-glacial isostatic emergence of the Scottish landmass and sea level increases attributable to global temperature increases predict an accelerating rise in relative sea level, with the Orkney archipelago among the areas most likely to be affected (Rennie and Hansom 2011).

The western shores of Mainland, Orkney, provide an ideal field laboratory to develop methodologies for long-term monitoring of rocky shores, specifically where changes relate to alterations in wave-energy profile. There is a richness of candidate sentinel species in Orkney, inclusive of southern and northern distribution ranges, and, as the major focus of WEC deployment in the foreseeable future, biological studies here are well placed to help inform these developments. Although previous monitoring studies in Orkney, largely associated with the expansion of the oil and gas industries in the 1970s, examined several habitats, they were monitored mainly on an annual basis, with a few seasonal studies made quarterly (Baxter et al. 1985). The ICIT monitoring programme described here collects data year-round, with greater regularity, and is the first to address specifically the nature of wave-energy effects on the rocky shore community in Orkney.

The two greatest challenges for the programme are to identify observable long-term changes and to distinguish between the relative roles that different environmental variables play in the changes. The development of paired-species monitoring and the use of multiple indicators may provide greater ability to dissect out the relative roles of energy extraction from climate change or other environmental fluctuations. The use of proportions for species pairs offers four main advantages over single species responses: (i) it is necessary only to know the relative preferences of the two species to an environmental variable rather than to have a detailed knowledge of their response curves—the precise adaptations of rocky shore organisms to energy conditions has not been explored fully; (ii) changes in proportions are easily interpreted in terms of environmental changes, because the response is likely to be highly directional; (iii) measurement of proportions is not dependent on being able to quantify habitat areas on rocky shores; and (iv) proportions may offer greater statistical power to detect changes than single-species responses. Monitoring a suite of sentinel organisms differentially responsive to energy extraction and climate change may provide greater confidence in determining the relative roles of each environmental variable.

Comprehensive surveying of Orkney's West Mainland littoral communities, including areas where access issues have prevented previous evaluations, has helped identify areas of potential environmental sensitivity. Although long-term monitoring may be necessary in identifying community population dynamics, government and business decision-making typically requires data delivery in a shorter time-frame; the inclusion of spatial monitoring helps to mitigate this issue. Methodologies such as barnacle clearance and patellid removal from rock pools may short-circuit normal community responses, potentially providing a more instant observable change. Although differences in limpet shell morphology related to wave-energy conditions are already established (Cabral and Jorge 2007), statistical analysis of preliminary data examining shell indices from midshore, high-energy rock pools suggests that shell shape in patellid limpets is more indicative of microhabitat differences (i. e. sheltering provided by rock pool immersion) than species differences. At least in high-energy environments, phenotypic adaptations appear to override genotypic differences, suggesting not only that patellid species determination cannot be based on external shell morphometrics alone but that small-scale differences in exposure can be detected in biometrics.

We have selected species and established methodologies that can inform the marine renewable energy sector. The work is, we hope, establishing essential baseline data necessary for longer term monitoring strategies, and the results may be transferable to other localities for similar programmes in future. The hope is to continue long-term monitoring and development of methodologies for these and other sentinel species, but more research is needed to determine specific high-energy adaptations and responses of many of the species, as well as the potential competitive relationships between them and other members in the community. Future detailed research strands for which the groundwork is being laid include studying shell morphometric adaptations on high-energy shorelines, evaluating the habitat and distribution of *F. distichus anceps*, more comprehensively quantifying pre-development littoral wave-energy exposure, optimizing semi-automatic image-analysis software and developing image-recognition software applications for barnacles. Continued long-term research is necessary, however, to understand the complex relationship between energy extraction, climate change and the rocky shore community.

References

Ang PO (1999) Natural dynamics of a *Fucus distichus* (Phaeophyceae, Fucales) population: reproduction and recruitment. Mar Ecol Prog Ser 78:71–85

Angus Council (2010). Wind, wave and tidal characteristics, coastal processes, shoreline management plan. http://www.angus.gov.uk/ac/documents/roads/SMP/default.html. Accessed 2012

Baardseth E (1970) A square-scanning, two-stage sampling method of estimating seaweed quantities. Norwegian Institute for Seaweed Research 33:1–41

Ballantine W J (1961) A biologically defined exposure scale for the comparative description of rocky shores. Field Stud 1:1–19

Barnes H (1956) The growth rate of *Chthamalus stellatus* (Poli). J Mar Biol Assoc UK 35:355–361

Barry J P, Baxter CH, Sagarin RD, Gilman SE (1995) Climate-related, long-term faunal changes in a California rocky intertidal community. Science 267:672–675

Baxter JM (1983) Allometric relationships of *Patella vulgata* L. shells characters at three adjacent sites at Sandwick Bay in Orkney. J Nat Hist 17:743–755

Baxter JM, Jones AM, Simpson JA (1985) A study of long-term changes in some rocky shore communities in Orkney. Proc Roy Soc Edinb 87B:47–63

Bekkby T, Rinde E, Erikstad L, Bakkestuen V (2009). Spatial predictive distribution modelling of the kelp species *Laminaria hyperborea*. ICES J Mar Sci 66:2106–2115

Bell EC, Denny MW (1994) Quantifying "wave exposure": a simple device for recording maximum velocity and results of its use at several field sites. J Exp Mar Biol Ecol 181:9–29

Broitman BR, Mieszkowska N, Helmuth B, Blanchette CA (2008) Climate and recruitment of rocky shore intertidal invertebrates in the eastern North Atlantic. Ecology 89:S81–S90

Burrows MT, Harvey R, Robb L (2008) Wave exposure indices from digital coastlines and the prediction of rocky shore community structure. Mar Ecol Prog Ser 353:1–12

Burrows MT, Jenkins SR, Robb L, Harvey R (2010) Spatial variation in size and density of adult and post-settlement *Semibalanus balanoides*: effects of oceanographic and local conditions. Mar Ecol Prog Ser 398:207–219

Burrows MT, Moore JJ, James B (2002) Spatial synchrony of population changes in rocky shore communities in Shetland. Mar Ecol Prog Ser 240:39–48

Cabral JP, Jorge RMN (2007) Compressibility and shell failure in the European Atlantic *Patella* limpets. Mar Biol 150:585–597

Castilla JC, Steinmiller DK, Pacheco CJ (1998) Quantifying wave exposure daily and hourly on the intertidal rocky shore of central Chile. Rev Chil Hist Nat 71:19–25

Connor D, Allen J, Golding N, Howell K, Liebernecht L, Noerten K, Reker J (2004) The marine habitat classification for Britain and Ireland (v04.05). http://jncc.defra.gov.uk/MarineHabitatClassification. Accessed 2012

Crisp DJ, Southward AJ (1953) Isolation of intertidal animals by sea barriers. Nature 172:208–209

Crisp DJ, Southward AJ (1958) The distribution of intertidal organisms along the coasts of the English Channel. J Mar Biol Assoc UK 37:157–203

Crisp DJ, Southward AJ, Southward EC (1981) On the distribution of the intertidal barnacles *Chthamalus stellatus, Chthamalus montagui and Euraphia depressa.* J Mar Biol Assoc UK 61:359–380
Delany J, Myers AA, McGrath D (1998) Recruitment, immigration and population structure of two coexisting limpet species in mid-shore tidepools, on the west coast of Ireland. J Exp Mar Biol Ecol 221:221–230
Denny MW (1983) A simple device for recording the maximum force exerted on intertidal organisms. Limnol Oceanogr 28:1269–1274
Denny M (1995) Predicting physical disturbance: mechanistic approaches to the study of survivorship on wave-swept shores. Ecol Monogr 64:371–418
Denny MW (1987) Life in the maelstrom: the biomechanics of wave-swept rocky shores. Trends Ecol Evol 2:61–66
Denny MW (2000) Limits to optimization: fluid dynamics, adhesive strength and the evolution of shape in limpet shells. J Exp Biol 203:2603–2622
Digby PGN, Kempton RA (1987) Multivariate Analysis of Ecological Communities. Chapman and Hall, London
EC (2012) Natura 2000 network. http://ec.europa.eu/environment/nature/natura2000/index_en.htm. Accessed May 2012
Edelstein T, McLachlan J (1975) Autecology of *Fucus distichus ssp distichus* (Phaeophyceae: Fucales) in Nova Scotia, Canada. Mar Biol 30:305–324
EEA (2008) About the EUNIS database. European environment agency, Copenhagen, Denmark. http://eunis.eea.europa.eu/about.jsp. Accessed 2012
EMEC (European Marine Energy Centre) (2006) Monthly Report January 2006. European Marine Energy Centre, Orkney
Firth LB, Crowe TP (2010) Competition and habitat suitability: small-scale segregation underpins large-scale coexistence of key species on temperate rocky shores. Oecologia 162:163–174
Fish JD, Fish S (1989) A Student's Guide to the Seashore. Unwin Hyman, London
Frid C, Andogeni E, Depestele J, Judd A, Rihan D, Rogers SI, Kenchington E (2012) The environmental interactions of tidal and wave energy generation devices. Environ Impact Assess 32:133–139
Fuji A (1988) Measuring wave force on a rocky intertidal shore. Bull Faculty Fisheries, Hokkaido Univ 39:257–264
Glanville J (2001) Littoral monitoring using fixed quadrat photography. In: Davies J, Baxter J, Bradley M, Conner D, Kahn J, Murray E, Sanderson W et al (eds) Joint Nature Conservation Committee, Peterborough. Marine Monitoring Handbook pp 307–314
Hawkins SJ, Jones HD (1992) Rocky shores. Immel, London
Hiscock K, Southward A, Tittley I, Jory A, Hawkins S (2001) The impact of climate change on subtidal and intertidal benthic species in Scotland. Scottish Natural Heritage (Survey and Monitoring Series), Edinburgh
Jones W, Demetropoulos A (1968) Exposure to wave action: measurements of an important ecological parameter on rocky shores in Anglesey. J Exp Mar Biol Ecol 2:46–63
Kendall MA, Lewis JR (1986) Temporal and spatial patterns in the recruitment of *Gibbula umbilicalis*. Hydrobiologia 142:15–22
Kendall MA, Williamson P, Garwood PR (1987) Annual variation in recruitment and population structure of *Monodonta lineata and Gibbula umbilicalis* populations at Aberaeron, mid-Wales. Estuar Coast Shelf Sci 24:499–511
Lewis JR (1964) The ecology of rocky shores. Hodder and Stoughton, London
Lewis JR (1986) Latitudinal trends in reproduction, recruitment and population characteristics of some rocky littoral molluscs and cirripedes. Hydrobiologia 142:1–13
Lewis JR (1999) Coastal zone conservation and management: a biological indicator of climatic influences. Aquatic Conserv 9:401–405
Lindegarth M, Gamfeldt L (2005) Comparing categorical and continuous ecological analyses: effects of "wave exposure" on rocky shores. Ecology 86:1346–1357
Lohse DP, Gaddam RN, Raimondi PT (2008). Predicted effects of wave energy conversion on communities in the nearshore environment. In: Nelson PA, Behrens D, Castle J, Crawford G, Gaddam RN, Hackett SC, Largier J et al (eds) Developing wave energy in coastal California: potential socio-economic and environmental effects by California Energy Commission, pp 75–98. http://www.opc.ca./gov/webmaster/ftp/project_pages/energy/CA_WEC_Effects.pdf. Accessed 2012
Lowell RB (1986) Crab predation on limpets: predator behaviours and defensive features of the shell morphology of the prey. Biol Bull 171:577–596
Mieszkowska N, Kendall MA, Hawkins SJ, Leaper R, Williamson P, Hardman-Mountford NJ, Southward AJ (2006) Changes in the range of some common rocky shore species in Britain—a response to climate change? Hydrobiologia 555:241–251
Moore J (2001) Fixed view-point photography. In: Davies J, Baxter J, Bradley M, Conner D, Kahn J, Murray E, Sanderson W et al (eds) Marine monitoring handbook. Joint Nature Conservation Committee Peterborough, pp 179–182
Mumby PJ, Green EP, Edwards AJ, Clark CD (1997) Coral reef habitat mapping: how much detail can remote sensing provide? Mar Biol 130:193–202
Murray E, Dalkin MJ, Fortune F, Begg, K (1999) Marine nature conservation review, Sector 2. Orkney: area summaries. Joint nature conservation committee, Peterborough
Neal K, Skewes M (2004) *Patella ulyssiponensis*. China limpet. Marine life information network: biology and sensitivity key information sub-programme [on-line]. Mar Biol Assoc UK Plymouth. Accessed 2012
O'Riordan RM, Power RM, Myers RM (2010) Factors, at different scales, affecting the distribution of species of the genus *Chthamalus* Ranzani (Cirripedia, Balanomorpha, Chthamaloidea). J Exp Mar Biol Ecol 392:46–64
Powell HT (1957) Studies in the genus *Fucus* L. J Mar Biol Assoc UK 36:663–693
Powell HT (1963) New records of *Fucus distichus* subspecies for the Shetland and Orkney Islands. British Phycol Bull 2:247–254
Power AM, McKrann K, McGrath D, O'Riordan RM, Simkanin C, Myers AA (2011) Physiological tolerance predicts species composition at different scales in a barnacle guild. Mar Biol 158:2149–2160
Rennie AF, Hansom JD (2011) Sea level trend reversal: land uplift outpaced by sea level rise on Scotland's coast. Geomorphology 125:193–202
Russell FS (1935) On the value of certain plankton animals as indicators of water movements in the English Channel and North Sea. J Mar Biol Assoc UK 20:309–322
Scottish Government (2009) Marine Renewables Masterplan. http://www.scotland.gov.uk/News/Releases/2009/01/2809505sc2. Accessed 2012
Shields MA, Woolf DK, Grist EPM, Kerr SA, Jackson AC, Harris RE, Bell MC et al (2011) Marine renewable energy: the ecological implications of altering the hydrodynamics of the marine environment. Ocean Coast Manage 54:2–9
Sjøtun K, Fredriksen S, Rueness J (1998) Effect of canopy biomass and wave exposure on growth in *Laminaria hyperborea* (Laminariaceae: Phaeophyta). Eur J Phycol 33:337–343
Solomon S, Qin D, Manning M, Chen Z, Marquis M, Averyt KB, Tignor M et al (eds) (2007) IPCC: climate change 2007: the physical science basis. Contribution of working group I to the fourth assessment report of the intergovernmental panel on climate change. Cambridge University Press, Cambridge
Southward AJ (1991) Forty years of changes in species composition and population density of barnacles on a rocky shore near Plymouth. J Mar Biol Assoc UK 71:495–513
Southward AJ, Hawkins SJ, Burrows MT (1995) Seventy years' observations of changes in distribution and abundance of zooplankton

and intertidal organisms in the western English channel in relation to rising sea temperature. J Therm Biol 20:127–155
Southwood TRE, Henderson PA (2000) Ecological methods. Blackwell, Oxford
Stephenson TA, Stephenson A (1949) The universal features of zonation between tide-marks on rocky coasts. J Ecol 37:289–305
Thomas MLH (1986). A physically derived exposure index for marine shorelines. Ophelia 25:1–13
Wells E, Wilkinson M, Tittley I, Scanlan C (2003) Intertidal seaweed biodiversity around Orkney. Coast Zone Topics 5:25–30
Wernberg T, Thomsen MS (2005) The effects of wave exposure on the morphology of *Eklonia radiata*. Aquat Bot 83:61–70
Westerbom M, Jattu S (2006) Effects of wave exposure on sublittoral distribution of blue mussels *Mytilus edulis* in a heterogeneous archipelago. Mar Biol 306:191–200
Wilkinson M (1975) The marine algae of Orkney. Eur J Phycol 10:387–397

Assessing the Impact of Windfarms in Subtidal, Exposed Marine Areas

4

Thomas G. Dahlgren, Marie-Lise Schläppy, Aleksej Šaškov, Mathias H. Andersson, Yuri Rzhanov and Ilker Fer

Abstract

Marine renewable energy conversion typically takes place at locations characterized by harsh physical parameters that challenge monitoring of the marine environment. These challenges are caused both by the lack of experience on what to expect in terms of impact, but also by a general lack of methods proven suitable for the monitoring of high-energy subtidal marine habitats. Here, the first offshore windfarm to be built in Norwegian waters, a project called Havsul I, is used as a model to provide (i) an overview contrasting the known effects and monitoring methods used at more sheltered offshore windfarms with those expected at a rocky, high energy site; (ii) a description and short assessment of the physical environment (bathymetry, current, wave and wind data) and marine assemblages at the site, (iii) an assessment of five methods used during the baseline study at Havsul I, including sediment grabs, sampling of assemblages from kelp stipes, video mosaics for rocky bottom benthic assemblages, traditional fishing gear for fish community evaluation, and C-PODs for harbour porpoise presence.

Keywords

Kelp · Marine renewable energy · Monitoring · Rocky seabed · Video mosaic

Introduction

Siting of offshore renewable energy devices has tended to move from nearshore, shallow waters (in the late 1990s), to offshore, deeper water (EWEA 2012). One of the drivers of this development is the lack of space on land and conflict with property owners claiming visual disturbance from onshore and nearshore windfarms (Esteban et al. 2011). Other potential conflicts are with shipping routes or alternative uses of the seabed, such as fishing or pipelines and cables (Burkhard et al. 2011). One could also add the increase in the quality of the wind and wave resource farther from shore. All these incentives apply to areas that are highly exposed to oceanic wind and wave energies, such as the steep and energetic seabeds off the coasts of Portugal, Ireland, Scotland and Norway in Europe, and in areas elsewhere such as Chile and California (Dvorak et al. 2010).

With few exceptions, offshore windfarms have to date been placed in relatively shallow seas on flat seabeds in the

T. G. Dahlgren (✉) · M-L. Schläppy
Uni Research, Thormøhlensgate 55, 5020 Bergen, Norway
e-mail: Thomas.Dahlgren@uni.no

M-L. Schläppy
EPFL, Lausanne, Switzerland

A. Šaškov
Coastal Research and Planning Institute, Klaipeda University, H. Manto 84, 92294 Klaipeda, Lithuania

M. H. Andersson
Department of Underwater Research, Swedish Defence Research Agency, 164 90 Stockholm, Sweden

Y. Rzhanov
Chase Ocean Engineering Laboratory, Center for Coastal and Ocean Mapping Joint Hydrographic Center, 24 Colovos Road, Durham, NH, 03824, USA

I. Fer
Geophysical Institute, University of Bergen, Postbox 7803, Bergen, Norway

M. A. Shields, A. I. L. Payne (eds), *Marine Renewable Energy Technology and Environmental Interactions,* Humanity and the Sea,
DOI 10.1007/978-94-017-8002-5_4,

southern Baltic and North Sea. The main environmental concerns about impacts on marine life in these areas relate to noise and sedimentation during the construction phase, and habitat change and noise during operation (Gill 2005; Wilhelmsson et al. 2010). The few studies of effects from the operation phase of a windfarm that have been published in peer-reviewed journals suggest that monitoring programmes have not detected any significant changes (e.g. Wilhelmsson et al. 2006; Lindeboom et al. 2011; Scheidat et al. 2011). A large volume of recently published reports from government agencies, research programmes and developers also indicate an absence of significant changes in community structure, species abundance and diversity after a few years of windfarm operation (Degrær et al. 2011; Stenberg et al. 2011; Bergström et al. 2012a, b). However, physical and biological conditions are dramatically different in more energetic coastal areas such as the Norwegian Sea (Shields et al. 2009). Bathymetry along the Norwegian coastal zone is typically steep, allowing little room for offshore wind developments (Fig. 4.1). These types of Norwegian offshore "banks" usually consist of pre-Cambrian crystalline rock with a rugged shape caused by glacial erosion. The resulting bathymetry is complex, giving such banks a mosaic of different benthic habitats. In the top 10–15 m, dense populations of kelp dominated by the species *Laminaria hyperboria* form a productive, diverse community (Mann 1972; Moore 1973), but below ~25 m, light intensities are too low to sustain the growth of brown algae and the wave action is too powerful to allow accumulation of sediment. With significant wave heights occasionally but annually in excess of 15 m, a highly eroded seabed extends down to around 70 m deep, forming a diverse habitat dominated by crust-forming algae and sessile invertebrates such as hydrozoans (Paine 1966). Fine sediment accumulates in deeper trenches (>100 m deep), where hydrodynamic forces are less, and because of the highly productive kelp community in the vicinity, the deeper trenches are organically rich and sustain an abundant and often diverse infaunal assemblage.

Some of the largest of these areas in Norway are found off the coasts of Møre and Romsdal county and have been subject to applications for offshore windfarm consents (Havsul I–IV). One project (Havsul I) was granted consent in 2009 and extended investigations were undertaken of bathymetry, geology, oceanography, wind resources and biology. The consent was given for a set of installations capable of producing 350 MW, covering an area of 49 km^2 centred on 62°49'37"N 06°18'29"E and situated 8 km from the closest inhabited island, Harøya (Fig. 4.1). The type of foundations or the size of turbines used had not been decided at the time of writing this chapter but will, because of the domination of rocky seabed, exclude monopiles. As the noise generated from pile-driving of monopiles has been the most important source of environmental concern during the construction phase of a windfarm (Wilhelmsson et al. 2010), disturbance effects during the construction phase will not be addressed further here, but we do discuss the challenges associated with planning and conducting environmental baseline studies and monitoring programmes suitable for marine renewable energy conversion projects in areas of high hydrodynamic forces. Calculated annual wave energy off the Møre coast is among the highest in the world, with an average of 438 MW m^{-1} year^{-1} (Golmen 2007). So-called extreme events are common there, with an average annual maximum significant wave height of 10.5 m for the period 1980–2006 (Golmen 2007), and with two events in excess of 12 m significant wave height during the last three months of 2011 (Fig. 4.2).

The environment there is, therefore, extremely harsh on any type of instrumentation left *in situ* to collect data over a period of time. This applies to instruments collecting physical data, such as current speed, temperature and salinity, and also to those collecting biological data, such as cetacean noise. Since the start of the project at Havsul I, no fewer than ten oceanographic, meteorological and biological instruments have been damaged or lost. The opportunities of calm weather available for fieldwork are also limited because of the high average wind speeds and exposure to oceanic swell breaking over shallow sites. The average mean wind speed at Ona Lighthouse (Fig. 4.1) between 15 August 2011 and 14 August 2012 was 8.8 m s^{-1} (data from www.yr.no). Navigation in the area by larger, less-weather-sensitive research vessels that would allow for more productive fieldwork from a stable and safe platform is also limited because of the narrow channels and shallow water.

Deeper offshore marine habitats in the Norwegian and North Sea have been subject, for some 30 years, to intense environmental monitoring warranted by petroleum extraction activities (e.g. Kingston 1992), but monitoring at exposed offshore rocky banks is not routinely conducted and standard methods are lacking (Shields et al. 2009). In fact, the limiting factors for researchers to work in such areas render them as *de facto* remote, and not very different from polar regions. For monitoring programmes, this means that systematically collected baseline data are not available and our understanding of the ecological responses to new stressors is limited (Shields et al. 2009, 2011). The poor knowledge of these habitats is also reflected by the absence of comprehensive species lists and a relatively large number of species newly discovered in recent years.

A major challenge for programmes trying to quantify environmental change at energy conversion structures placed in high-energy sites such as Havsul I is that hypotheses on expected impact and the resulting effect on ecosystems are not well developed. In contrast to the relatively shallow, low energy, soft-sediment marine environments where offshore windfarms have been operating for up to as much as a decade already, little is known about what to expect for high-

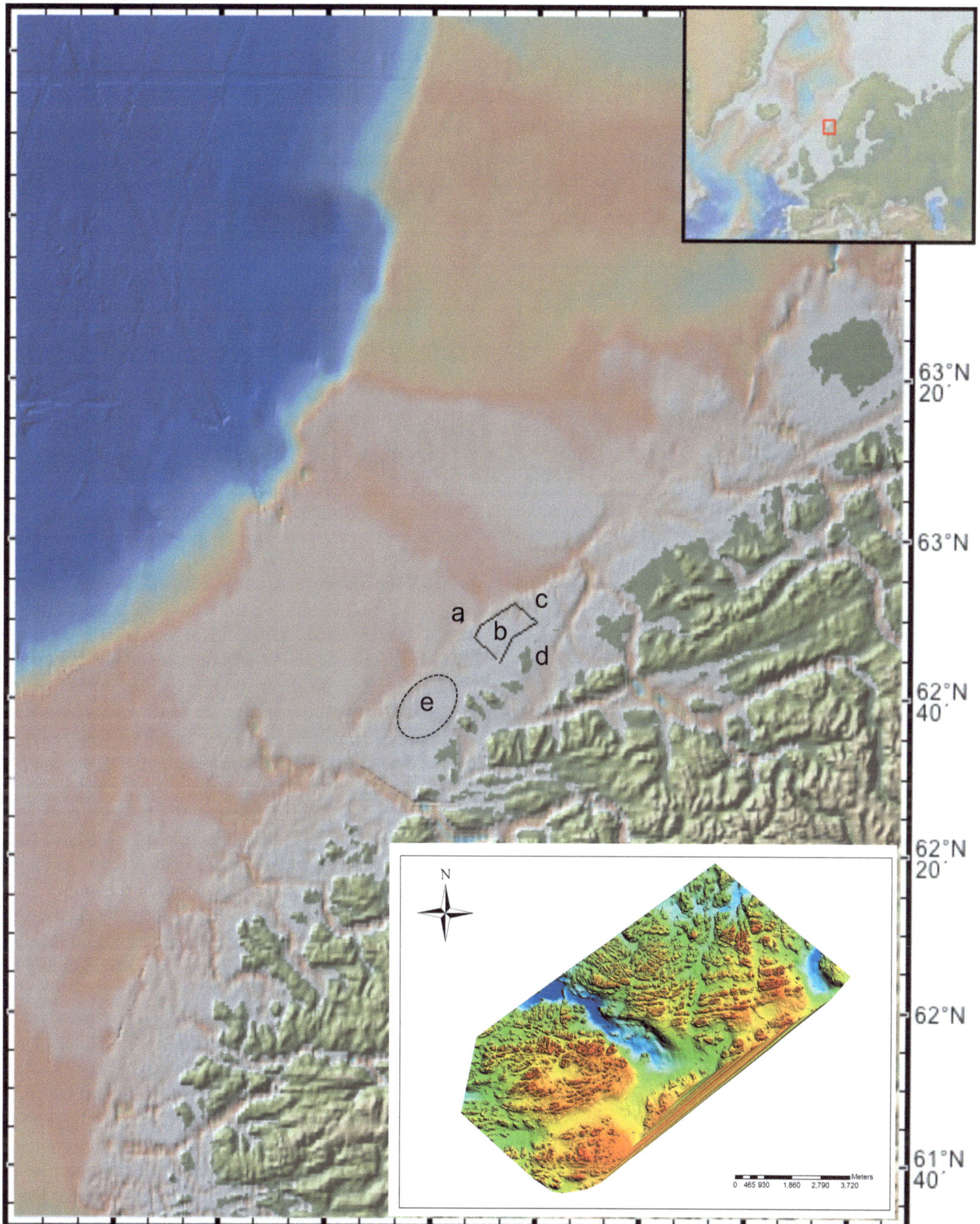

Fig. 4.1 Map of coastal Møre and Romsdal area in western Norway showing **a** the position of the oceanographic mooring, **b** an outline of the consent area (Havsul I), and **c** the position of Ona Lighthouse, **d** Harøya, and **e** the reference area. The insert at the *lower right* is a multibeam bathymetric map of the consented area. The map was created using GeoMapApp. (http://www.GeoMapApp.org; Ryan et al. 2009)

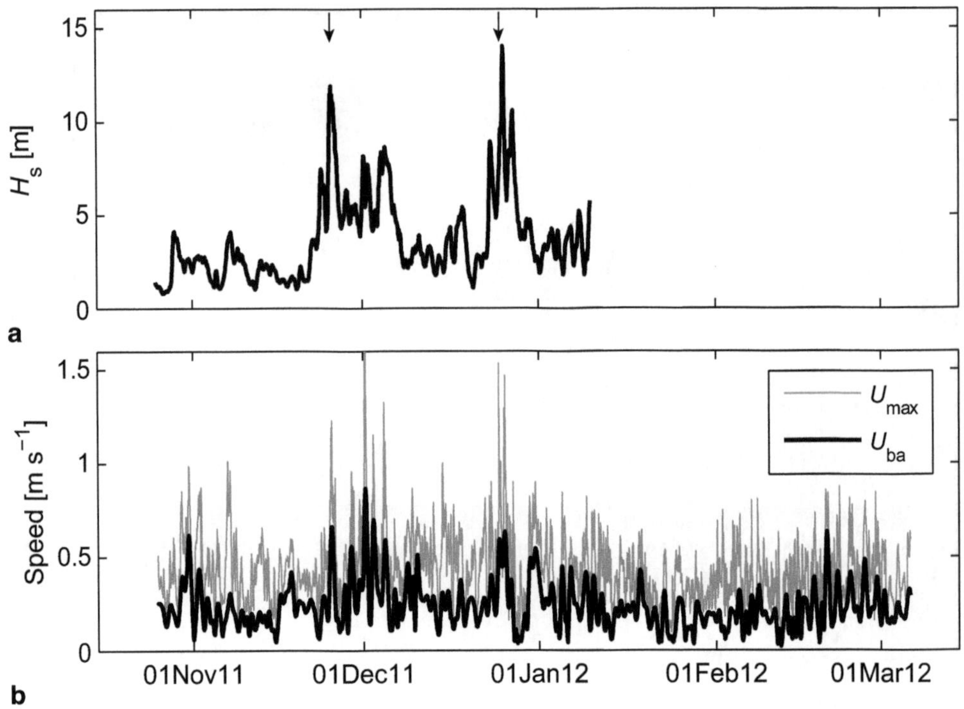

Fig. 4.2 Havsul I offshore windfarm. Position 62°50'07"N 06°08'14"E. Time-series of **a** significant wave height (H_s), and **b** magnitude of the maximum hourly velocity (U_{max}, *grey*) and the depth-averaged 25-h low-pass velocity (U_{ba}, *black*) in the Havsul I area measured at approximately the 130 m isobath. The *arrows* in **a** mark storms *Berit* and *Dagmar* in late November and December 2011, respectively

energy seabeds. The highly energetic offshore areas of the Norwegian coast can be regarded as less affected by the most serious threats to European marine communities, so arguably also more vulnerable to low levels of disturbance. Compared with coastal sediment habitats, energetic hard seabed communities are more often regarded as less affected by accumulation of contaminants, less affected by habitat-degrading fishing activities such as bottom trawls, and less affected by eutrophication (e.g. Gray 1997, but see Piola and Johnston 2008). Following the beliefs of Foley et al. (2011), we should strive to reduce the environmental footprint from energy and food production by focusing on halting the expansion of the area used for such activities. A more efficient use of areas could be achieved in coastal and offshore regions by placing windfarms in already impacted areas and combining them with, for example, aquaculture (Buck et al. 2008).

Direct impacts from offshore windfarms, such as the addition of habitats with the introduction of hard substrata in areas otherwise devoid of them, habitat loss from excavation of sand or replacement of soft sediments with hard blocks for scour protection, are not easily discernible at high-energy sites. If turbine foundations are placed at more exposed sites, one may expect increased drag causing decelerations, wakes and a sheltering effect. This, in turn, can increase the number of available microhabitats for fauna such as crabs (Langhamer and Wilhelmsson 2009). Parts of the planned windfarm Havsul I overlap with an area where kelp is harvested for the alginate industry, harvesting that can be compared with the bottom trawl fisheries excluded from some windfarms in sedimentary seabeds. By removing kelp from part of the Havsul area on a regular five-year cycle, the practice has been shown to increase net kelp productivity but to decrease the diversity of associated fauna (Steneck et al. 2002; Lorentsen et al. 2010). Hence, cessation of kelp removal by banning harvesting within the windfarm would probably increase the diversity of associated fauna. An increased diversity of fish species was observed at Horns Rev windfarm in Denmark, probably in response to an increase in habitat heterogeneity (Stenberg et al. 2011). The end (or reduction) of kelp trawling at Havsul is expected to result in greater species richness because of a changed demography of the kelp population with increased longevity of kelp plants (Christie et al. 2003). The diversity of fauna and flora associated with kelp stipes and holdfasts increases with the age of the plants, and the recovery of the associated fauna from regular kelp removal by trawling depends on the dispersal capabilities and assemblage structure of the surrounding kelp forests (Christie et al. 1998).

Whereas sessile benthic fauna will be impacted directly by all phases of the Havsul I windfarm construction, operation and decommissioning, mobile fauna such as fish and mammals have the choice of entering or leaving the area. Laboratory simulations have suggested that harbour porpoises (*Phocoena phocoena*) and common seals (*Phoca vitulina*) can detect the noise generated by a 2 MW wind turbine at sea (Koschinski et al. 2003). The Harøy archipelago has a large population of common seals that frequently use the Havsul I area for foraging and haul-out (Bjørge et al. 2002). Harbour porpoises are found in fjord systems right along the Norwegian coast, but little is known about their abundance offshore outside the North Sea, although a regional census was undertaken in 1994 (Bjørge and Øien 1995; Hammond et al. 2002). Current understanding of the impact on seal and porpoise populations from operational offshore windfarms is limited, but suggests that if the area is important for foraging, the long-term abundance of seals and porpoises within the farm will not be altered significantly (Tougaard et al. 2003, 2006). One study suggests that the abundance of porpoises may actually increase, possibly as a consequence of lessened disturbance from fishing vessels and the greater patchiness in fish abundance increasing foraging success (Petersen and Malm 2006; Scheidat et al. 2011).

To monitor environmental change in a mosaic of different habitats with limited access to evidence-based impact hypotheses, a diverse set of methods is required. Infaunal diversity and abundance of deeper areas with soft sediment can be monitored successfully using established grab methods, but other less proven methods are required for rocky seabeds and in kelp forests. Below, some of these methods are outlined and experiences from assessing them during a baseline study at the Havsul I windfarm site discussed.

Monitoring at Havsul I Offshore Windfarm

Here, we limit ourselves to the methods being considered by the developing company and the responsible authorities for baseline studies at Havsul I and an associated reference area. In particular, we look at technical challenges, limitations and potential sensitivity specific to the extreme physical environment (chaotic bathymetry, currents, wind and wave action) experienced at this offshore high-energy site. The following methods have been adopted:

i. physical oceanography;
ii. traditional van Veen grabs to sample the biota of sediments;
iii. video mosaics to map rocky seabed habitats;
iv. traditional kelp dredges to sample the diversity and demography of kelp forests;
v. traditional bottom-set longlines, gillnets, traps and fykenets to sample assemblages of benthic fish;
vi. C-PODs to sample porpoise abundance.

Physical Oceanography

On 25 October 2011, an oceanographic mooring consisting of instrumentation to measure the vertical distribution of ocean currents, temperature and salinity was deployed approximately 6 km offshore of the Havsul I area off the coast of Ålesund (Fig. 4.1). The water depth at the mooring site was ~130 m and the hourly averaged time-series for currents was obtained between 10 and 120 m, and for other parameters between 25 and 115 m. The mooring was recovered on 6 March 2012. An additional subsurface buoy (at ~10 m over the 130 m isobath) equipped with high-resolution pressure and motion sensors was deployed to infer surface wave parameters. The wave spectra and the corresponding wave parameters were obtained using 15-min segments of data, after applying the appropriate corrections for vertical acceleration and pitch of the platform and the transfer function for the attenuation of surface wave pressure signal with depth. Wave data were collected from 25 October 2011 to 10 January 2012. Current measurements were made by an RD-Instruments 300 kHz acoustic Doppler current profiler (ADCP) and a pair of Nortek Aquadopp current meters fixed at the bottom on the same mooring. Temperature and salinity measurements were made with Sea-Bird Electronics (SBE) loggers (6 Microcats and 2 Seacat) distributed evenly in the vertical. The measurement period covered two storms with wind speeds in excess of 20 m s^{-1} and 30 m s^{-1} (storms *Berit* and *Dagmar*, respectively), as measured at the nearby Vigra Airport meteorological station on 25 November and 25 December 2011. The site is highly energetic (Fig. 4.2), and the significant wave height, H_s, typical of the region varied between 1 and 5 m, increasing to >12 m during storms. Although the hourly maximum velocity in the water column typically varied between 0.2 and 1 m s^{-1}, it did reach ~1.5 m s^{-1} during storms. When tidal variability is removed (using a 25-h low-pass filter), depth-averaged currents there vary between 0.1 and 0.5 m s^{-1}, occasionally reaching values >0.6 m s^{-1}.

Sediment Habitat

The deeper trenches in the area are filled with soft sediment. The hypothesis behind monitoring the soft-sediment community is that any change in productivity at shallower depths caused by the windfarm (Wilhelmsson and Malm 2008) would lead to changed flux of organic carbon to the surrounding sediments. Changes in hydrodynamics of the area from, for instance, the presence of turbine foundations or wake gradients, could also lead to changes in organic carbon flux (Broström 2008). A common method of monitoring infaunal organisms uses a 0.1 m^2 van Veen grab (Norsk Standard 2005). In addition, samples for sediment characteristics and organic content were also collected. Biological samples were collected on a sieve of 1-mm diameter holes, fixed in formaldehyde, then rinsed in seawater and preserved in ethanol. During the first year of baseline data collection, a number of the randomized sample stations were dominated by sediment and gravel too coarse for the van Veen grab to close properly, and new replacement positions had to be selected and new samples taken. The feasibility of the method in areas with chaotic bathymetry and large hydrodynamic forces is limited by the heterogeneity of the seabed characteristics. In such rough seas, large vessels are normally used to withstand the

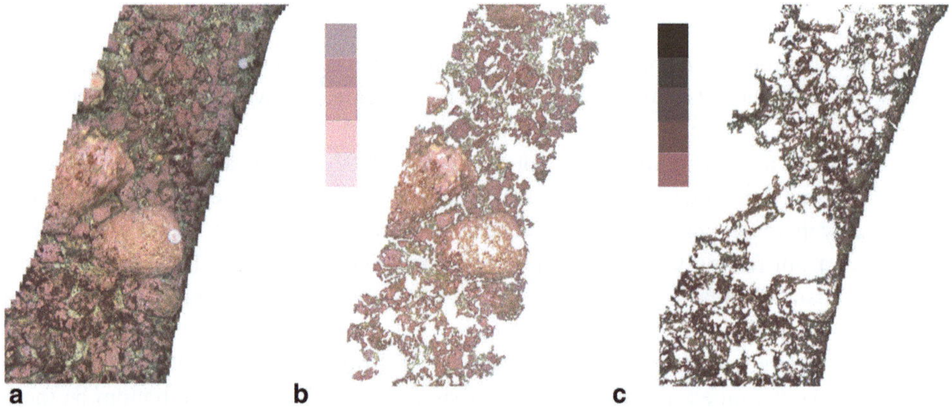

Fig. 4.3 Example of the bottom video mosaic. Two preliminarily named biological features, "*Lithothamnion*" and black crust, are extracted from the initial mosaic using selected training colours (shown next to each layer). The coverage is calculated as a proportion of pixel count. **a** Initial mosaic. **b** *Lithothamnion* sp. **c** Crust-forming algae

bad weather, allowing for a stable working platform, access to powerful winches, plenty of deck space and storage space, repair workshops, and well-ventilated areas and cabins. With narrow channels, limited depth and hence limited possibilities to manoeuvre large vessels, smaller, less optimal boats have to be used in combination with the use of various support facilities on shore.

Rocky Seabed

Traditional benthic sampling techniques are not feasible on hard substrata. The use of a SCUBA-based monitoring method is also limited by cost and safety issues in this highly energetic offshore area (Sisson et al. 2002). Therefore, we used a camera-based approach, with data collected as video imagery (Sheehan et al. 2010). Three types of platform can be used to collect the data, autonomous, towed or remotely operated. The last of these allow for better compensatory manoeuvrability in high energy situations (Sheehan et al. 2010). A work-class remotely operated vehicle (ROV) was used to collect data at Havsul I, the system equipped with powerful xenon lights (total power 600 W), colour HD camera (resolution 1920 × 1020 pixels) and two laser-line pointers for image-scaling. Video data were collected in transects with an average length of ~200 m. To optimize the video footage for mosaic construction, the camera was orientated vertically, and ROV altitude was kept as constant as possible. This was done as consistently as possible although water movement in the area is very dynamic, and some variations in camera altitude and angle to the seafloor were unavoidable. The optimal ROV altitude is dictated by illumination of the seafloor; when the ROV is too close to the seabed, illumination is excessive and there is image brightness saturation, but when it is too far from the seabed, images are dark through insufficient lighting and there is strong distortion of colour attributable to wavelength-dependent light absorption.

Data acquired from a moving camera are difficult to analyse using simple computer algorithms, so video mosaics were created using software developed at the Center for Coastal and Ocean Mapping (Rzhanov et al. 2004). Combining overlapping frames into a single picture allowed consideration of all the data collected (i.e. omitting no frames containing unique visual information). At the same time, overcounting features present in several video frames was avoided, because they appeared only once in a mosaic picture. To construct video mosaics of manageable size, all videos were segmented into 30-s clips, each corresponding to ~10 m of transect length.

Visually, it proved possible to count megabenthos diversity and abundance and to estimate the coverage of different algal species. Although the manual counts of megafauna species were fairly reliable (Jones et al. 2006), the estimation of percentage cover is more challenging, because benthic microhabitat types in the area were extremely patchy and diverse. Use of the video mosaic approach for manually counting megafauna was faster, less time-consuming and more accurate because it was easier to handle the still pictures than the raw video, allowing the operator to zoom in and out and to scroll in any direction.

For the computer-aided coverage estimation, we used a colour-based approach. For each feature, a set of training colours was selected, and features were assigned a value (a microhabitat) on the basis of this set (Fig. 4.3), allowing fast, reproducible extraction of features from mosaics. Once appropriate training sets had been selected, there was no need for an expert to do the balance of the analysis. The quality of the output at this stage is operator-independent, and after a small amount of training, any technician could process the data. The final results depend upon the training sets of colours and can vary. To evaluate possible errors, three mosaics were selected for testing, and for each feature in each, an expert picked seven different training colour sets. To compare the method with manual analysis, the same video segments were analysed manually using point-based feature selection (Carleton and Done 1995). Comparison of the results obtained with different training colours and between computer-aided and manual analyses revealed that devia-

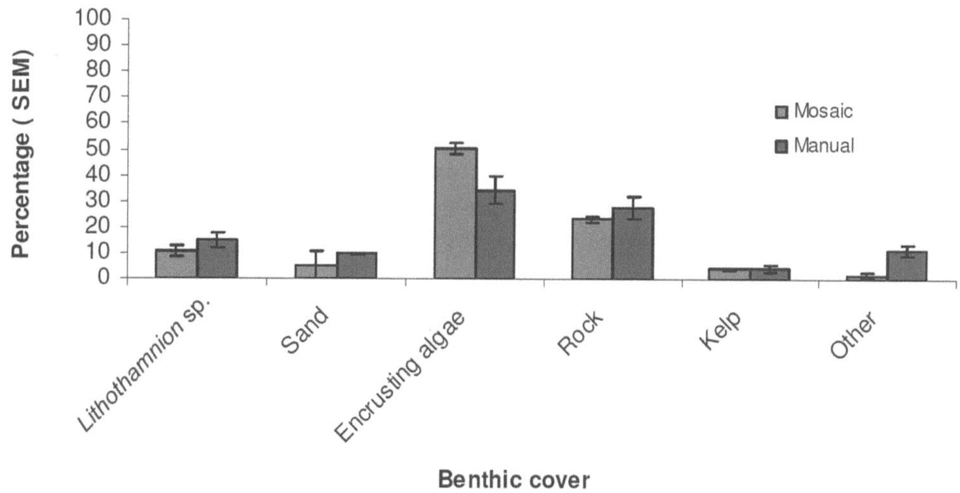

Fig. 4.4 Comparison of mosaic-based analyses of benthic cover and a manual method for six features encountered at Havsul I

tions attributable to a different choice of training colour sets were minimal (<5%, and for some features <2%), so the results were comparable with manual analysis performed by a trained marine benthic ecologist (Fig. 4.4).

Kelp Ecosystem

The area where Havsul I is planned overlaps with a key area for kelp harvesting along the Norwegian coast. Water 10–15 m deep is dominated by dense kelp, mainly *Laminaria hyperboria*, a species that is harvested regularly in some areas of Norway (Vea and Ask 2011). The presence of turbines and cable trenches will affect kelp harvesting inside a planned windfarm area, but mitigate any negative effects on the habitat caused by trawling. A baseline study is crucial to the quantification of any impacts, so to assess the impact on the kelp forest and the associated community of plants and animals, we collected samples of kelp stipes with a small commercial kelp trawl (Vea and Ask 2011). The associated species were removed from the kelp stipes, then fixed in formaldehyde, and sections of the stipes were made to estimate kelp age structure (Kain and Jones 1964). The diversity and abundance of associated animals can be enormous, with up to 80,000 individuals from up to 238 species on a single stipe (Christie et al. 2003). To render monitoring of this diversity and abundance feasible, a subsample of representative taxa is required. Following work by Kongsrud (2000), the diversity and abundance of crustaceans and annelids was sampled in a semi-quantitative design in which kelp stipes were cleared of all associated fauna. All samples from the baseline study were preserved and stored for future reference. One reason for potential later use would be the need to re-examine the baseline samples using broader taxonomic sampling, if changes in diversity are suggested from the more restricted sample.

Fish Community

The Havsul I area is normally avoided by larger commercial fishing vessels because it is relatively shallow and subject to complex and large waves. Only local fishers conduct a restricted small-scale fishery within the area, using passive gear and small craft able to deploy quickly when weather conditions turn favourable. Our monitoring of the fish community was based on the same type of gear and conducted by the same local fishers, using three types of bottom-set gillnet, bottom-set longlines and crab pots. In addition, during the first year of baseline sampling, cod fykenets were deployed, but the massive catches of crabs (*Cancer pagurus*) and low catch rates of fish led to the abandonment of such fishing activity. The catch was identified where possible to species level, measured and weighed on board, although it was impossible to identify with this sampling method changes in the abundance of pelagic species such as herring, for which ship-borne acoustic estimation is most commonly used today (MacLennan and Simmonds 1992). However, ship-borne acoustic monitoring is of limited value for monitoring around windfarms (Bergström et al. 2012a) and cannot be used to estimate groundfish abundance around rocky banks (Starr et al. 1996). Upwardly directed bottom-set acoustic monitoring of pelagic species is considered to be too expensive (Axenroth et al. 2004), so given the frequent loss of moored equipment in the Havsul I area, it was not attempted. Also, although the use of baited underwater cameras is promising for non-destructive monitoring in a rocky, energetic habitat (Harvey et al. 2007), deployment using landers can also be difficult with such strong currents. Once the windfarm is constructed, web-based underwater visual observations relayed to land via the control system of the windfarm may provide a means of monitoring remotely any changes in fish behaviour and abundance attributable to, for example, turbine load (Glover et al. 2010).

Porpoise and Seal Abundance

We opted for annual estimation of harbour seal abundance by aerial survey in August at known haul-out sites within the area (Bjørge et al. 2002); in August, the seals are moulting and are more predictably out of water. The average ratios of seals at haul-out sites in relation to the total population size have been calculated for the moulting period at different areas along the Norwegian coast (Bjørge et al. 2007), so using a correction factor of 1.35 for Møre and Romsdal, the total population size can be estimated from the number of seals at the haul-out sites (Bjørge et al. 2007). Seals are counted from photographs taken from a light aircraft collecting, to reduce costs and environmental impact, material for baseline studies of seabird abundance.

Harbour porpoises are the most common cetacean in many north European waters and are frequently monitored when offshore windfarms are being built, so as to better understand and minimize the impact of offshore windfarm construction on their population size. Although there are no other data on their abundance in the Havsul area, data from bycatch and other studies in Norwegian waters indicate that they are common year-round, peaking in coastal areas between July and October (Bjørge et al. 2011). The aim was to monitor harbour porpoises acoustically using autonomous underwater echolocation click detectors, called C-PODs (Chelonia Limited; www.chelonia.co.uk), for abundance and habitat use at both the planned windfarm site and a control site. The hypothesis was that neither construction nor operation of the proposed windfarm site would have an impact (negative or positive) on porpoise abundance. The C-PODs were deployed from August to October 2011 and from July to September 2012, and deployments are planned for four more periods during the operational phase. The Havsul I region is a very challenging area to work in weather-wise and particularly in winter, so to take advantage of short windows of good weather during summer, a small rib boat was used to deploy and retrieve the sensors quickly.

For the first two periods of deployment of the sensors, the location was selected using available data on bottom substrata and oceanographic conditions. To allow the recorded data to be analysed separately, the sensors were positioned in positions of similar environmental conditions. The three deployment positions in each of the impact and reference areas were selected to be of similar depth, topography and, at the future construction site, 100 m from the planned wind turbine position. No data exist for porpoise habitat usage or behaviour for this area or any Norwegian offshore site, so only one type of habitat was chosen, a plateau 30 m deep at the edge of a much deeper (>50 m) area. The rigs were bottom-mounted and without a surface buoy, to reduce the risk of theft and impact from wave motion, advice being offered by scientific groups that had experience of the deployment of C-PODs in various types of water body. The rigs each contained a C-POD, an acoustic release (pop-up), ballast weights of jute bags containing 35 kg of gravel, and buoys for buoyancy (Fig. 4.5). The choice of size and weight of the ballast was also constrained in that it had to be managed by two people in a rib boat. Each bag was tied with ropes to a shackle and 2-mm stainless steel wire to the acoustic release. For buoyancy, two hard Nokalon trawling buoys were used, each with a lifting force of 4 kg. Both C-POD and acoustic release are durable and reliable in rough sea conditions.

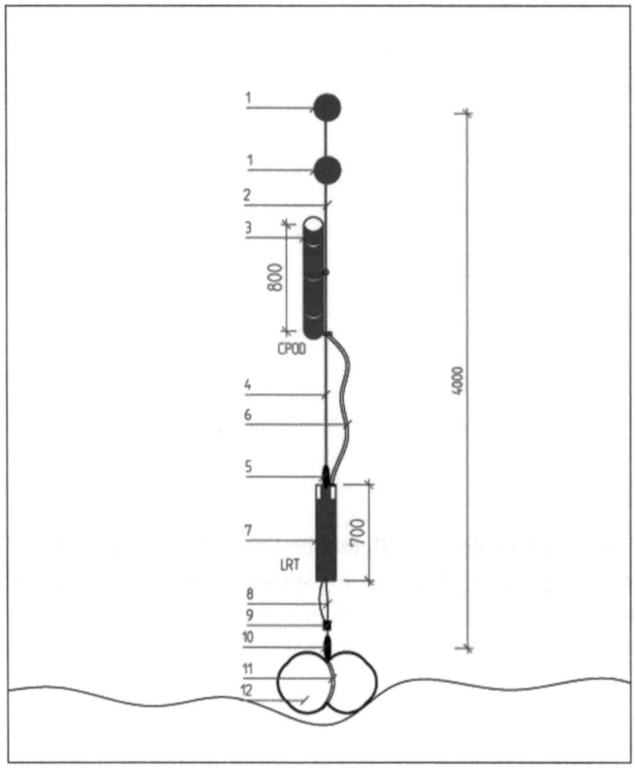

Fig. 4.5 Sketch of a C-POD sensor rig. *1* Nokalon trawling buoys, diameter (Ø) 20 cm, lifting force 4 kg. *2* Attachment rope (nylon), Ø = 10 mm, 2 m long. *3* C-POD. *4* Attachment rope (polypropylene), Ø = 20 mm, 1 m long. *5* Shackle (stainless steel), Ø = 4 mm. *6* Security rope (nylon), Ø = 5 mm, 1.5 m long. *7* Acoustic release unit (LRT). *8* Wire (stainless steel), Ø = 2 mm, 30 mm long. *9* Wirelock (stainless steel). *10* Shackle (galvanized), Ø = 10 mm with a breaking strength of 1200 kg. *11* Ballast rope (nylon), Ø = 10 mm, 2 m long. *12* Ballast weight, jute bag with 35 kg of stones and gravel

During the first year of the baseline studies (2011), harbour porpoises were sighted on the surface during deployment, in both impact and reference areas. In early October 2011, before the planned C-POD retrieval, however, the Havsul I region was hit by several severe storms, resulting in delayed recovery and the breaking loose of several of the C-PODs, probably as a result of failure of the stainless steel wire and wire lock (Fig. 4.5). Local fishers up to 130 km north of the deployment site found some of the lost sensors, but two sensors were never found and the data therefore lost. Of the other four, two were originally deployed at the

Havsul I site and two at the control site, allowing balanced comparative analysis. A common way of analysing click detector data is to calculate Detection Positive Minutes (DPM; see Leeney et al. 2007), i.e. at least one porpoise echolocation train detected during 1 min of the total logging time. In the analysis, only data recorded while the C-PODs were at the mooring site were used, although clicks were recorded also while sensors were afloat on the surface after being torn from their mooring. Preliminary analysis of that first year's data showed considerable porpoise activity, though relatively unevenly distributed among sensors, and that the overall estimates of porpoise presence in the Havsul I area and the control area were similar, indicating even usage of the area as a whole. For the second baseline year, the rig was modified according to lessons learned during the first year, and final analyses including both years of baseline data will include an evaluation of porpoise habitat use as well as an estimate of the possible impact of weather on the results.

Conclusions

There are clearly multiple, complex challenges associated with environmental monitoring at installations in an energetic marine environment that is a mix of habitats with differing characteristics. The impacts on these different habitats are often not well understood, in part because these harsh coastal environments have historically been avoided by scientists, resulting in a lack of baseline data in terms of both community structure and ecosystem function. The importance of this baseline study is clear, therefore, and so too is it that careful analysis is carried out before reaching any firm conclusions. Monitoring work involving boats is always weather-dependent, and the practical use of permanent or long-term installations such as bottom-mounted instruments or oceanographic buoys is limited by the higher-than-normal frequency of extreme weather events. The monitoring methodology also has to be adjusted to facilitate deployment of gear from small boats, because shallow water and narrow channels between underwater obstacles or topographic features generally preclude the use of larger research vessels. Moreover fieldwork has to be planned with weather in mind, possibly combining several methods to cover possible effects in different habitats and under different conditions.

Acknowledgements The work was conducted within Work Package 5 of the Norwegian Centre for Offshore Wind Energy (NORCOWE). We acknowledge the support at marine operations provided by Halvor Mohn, Argus AS and the backing of Vestavind Offshore AS and their representative Dag Breistein. Svein Winther, Sergei Olenin and Erling Heggøy initiated parts of the project, and the captain and crew of RV "Hakon Mosby" provided encouragement and support throughout the physical oceanography cruises.

References

Axenroth T, Didrikas T, Danielsson C, Hansson S (2004) Diel patterns in pelagic fish behaviour and distribution observed from a stationary, bottom-mounted, and upward-facing transducer. ICES J Mar Sci 61:1100–1104

Bergström L, Kautsky L, Malm T, Ohlsson H, Wahlberg M, Rosenberg R, Åstrand Capetillo N (2012a) Vindkraftens effekter på marint liv—en syntesrapport. Naturvårdsverket Rapport 6488, 94 pp. ISBN 978-91-620-6488-4 (in Swedish with English summary)

Bergström L, Sundqvist F, Bergström U (2012b) Effekter av en havsbaserad vindkraftpark på fördelningen av bottennära fisk. En studie vid Lillgrunds vindkraftpark i Öresund. Vindval. Naturvårdsverket Rapport 6485, 37 pp. ISBN 978-91-620-6485-3 (in Swedish with English summary)

Bjørge A, Bekkby T, Bakkestuen V, Framstad E (2002) Interactions between harbour seals, *Phoca vitulina*, and fisheries in complex coastal waters explored by combined Geographic Information System (GIS) and energetics modelling. ICES J Mar Sci 59:29–42

Bjørge A, Godøy H, Skern-Mauritzen M (2011) Estimated bycatch of harbour porpoise *Phocoena phocoena* in two coastal gillnet fisheries in Norway. Report of the International Whaling Commission, IWC SC63/SM18

Bjørge A, Øien N (1995) Distribution and abundance of harbour porpoise, *Phocoena phocoena*, in Norwegian waters. Reports of the International Whaling Commission, Special Issue 16:89–98

Bjørge A, Øien N, Fagerheim KA (2007) Abundance of harbour seals (*Phoca vitulina*) in Norway based on aerial surveys and photographic documentation of hauled-out seals during the moulting season, 1996 to 1999. Aquat Mamm 33:269–275

Broström G (2008) On the influence of large wind farms on the upper ocean circulation. J Mar Syst 74:585–591

Buck BH, Krause G, Michler-Cieluch T, Brenner M, Buchholz CM, Busch JA, Fisch R et al (2008) Meeting the quest for spatial efficiency: progress and prospects of extensive aquaculture within offshore wind farms. Helgoland Mar Res 62:269–281

Burkhard B, Opitz S, Lenhart H, Ahrendt K, Garthe S, Mendel B, Windhorst W (2011) Ecosystem based modeling and indication of ecological integrity in the German North Sea—case study offshore wind parks. Ecol Indic 11:168–174

Carleton JH, Done TJ (1995) Quantitative video sampling of coral reef benthos: large-scale application. Coral Reefs 14:35–46

Christie H, Fredriksen S, Rinde E (1998) Regrowth of kelp and colonization of epiphyte and fauna community after kelp trawling at the coast of Norway. Hydrobiologia 375:49–58

Christie H, Jorgensen NM, Norderhaug KM, Waage-Nielsen E (2003) Species distribution and habitat exploitation of fauna associated with kelp (*Laminaria hyperborea*) along the Norwegian coast. J Mar Biol Assoc UK 83:687–699

Degrær S, Brabant R, Rumes B (2011) Offshore wind farms in the Belgian part of the North Sea. Royal Belgian Institute of Natural Sciences Management. Unit of the North Sea Mathematical Models, Marine Ecosystem Management Section

Dvorak MJ, Archer CL, Jacobson MZ (2010) California offshore wind energy potential. Renew Energy 35:1244–1254

Esteban MD, Diez JJ, López JS, Negro V (2011) Why offshore wind energy? Renew Energy 36:444–450

EWEA (2012) The European offshore wind industry—key 2011 trends and statistics. A report by the European Wind Energy Association. http://www.ewea.org/fileadmin/files/library/publications/statistics/EWEA_stats_offshore_2011_02.pdf. Accessed 5 Dec 2012

Foley JA, Ramankutty N, Brauman KA, Cassidy ES, Gerber JS, Johnston M, Mueller ND et al (2011) Solutions for a cultivated planet. Nature 478:337–342

Gill A (2005) Offshore renewable energy: ecological implications of generating electricity in the coastal zone. J Appl Ecol 42:605–615

Glover AG, Higgs ND, Bagley PM, Carlsson R, Davies AJ, Kemp KM, Last KS et al (2010) A live video observatory reveals temporal processes at a shelf-depth whale-fall. Cahiers Biologie Mar 51:375–381

Golmen LG (2007) Potensiale for havenergiproduksjon i Møre og Romsdal. Runde Miljøsenter, Rapport 04/2007, 53 pp

Gray JS (1997) Marine biodiversity: patterns, threats and conservation needs. Biodivers Conserv 6:153–175

Hammond PS, Berggren P, Benke H, Borchers DL, Collet A, Heide-Jørgensen MP, Heimlich S et al (2002) Abundance of harbour porpoise and other cetaceans in the North Sea and adjacent waters. J Appl Ecol 39:361–376

Harvey ES, Cappo M, Butler JJ, Hall N, Kendrick GA (2007) Bait attraction affects the performance of remote underwater video stations in assessment of demersal fish community structure. Mar Ecol Prog Ser 350:245–254

Jones DOB, Hudson IR, Bett BJ (2006) Effects of physical disturbance on the cold-water megafaunal communities of the Faroe–Shetland Channel. Mar Ecol Prog Ser 319:43–54

Kain JM, Jones NS (1964) Aspects of the biology of *Laminaria hyperborea*. 3. Survival and growth of gametophytes. J Mar Biol Assoc UK 44:415–433

Kingston PF (1992) Impact of offshore oil production installations on the benthos of the North Sea. ICES J Mar Sci 49:45–53

Kongsrud JA (2000) Flora og fauna tilknyttet stortarestipes (*Laminaria hyperborea* (Gunnerus) Foslie) ved Færøyene. Hovedfagsoppgave i marinbiologi. Institutt for fiskeri- og marinbiologi, Universitetet i Bergen (in Norwegian)

Koschinsky S, Culik BM, Henriksen OD, Tregenza N, Ellis G, Jansen C, Kathe G (2003) Behavioural reactions of free-ranging porpoises and seals to the noise of simulated 2 MW windpower generator. Mar Ecol Prog Ser 265:263–273

Langhamer O, Wilhelmsson D (2009) Colonisation of fish and crabs of wave energy foundations and the effects of manufactured holes—a field experiment. Mar Environ Res 68:151–157

Leeney RH, Berrow S, McGrath D, O'Brien J, Cosgrove R, Godley BJ (2007) Effects of pingers on the behaviour of bottlenose dolphins. J Mar Biol Assoc UK 87:129–133

Lindeboom H, Kouwenhoven H, Bergman M, Bouma S, Brasseur S, Daan R, Fijn R et al (2011) Short-term ecological effects of an offshore wind farm in the Dutch coastal zone; a compilation. Environ Res Lett 6:1–13

Lorentsen S-H, Sjøtun K, Grémillet D (2010) Multi-trophic consequences of kelp harvest. Biol Conserv 143:2054–2062

MacLennan DN, Simmonds EJ (1992) Fisheries acoustics. Chapman and Hall, London

Mann KH (1972) Ecological energetics of the sea-weed zone in a marine bay on the Atlantic coast of Canada. 2. Productivity of the seaweeds. Mar Biol 14:199–209

Moore PG (1973) The kelp fauna of northeast Britain. 2. Multivariate classification: turbidity as an ecological factor. J Exp Mar Biol Ecol 13:127–154

Norsk Standard (2005) NS-EN ISO 16665. Vannundersøkelse—Retningslinjer for kvantitativ prøvetaking og bearbeiding av marin bløtbunnsfauna (ISO 16665:2005)

Paine RT (1966) Food web complexity and species diversity. Am Nat 100:65–75

Petersen JK, Malm T (2006) Offshore windmill farms: threats to or possibilities for the marine environment. Ambio 35:75–80

Piola RF, Johnston EL (2008) Pollution reduces native diversity and increases invader dominance in marine hard-substrate communities. Divers Distrib 14:329–342

Ryan WBF, Carbotte SM, Coplan JO, O'Hara S, Melkonian A, Arko R, Weissel RA et al (2009) Global multi-resolution topography synthesis. Geochem Geophy Geosys 10:Q03014. doi:10.1029/2008GC002332

Rzhanov Y, Mayer L, Fornari D (2004) Deep-sea image processing. Proceedings of Oceans'04, Kobe, pp 647–652

Scheidat M, Tougaard J, Brasseur S, Carstensen J, Van Polanen Petel T, Teilmann J, Reijnders P (2011) Harbour porpoises (*Phocoena phocoena*) and wind farms: a case study in the Dutch North Sea. Environ Res Lett 6:1–10

Sheehan EV, Stevens TF, Attrill MJ (2010) A quantitative, non-destructive methodology for habitat characterisation and benthic monitoring at offshore renewable energy developments. PLoS ONE 5(12):e14461. doi:10.1371/journal.pone.0014461

Shields MA, Dillon LJ, Woolf DK, Ford AT (2009) Strategic priorities for assessing ecological impacts of marine renewable energy devices in the Pentland Firth (Scotland, UK). Mar Policy 33:635–642

Shields MA, Woolf DK, Grist EPM, Kerr SA, Jackson AC, Harris RE, Bell MC et al (2011) Marine renewable energy: the ecological implications of altering the hydrodynamics of the marine environment. Ocean Coast Manag 54:2–9

Sisson JD, Shimeta J, Zimmer CA, Traykovski P (2002) Mapping epibenthic assemblages and their relations to sedimentary features in shallow-water, high-energy environments. Cont Shelf Res 22:565–583

Starr RM, Fox DS, Hixon MA, Tissot BN, Johnson GE, Barss WH (1996) Comparison of submersible-survey and hydroacoustic-survey estimates of fish density on a rocky bank. Fish B-NOAA 94:113–123

Stenberg C, van Deurs M, Støttrup J, Mosegaard H, Grome T, Dinesen G, Christensen A et al (2011) Effect of the Horns Rev 1 offshore wind farm on fish communities—Follow-up seven years after construction. In: Leonard SB, Stenberg C, Støttrup J (eds). DTU Aqua Report, 246-2011, 99 pp

Steneck RS, Graham MH, Bourque BJ, Corbett D, Erlandson JM, Estes JA, Tegner MJ (2002) Kelp forest ecosystems: biodiversity, stability, resilience and future. Env Conserv 29:1–24

Tougaard J, Carstensen J, Wisz MS, Teilmann J, Bech NI, Skov H (2006) Harbour porpoises on Horns Reef—effects of the Horns Reef Wind Farm. Final report to Elsam Engineering A/S. NERI Technical Report, Roskilde, Denmark

Tougaard J, Ebbesen I, Tougaard S, Jensen T, Teilmann J (2003) Satellite tracking of harbour seals on Horns Reef. Use of the Horns Reef wind farm area and the North Sea. National Environmental Research Institute, Roskilde, Denmark, 43 pp

Vea J, Ask E (2011) Creating a sustainable commercial harvest of *Laminaria hyperborea*, in Norway. J Appl Phycol 23:489–494

Wilhelmsson D, Malm T (2008) Fouling assemblages on offshore wind power plants and adjacent substrata. Estuar Coast Shelf Sci 79:459–466

Wilhelmsson D, Malm T, Öhman M (2006) The influence of offshore windpower on demersal fish. ICES J Mar Sci 63:775–784

Wilhelmsson D, Malm T, Thompson R, Tchou J, Sarantakos G, McCormick N, Luitjens S et al (2010) Greening blue energy. identifying and managing the biodiversity risks and opportunities of offshore renewable energy. IUCN, Gland, Switzerland, 102 pp

5 The Influence of Fisheries Exclusion and Addition of Hard Substrata on Fish and Crustaceans

Dan Wilhelmsson and Olivia Langhamer

Abstract

Offshore renewable energy development (ORED) could induce local ecological changes and put species assemblages of conservation interest at risk. If well planned and coordinated, however, ORED could be beneficial to the local subsurface marine environment in several aspects. Acknowledging the scale of ORED, there is increasing interest in the opportunities offered by the resulting changes in fishing patterns, such as exclusion or limitation of bottom trawling, in wind and wave farms. Areas encompassing several square kilometres may in some important aspects resemble Marine Protected Areas, and wind and wave-energy foundations and other associated structures can function as artificial reef modules and enhance the local abundance of marine organisms, including commercially important fish and crustaceans. It is also possible that floating offshore energy devices can function as fish aggregation devices for pelagic fish. Here, the potential influence of offshore wind and wave farms on fish and commercially important crustaceans is described, mentioning the uncertainties with regard to positive and negative effects on benthic and pelagic assemblages and specific species.

Keywords

Artificial reefs · Crustaceans · Fish · Marine protected areas · Wave energy · Wind power

Introduction

Several countries are planning for massive offshore renewable energy development (ORED; wind, wave, tidal and marine current). For offshore wind power within just the EU, plans for development contain nearly 40 GW until year 2020, and another 100 GW between 2020 and 2030 (EWEA 2009). This is equivalent to >25,000 wind turbines in offshore wind farms covering 15,000–25,000 km^2 of Europe's continental shelf. In addition, the technological development of wave power is progressing rapidly, and the first units are already in operation. The influence on already stressed marine environments of this large-scale development is uncertain. Indeed, ORED could induce local ecological changes and put species assemblages of conservation interest at risk. Discussions and research commonly centre on the effects of noise, flickering, electromagnetic fields, and changed hydrodynamic conditions, on benthic communities, fish, mammals and birds (Wilhelmsson et al. 2010).

However, once the construction phase is past, ORED may have a positive effect on the abundance of many species. The introduction of turbines will alter fishing patterns because bottom trawling will in most cases be limited or excluded.

D. Wilhelmsson (✉)
Swedish Secretariat for Environmental Earth System Sciences, Royal Swedish Academy of Sciences, Stockholm 104 05, Sweden
e-mail: dan.wilhelmsson@sseess.kva.se

O. Langhamer
Department of Biology, Norwegian University of Science and Technology, Høgskoleringen 5, Trondheim N-7491, Norway

M. A. Shields, A. I. L. Payne (eds), *Marine Renewable Energy Technology and Environmental Interactions,* Humanity and the Sea,
DOI 10.1007/978-94-017-8002-5_5,

Floating devices and/or artificial reef effects around foundations may have positive impacts on both benthic and pelagic species (Wilhelmsson et al. 2006; Langhamer et al. 2009). Aggregations of certain species may at the same time adversely influence the abundance of some species through increased predation pressure and the depletion of food resources. Here, we describe the potential influence of fishery closures and the introduction of artificial substrata in relation to ORED on fish and commercially important crustaceans. The scope is limited to offshore wind power and wave power, although the reasoning can be applied largely to tidal and ocean current energy too. Potential negative impacts on the marine environment during construction of the wind and wave farms are not addressed, and the underlying assumptions are that noise, electromagnetic fields and human activities during operation of the offshore wind and wave farms (OWFs) do not cause major disturbances to or avoidance of the area by fish and crustaceans.

Fishing Exclusion and Limitation Effects in Wind and Wave Farms

Wave-energy farms will primarily be developed in coastal waters in waters up to 200 m deep. Current wind power technologies, including monopiles, tripods, and jacket and gravity foundations limit offshore, non-floating wind turbines to coastal areas no deeper than 40 m (Wilhelmsson et al. 2010). Seabeds of muddy sand, sand or gravel beds with only scattered boulders are often preferred for technical and economic reasons (Wilhelmsson et al. 2010). Infaunal assemblages that are important sources of food for birds and fish usually dominate the seabed in these habitats. Offshore wind power is often developed or prospected on offshore banks, which can constitute important habitats for fish, mammals and seabirds (Frederiksen et al. 2006; Inger et al. 2009).

Trawling, which is one of the most severe threats to the marine environment (Thrush and Dayton 2002), and gillnetting is either prohibited or inevitably inhibited inside OWFs and safety zones; note that the latter, 50–500 m in extent, are always required around OWFs, to protect lives, property and environment. Limiting fisheries in these areas should to some degree be acting in a manner similar to the functions of marine protected areas (MPAs) or so-called "no-take zones" (NTZs). An MPA is an umbrella term defined by IUCN in 1988 as "any area of intertidal or subtidal terrain, together with its overlying water and associated flora, fauna, historical and cultural features, which has been reserved by law or other effective means to protect part of or the entire enclosed environment". NTZs are the most restrictive form of MPAs, and are areas that have temporarily or permanently been closed to all (not only some gear types) fishing and other extractive activities to protect and restore fish stocks and natural habitats (Roberts et al. 2005). MPAs are established primarily to protect certain species, ecosystems and habitats, as well as spawning and nursery grounds for fish and commercially important shellfish (Côté et al. 2001; Walmsley and White 2003). MPAs often also provide opportunities for research, education and recreational activities, and can be established to protect historical sites such as areas of archaeological interest such as shipwrecks and important cultural sites (IUCN 1988).

A comparative analysis of a large set of MPAs suggested that the establishment of an MPA may on average result in doubled species density, tripled biomass and increased size of individuals and species diversity relative to unprotected areas (Halpern 2003), though these averaged estimates are not applicable to site-specific predictions and in relation to certain species or species assemblages. A meta-analysis of 58 raw datasets from 19 European MPAs suggested that for every tenfold increase in the size of the protected areas, the density of commercial fish increased by 35 % (Claudet and Pelletier 2004). The subsequent increase in fish biomass may spill over into neighbouring areas (Christie et al. 2010; Russ and Alcala 2011). Larger parks are more likely to contain more species of fish and crustaceans, including rare species, and smaller ones are less likely to produce spill-over effects and to serve as larval sources (Halpern 2003). Smaller areas are also more vulnerable to disturbance, e.g. oil spills.

Fishers may benefit from increased catches of fish close to zone boundaries (Guénette et al. 2000). The extent of this depends on the life history and dynamics of the target stock as well as on the size of the protected area. For example, plaice (*Pleuronectes platessa*) in the North Sea spend their first few years in relatively restricted habitats (Deveen 1978), so a protected area of 38,000 km^2 was designated (the "plaice box") as part of that stock's management strategy, and a considerable reduction in juvenile fishing mortality was achieved within it (Piet and Rijnsdorp 1998). As they mature, the plaice migrate out of the zone, boosting fishery yields, despite derogations for small beam trawlers to enter the box to fish for shrimp. It is an open question, of course, whether yields to the plaice fishery might be greater if the area was to become a NTZ where all fishing is prohibited.

For highly migratory species, however, MPAs may be of limited use because the species may be subject to displaced fishing efforts outside the area protected (Claudet et al. 2008). Also, for more stationary fish, experience in the North Sea has shown that the fishing effort that would have taken place within an area that was closed moved to adjacent waters, and this more or less cancelled out the beneficial effects of the closed area (Sanchirico et al. 2006). To realize an MPA's full potential, therefore, an overall fishing effort reduction is generally advocated.

Another example, including invertebrates, is the fishery exclusions enforced in Strangford Lough, Northern Ireland, primarily to protect parts of the queen scallop (*Aequipecten opercularis*) population from a dredge fishery (Rogers 1997). Also, in the Isle of Man, great/king scallop (*Pecten maximus*) densities are much greater within the closed area at Port Erin than in surrounding areas, and these scallops may constitute an important spawning stock (Beukers-Stewart et al. 2005).

There are almost 6,000 MPAs designated around the world, but they encompass just over 1% of the world's oceans, and many lack actual protection (being so-called paper parks; Spalding et al. 2010). In terms of protection of fishery resources through MPAs, around 13% of the world's MPA areas are currently designated as NTZs (Spalding et al. 2010). As an example of regional initiatives, contracting parties of the Oslo–Paris (OSPAR) Commission for the protection of the Northeast Atlantic have agreed to identify and select MPAs within their waters in accordance with OSPAR Guidelines (OSPAR 2006). The main criteria for selection of sites include the presence of a species or habitat in need of protection (as identified on the initial OSPAR list of threatened and/or declining species and habitats), ecological significance, high natural biodiversity, representativeness, sensitivity and naturalness. The bio-geographic representation of MPAs nominated for the OSPAR network makes up ~15% of the whole European coast (OSPAR 2006).

The effects of decreased fishing pressure and habitat disturbance primarily by trawling exclusion in OWFs may resemble the functions of MPAs and lead to average increases in the biomass of motile organisms for the area as a whole. The empirical evidence for this is, however, weak. Significant and long-term survey effort is often required to isolate any effects of the limitations of fishing within OWFs from seasonal and interannual variability at the levels of both ecosystem and species. Other factors related to the construction of turbines in an area also need to be considered, such as habitat alterations by the presence of turbines and associated devices that can make the spatial distribution of fish biomass at post-construction levels more heterogeneous, as discussed in detail below. Results from surveys targeting fish assemblages within an OWF area as a whole (i.e. not designed to capture potential aggregations of fish around turbines) in Denmark, Holland and Sweden basically indicate either increased abundance of some species (e.g. sandeels Ammodytidae, cod *Gadus morhua*, whiting *Merlangius merlangus*, sole *Solea solea*) or no effects (Leonhard et al. 2011; Lindeboom et al. 2011; Bergström et al. 2012).

For example, the effect of offshore wind power on fish assemblages was monitored in the Belgian C-Power project on Thornton Bank, 27 km off the Belgian coast (Degraer et al. 2011). Six turbines have been in operation since 2009, and the installed capacity of the entire windfarm is now 325 MW (54 turbines). Water depth in the concession area ranges between 18 and 24 m. Belgian beam trawlers were excluded from the windfarm, which may have had positive impact on the soft bottom benthos within the area. Certainly, larger individuals and greater densities of swimming crabs (*Liocarcinus holsatus*) and of commercially important brown shrimp (*Crangon crangon*), pouting (*Trisopterus luscus*) and whiting have been recorded within the area since windfarm construction (Degraer et al. 2011; Reubens et al. 2011a). Further, monitoring of fish assemblages before (2001) and after (2010) construction of the Horns Rev windfarm in Denmark revealed minor discernible changes in fish abundance after the construction of the windfarm, possibly resulting from exclusion of trawlers from the area (Leonhard et al. 2011, but see below for potential artificial reef effects). Other examples exist locally too. Before construction of the Dutch offshore windfarm Egmond aan Zee (OWEZ), the area was heavily fished by beam trawlers. The windfarm, situated 10–18 km from the Dutch coast in water depths of 17–21 m, was built in 2006 and became fully operational in early 2007 (Lindeboom et al. 2011). It consists of 36 turbines and its total surface area is ~40 km^2. Fish surveys in the windfarm area and two control areas were conducted before (2003) and after (2007) construction. Overall differences in fish catches between the surveys were similar in all areas investigated, and the cause for these changes cannot therefore be attributed to construction of the windfarm. At a species level, however, significant increases inside the windfarm area were recorded for sole, whiting and red mullet (*Mullus surmuletus*), and there was a decrease in lesser weever (*Echiichthys vipera*) abundance. Finally, at the Swedish windfarm Lillgrund, no changes in fish abundance considering the area as a whole were detected over 3 years of fishing surveys (but see the discussion on artificial reef effects at the same windfarm below). The windfarm area had not been trawled before the farm was constructed (Bergström et al. 2012).

Based on research into the effects of MPAs, potential positive effects of limitations of fisheries alone on fish and commercially important shellfish are likely to be most prominent for species that were heavily exploited in the area prior to OWF establishment. The largest OWFs have the potential to contain not only greater total biomass but also greater densities of commercial fish. The potential protective function of OWFs may depend too on their spatial configuration, based on evaluations of MPA set-ups in both temperate and tropical seas. Importantly, the location of an OWF depends primarily on the harvestable energy source, the factors influencing the costs of construction and maintenance (e.g. distance to a port) and the potential for cable connection to land. Although MPAs that have been designed and sited with no primary focus on conservation or fisheries enhancement can provide these functions to a varying extent, the protection from fishing may not necessarily be of significance for the habitats and species in question.

Deployment of Artificial Habitats

Artificial Reefs and Fish Aggregation Devices

The presence of submerged and floating parts of offshore renewable energy devices will influence marine organisms in different ways and to varying extents. For example, the structures will inevitably be colonized by fouling organisms and provide artificial habitats for fish and invertebrates. More formally, the term "artificial habitats" refers to structures deployed with the primary purpose of providing habitats for and/or enhancing the abundance of different fish and invertebrate species. Such structures include artificial reefs (ARs) placed on the seabed, and devices, such as buoys and rafts, placed on the surface or in midwater (i.e. fish aggregation devices, FADs). Artificial habitats can also be seaweed-like habitats placed on the seabed (Godoy and Coutinho 2002). Figure 5.1 shows some widely used artificial habitat types, including ARs and FADs.

ARs are generally defined as any man-made structure intentionally or unintentionally placed on the seafloor, although they are mostly deployed for a particular economic or natural resource management purpose. The construction and deployment of ARs in coastal waters is common worldwide in managing fisheries (Seaman 2004; Whitmarsh et al. 2008), mitigating damage to the environment (Powers et al. 2003), protecting (i.e. from trawling) and/or facilitating the rehabilitation of certain habitats (e.g. spawning sites) or water bodies (Clark and Edwards 1995; Sánchez-Jerez et al. 2002; Kaiser et al. 2006; Gao et al. 2008), for aquaculture purposes (Fabi and Fiorentini 1997), or in increasing the recreational value of an area (e.g. for diving, Wilhelmsson et al. 1998; for fishing, Milon 1989). Further, because artificial habitats often allow the control of different habitat factors, much knowledge on natural reefs has been generated through experiments with ARs (Bohnsack et al. 1991; Jensen 2002). The materials used for AR construction range from specially designed concrete or steel units to scrap materials such as car tyres, shipwrecks and railway cars (Baine 2001). The reefs are created by arranging modules in planned clusters or by deploying material haphazardly. Their sizes for applied (e.g. fisheries) purposes vary from tens of m^2 to nearly 1 km^2. The major incentives for artificial habitat construction vary between countries and regions. AR deployment for enhancement of commercial fisheries is most developed in Japan, and by the end of the 1980s, ARs had been created at some 6,000 locations, and man-made habitats influenced ~10 % of Japan's coastal seafloor down to 200 m (Polovina 1991).

Although some studies have revealed that ARs do not have a significant impact on fish assemblages, accumulated evidence suggests that they generally hold greater fish and decapod densities and biomass, and provide better catch rates, than surrounding soft bottom areas, and in many cases also in relation to adjacent natural reefs (Ambrose and Swarbrick 1989; Bohnsack 1989; Wilhelmsson et al. 1998). Reasons suggested for greater abundance and diversity of fish on and around ARs include enhanced protection (from predation and water movement) and food availability, and the use of the structures by fish as reference points for spatial orientation (Jessee et al. 1985; Nakamura 1985; Bohnsack 1989; Grove et al. 1991). A range of design and location factors may influence the fish community structure on ARs, such as height, size, inclination, protuberance, surface structure, void space and number of interior hollows, shade effects, distance between modules, isolation and composition of the surrounding seabed (Jessee et al. 1985; Stephan and Lindquist 1989; Bohnsack et al. 1994). The importance of the different factors varies between species, trophic groups and spatial use patterns of fish (Risk 1972; Luckhurst and Luckhurst 1978). Attempts to generalize the ecology of ARs need to be made cautiously, because ecological systems are complex, the comparably few data derived from rigorous studies are often restricted in geographic scope, and there is a general lack of standardization in survey methodology and aims.

Another category of artificial habitat is a FAD, which is a floating structure deliberately placed on the surface or suspended in the water column to attract fish and enhance fishing efficiency. At least 333 species belonging to 96 families have been described in the literature as aggregating near/under or to be more distantly associated with floating structures such as logs, drifting algae, jellied zooplankton, whale corpses, floats or other man-made structures (Castro et al. 2002). Fishers through the ages have taken advantage of such aggregating behaviour through, for example, "log fishing", which took place around such natural debris as logs.

With this perspective, it is not surprising that FADs have a proven record of aggregating pelagic fish, in particular tuna (*Thunnus* spp.), mackerel (*Scomber* spp.), scads (Carangidae), and dolphinfish (*Coryphaena hippurus*). The radius of influence of FADs may range from metres (damselfish) to kilometres (tunas, dolphinfish), and from the surface to 700 m deep, depending on species (Seaman and Sprague 1991; Relini et al. 1994; Castro et al. 2002). They can be manufactured buoys or be made, for example, of car tyres, synthetic mesh or palm leaves (Fig. 5.1) and are used widely in both artisanal and larger scale commercial fisheries, particularly in the tropics. For example, some 2,500 FADs are used at any given time in the western Indian Ocean, and 50–70 % of the purse-seine catch of tunas there is from fisheries employing FADs (Moreno et al. 2007). In Sicily too, 30 % of fish landings originate from FAD fisheries (Seaman and Sprague 1991).

Both ARs and FADs can have negative environmental and social effects if not properly planned and/or used. If

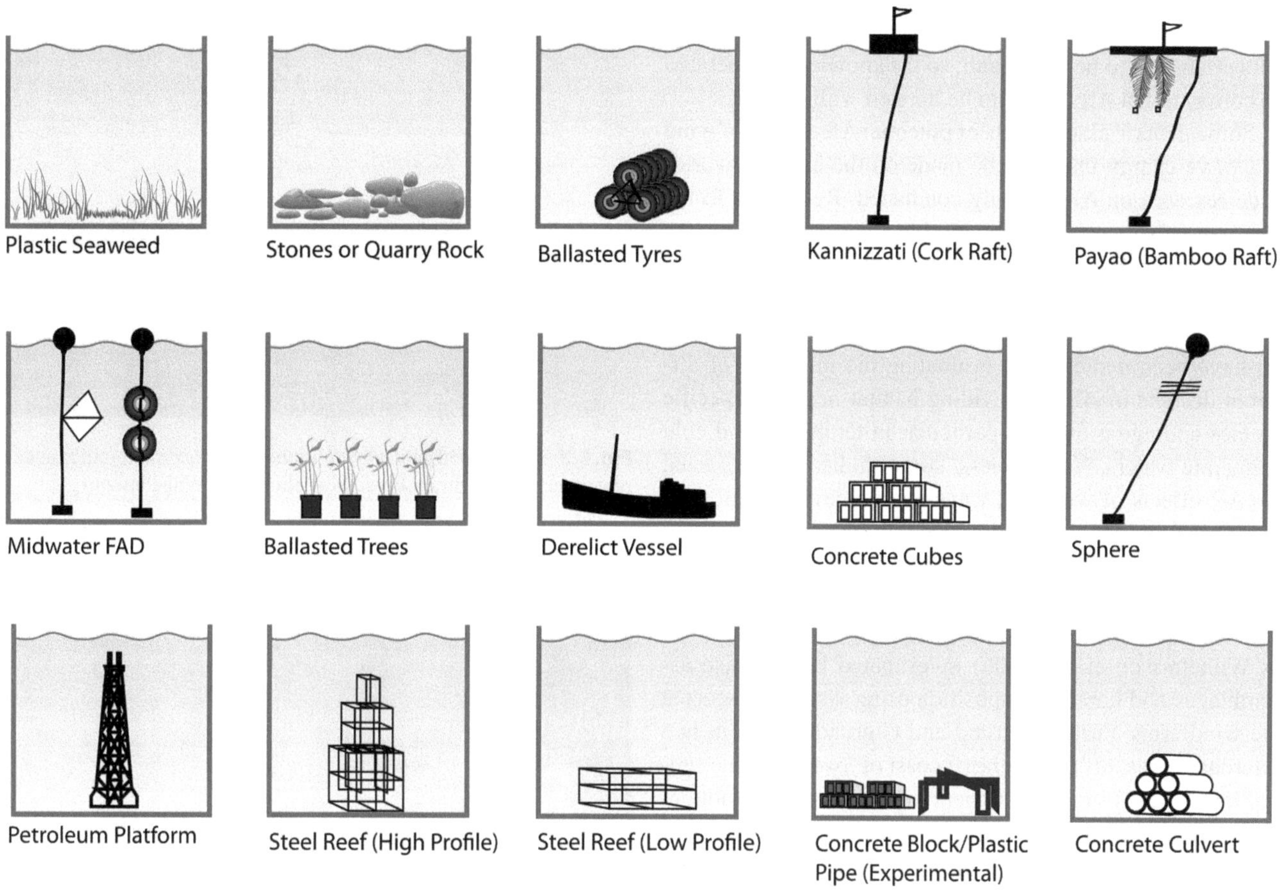

Fig. 5.1 Examples of widely used artificial habitats (after Seaman and Sprague 1991; © C. Wilhelmsson, reproduced with permission)

ARs only aggregate fish from surrounding areas and do not contribute to added production, enhanced fishing efficiency in the area may aggravate overfishing if the new circumstances are not managed with caution. Similarly, as increased catchability of fish is the main purpose of FADs and most commercial species are overfished worldwide, FADs contribute to worsening the problem. Recently, the Japanese Government temporarily banned of the use of FADs in the tuna fisheries to reduce fishing pressure. Conflicts over user rights between fisher groups, and between recreational divers and fishers, can also arise. Further, some studies have revealed that densities of benthic prey items decrease with proximity to an AR, which may be caused by predation pressures exerted by the fish resident on the structures, and it has even been suggested that ARs have the potential to cause the elimination of prey species in adjacent areas (Guichard et al. 2001; Danovaro et al. 2002). FADs have been suggested as potential "ecological traps", meaning that their proliferation could make fish stay too long around structures under non-optimal local feeding conditions, affecting physical condition and growth (Hallier and Gaertner 2008).

Offshore Renewable Energy Devices as Artificial Habitats

Different types of urban structure in the sea, constructed primarily for other purposes, such as oil platforms, breakwaters, pier pilings and pontoons, can also serve as habitats for dense fish and invertebrate assemblages (see Wilhelmsson et al. 2006, for further references). These have been referred to as secondary artificial reefs. Interestingly, Helvey (2002) argued that the potential for oilrigs in California to be essential fish habitat (EFH) should be considered in the environmental review process conducted before their decommissioning. It is reasonable to assume that offshore wind turbines, including the boulders that generally encircle the support structures of wind turbines for scour protection, as well as wave-energy devices anchored with gravity foundations, can function as ARs and potentially enhance the local biomass of sessile and motile organisms. Wind turbines anchored with gravity foundations or monopiles encircled with scour protection should have the greatest potential to give rise to AR effects on fish, crabs and lobsters. As noted earlier, however, the caveat to this would be that the fish do not avoid the area around an energy device because of, for example, noise or other disturbance. Further, these secondary ARs often tend to be very small, so the knowledge generated on conventional ARs needs to be applied with caution.

Some generic assessments of potential AR effects of wind and wave-energy units can be made on the basis of world-wide research on ARs already conducted. Responses to the deployment of artificial habitats, however, vary greatly by species (behaviour, habitat requirements, etc.) and local environmental and ecological conditions (species interactions, food availability, depth, etc.). Moreover, relatively few studies have been dedicated to evaluating the influence of different designs of AR in providing habitat needs of specific species and age groups, in particular in temperate and cold temperate waters. Nevertheless, some studies have targeted the AR effects of wind and wave-energy devices, and such work has provided some indication of how wind- and wave-energy foundations primarily deployed on soft bottoms, as well as their added structural components, can enhance local abundances of fish and invertebrates.

Wilhelmsson et al. (2006) investigated benthic fish assemblages and habitat composition using visual transects at the windfarms Yttre Stengrund and Utgrunden, and in two reference areas, off the southeast coast of Sweden in the central Baltic Sea. Total fish numbers were higher on the bottom in the vicinity of the wind turbines than in surrounding areas. On, or in the water column around, the turbines too, fish community structure was different and total fish abundance higher than on the surrounding seabed. The species contributing most to the differences was the two-spotted goby (*Gobiusculus flavescens*), juveniles of which were ~100×more abundant at the turbine structures than on the seabed 20 m away and in the control areas. The eelpout (*Zoarces viviparous*), the black goby (*Gobius niger*) and the sand goby (*Pomatoschistus minutus*) also contributed to spatial differences in overall fish distribution. At the turbines, bottom-dwelling fish were found mainly in the pockets of steel mouldings encircling the turbine piles, in association with protruding structures, and at the sheltered corner where the wall met the seabed. The turbines were particularly interesting study objects in terms of artificial reef effects, because they were monopiles driven into the seafloor with no scour protection (e.g. rock rubble/boulders) placed around the base of the turbines. Hence, relative to the surrounding seabed, they have little structural complexity at a relevant scale for most fish species, and it has been suggested that they will not have any noteworthy function as ARs, for which shelter is considered often to be the most important factor (Fig. 5.2).

Langhamer and Wilhelmsson (2009) and Langhamer et al. (2009) conducted detailed studies on the AR function of Seabased Ltd wave-energy foundations (Fig. 5.3) during the period 2006–2008. The footprints of the foundations of the Seabased structures were approximately 9 m^2, and the relatively non-complex structures reached 1 m above the seabed

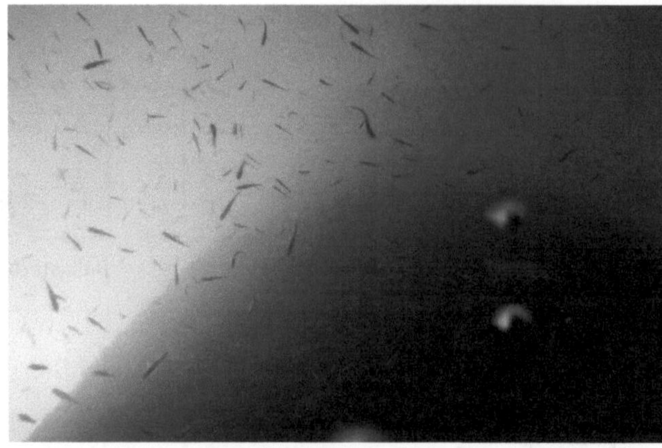

Fig. 5.2 Two-spotted gobies (*Gobiusculus flavescens*) around a monopile in the Kalmar Strait, Baltic Sea (photo Dan Wilhelmsson)

Fig. 5.3 Deployment of a Seabased Ltd wave-energy foundation perforated with 26 holes. Wave-absorbing buoys can be seen in the background (Langhamer and Wilhelmsson 2009, photo Olivia Langhamer)

and were located 25 m deep off Lysekil on the Swedish west coast. Visual surveys conducted during daylight, 2–3 months after the foundations were deployed, showed that the foundations did have an effect on the distribution patterns of fish. Typical observed fish species associated with the foundations were cod, rock gunnel (*Pholis gunnellus*), sand goby and flatfish (Pleuronectidae). Edible/brown crabs (*Cancer pagurus*) also responded postitively to the foundations, and a number of crabs associated with the steel bolts on the foundations.

The structural architecture of the foundations was also manipulated by making 26 holes on the side of 11 of 21 foundations, to investigate the potential to enhance the abundance and diversity of associated organisms, including fish (Langhamer et al. 2009; Langhamer and Wilhelmsson 2009). Primarily edible crabs used the holes, as well as the steel bolts, for shelter, the average number of crabs being five times greater on the foundations with holes than on foundations without holes. The crabs clearly favoured holes

Fig. 5.4 A European lobster (*Homarus gammarus*) residing under a wave-energy foundation (photo Olivia Langhamer)

located 0.5 m above the seabed over holes close to the bottom. Fish were mostly resident on the seabed along the sides of the foundations, and the holes had no apparent effect on fish abundance although abundance was significantly greater on the foundations than in the controls. In both studies, lobsters, both European (*Homarus gammarus*) and Norwegian (*Homarus norvegicus*), were recorded in cavities under some of the foundations (Fig. 5.4).

The studies mentioned above were conducted by visual census using SCUBA, which at least in these waters limits the species recorded to benthic, semi-pelagic and relatively stationary fish. As described above, several fishing surveys and monitoring programmes at windfarms have been conducted targeting larger and more mobile pelagic fish over larger spatial and temporal scales than what can be achieved by visual census. Most of the surveys, however, have not been designed to capture AR effects, at least in terms of survey design and methodology. Nevertheless, the Swedish Fisheries Board conducted fishing surveys for 3 years at Lillgrund windfarm in Öresund, Sweden, with focus on the nature and extent of the AR effect of the turbines (Bergström et al. 2012). Five species of fish clearly aggregated: cod, European eel (*Anguilla anguilla*), eelpout, short-horn sculpin (*Myoxycephalus scorpius*), and goldsinny wrasse (*Ctenolabrus rupestris*). An aggregation of shore crabs (*Carcinus maenas*) was also shown during the second and third year. Further, at windfarm Horns Rev 1, fish species diversity increased close to the turbines as a result of the presence of reef-dwelling species such as goldsinny wrasse, eelpout and lumpsucker (*Cyclopterus lumpus*), which are known to associate with turbines (Leonhard et al. 2011). It was suggested that there was a redistribution of fish towards the areas in the vicinity of wind turbines as a result of feeding opportunities provided by the benthic epifauna on the turbines.

Many species are likely to reside and aggregate in high density relatively close to turbines, but such assemblages would not be recorded with most survey techniques and designs used at windfarms today. Acknowledging this, Couperus et al. (2010) report in the grey literature on their use of high-resolution sonar around five turbines at the Dutch OWEZ windfarm. Aggregations of fish, primarily horse mackerel (*Trachurus trachurus*) and cod, were recorded within 20 m of the turbines at densities as much as 37× higher than in surrounding open waters. A similar pattern was supported by inventories of fish and crustaceans around single wind turbines. For instance, Reubens et al. (2011a) recorded a school of pouting (*Trisopterus luscus*) around a turbine at a windfarm in the Belgian part of the North Sea. The school, of some 22,000 fish with an estimated total biomass of 2.5 t, stayed associated with the structure for at least a year. Monitoring of tagged cod at the same windfarm revealed that also here, cod seemed to be attracted to turbine foundations including the scour protection (Reubens et al. 2011b). Further, at Nysted windfarm, Denmark, >2,000 shore crabs were recorded by Maar et al. (2009) associating with a single turbine.

Apart from the studies at the Seabased foundations described above, comparably little work has been done on the potential for wave-energy devices to function as ARs and FADs. Available data only allow for qualitative estimates to be drawn based on scientific speculation. In terms of AR effects of current technologies, only a few types of device may influence distribution patterns of fish to a noteworthy extent. In shallower water (<40–50 m), it is possible that protruding structures such as the generators standing on the Seabased foundations do affect pelagic fish (Wilhelmsson and Langhamer 2010). Semi-submerged elongated Pelamis wave-energy converters and point-absorbing Wavebob devices are not likely to function as ARs to the same extent because such technologies do not provide complex and/or voluminous structures on the seabed (Wilhemsson and Langhamer 2010). The same study showed that all the wave-energy devices assessed may, on the other hand, have the potential to function as FADs for certain species, though which technology would affect which species of fish most effectively remains uncertain. For example, according to some surveys on FADs in other parts of the world, neither the size nor the number of devices may influence the biomass of fish associated with or aggregated under floating structures (Castro et al. 2002; Soria et al. 2009). Still, for the fish species or life stages where the resources provided by FADs (e.g. food and shelter) are limiting factors, it is possible that biomass increases with the size of the floating parts. Moreover, although it may enhance the abundance of some species, the deployment of numerous floating devices in a single area may actually decrease the FAD function for species that use FADs as reference or meeting points (Dagorn and Fréon 1999).

A number of fish species forage in neighbouring habitats while using an AR mostly or solely as a refuge from predation and water currents/swells. However, many species benefit from the food resources on and around ARs. Stomach analyses by Reubens et al. (2011a) showed that pouting, for instance, feed on the macrobenthos produced on the wind turbine structures. In the study by Maar et al. (2009) cited above, for example, it was estimated that, unless blue mussels (*Mytilus edulis*) and other prey were produced on and around the turbines, the shore crabs, shrimps and fish associated with the turbines would have needed a food resource equivalent to what is naturally provided in a larger area than claimed by the entire windfarm.

Wind turbines appear to offer a suitable substratum and good feeding conditions for blue mussels (Wilhelmsson et al. 2006; Wilhelmsson and Malm 2008; Maar et al. 2009). They can harbour 10× more biomass per unit area of blue mussels than bridge pilings in the same region (Maar et al. 2009), so each turbine can support 1–2 t of mussels and a windfarm in total can double the biomass of filter-feeders in an area (Maar et al. 2009). Also the floating parts of wave-energy devices can provide suitable substrata for blue mussels (Langhamer et al. 2009). The submerged and floating parts of wind- and wave-energy devices and the adjacent seabed may therefore resemble a reef ecosystem, with a variety of interacting components. Fish and crustaceans feed on macroinvertebrates on the wind turbines, and on mussels subsequently dislodged from the devices (Langhamer and Wilhelmsson 2009; Maar et al. 2009; Reubens et al. 2011a). The large numbers of semipelagic fish (see Wilhelmsson et al. 2006) and sessile organisms associated with the monopiles may contribute to increased benthic production of food for fish and decapods on the surrounding seabed through the deposition of organic material such as faecal matter, organic litter and dead organisms, which may in turn attract benthos-feeding fish (e.g. Wilhelmsson et al. 2006).

The Management and Manipulation of AR and FAD Effects

It is clear that wind- and wave-energy devices may locally enhance the biomass of a number of fish and decapod species of commercial importance. The parts of offshore energy devices that are surface orientated or penetrate the whole water column (e.g. buoys and supporting structures) may function as FADs for pelagic fish, potentially providing additional fisheries management challenges (Fayram and de Risi 2007).

For some fish species limited by the availability of reef habitat for refuge, territory, food and behavioural requirements, and for heavily fished and vulnerable species, the habitat provided by primarily wind- but also wave-energy devices may serve as total stock size enhancers, although this may mostly be of local relevance (Bohnsack 1989). The availability of suitable habitat may, for example, be limiting during life stages such as the early benthic phase, moulting or spawning, and these demographic bottlenecks could be widened through provision of artificial habitats (Butler and Herrnkind 1997). In theory, some structures can cater particularly well for spawning, recruitment, survival and growth simply by providing better refuge from fishing and predation than natural habitats, or by enhancing foraging efficiency. For other species, or for the same species in other regions, however, such artificial habitats will only redistribute fish production and existing fish biomass.

By considering specific habitat preferences of marine organisms in the design of OWFs and devices, the abundance and diversity of associated species can clearly be enhanced. Commercially important or threatened species can be specially catered for where desired, and for both fish and decapods, the location of energy installations and the configuration of scour protection in terms of the density of boulders and void space can be of significance. The diversity of microhabitats can be further enhanced through structural modifications of the foundations themselves (Wilhelmsson et al. 2006; Langhamer and Wilhelmsson 2009). Overhangs, under which juvenile fish can reside, pockets and other protruding structures can be mounted at different depths on foundations to enhance abundance, diversity and recruitment rates of fish and decapods (Wilhelmsson et al. 2006). Holes can be made in foundations to improve their utilization, preferably with multiple openings because some fish avoid holes with just one opening as a result of limited escape options and water exchange. The single-entrance holes made in the experiments by Langhamer and Wilhelmsson (2009) and Langhamer et al. (2009) had, for example, no notable influence on associated fish numbers.

Frond mats, which can be used to prevent erosion around wind turbines, may function as artificial algae or seagrass beds, providing shelter for juvenile fish and serving as habitat for fish of conservation importance, such as pipefish and seahorses, as suggested by Linley et al. (2007) and Wilson and Elliot (2009). Habitat preferences of the European lobster at different life stages are often of interest when considering the AR aspects of wind- and wave-energy structures, owing to the commercial importance of the species and indications that the species may indeed be largely habitat-limited (Jensen et al. 1994). It is also possible to influence the composition of fouling assemblages on the footings of energy installations, to enhance feeding and shelter opportunities. Position in the water column and orientation of the substratum could, for example, be adapted specifically to cater for macroalgae or filter-feeding animals (e.g. mussels, Petersen and Malm 2006; Wilhelmsson and Malm 2008).

Once shelter and compartment have been provided in sufficiency, food supply on and in the vicinity of the artificial structures can become a limiting factor for the numbers of associated fish and crustaceans (Bohnsack et al. 1991). A threshold for when design features do not increase abundance of fish and decapods could therefore be reached. Fish and decapods may, further, take advantage of the combined resources on and around several energy devices. The space between the devices and the mobility of a species may influence whether predatory and grazing species can be sustained within or will visit an area (Overholtzer-Mcleod 2006). This would also influence the strength of potential cascade effects of increased predation, on for instance the spatial distribution of prey and competitive species. To seek to influence this, additional ARs can be deployed around turbines to create larger continuous areas of "reef".

Many fish and crustaceans associated with ARs forage largely on the seabed around the reefs (Ambrose and Anderson 1990; Kurz 1995; Einbinder et al. 2006). This behaviour may explain the low faunal biomass recorded on the seabed around wind turbines by Maar et al. (2009). Sheehan et al. (2008) showed that the addition of artificial habitats (tiles) can enhance the abundance of shore crabs throughout estuaries, despite heavy fishing pressure in them. This can increase the predation on juvenile fish (e.g. flounder, *Platichthys flesus*, and cod), and adversely affect biomass of fauna such as mussels and oysters (Sheehan et al. 2008). Moreover, in experiments with wave-energy foundations, the densities of spiny starfish (*Marthasterias glacialis*) were negatively influenced by the presence of holes, potentially because of increased predator abundance (e.g. edible crabs; Langhamer and Wilhelmsson 2009). There may, therefore, be situations and habitats where the reef effects of offshore energy installations, including predation, local habitat changes and altered energy and nutrient transports, are unwanted. Knowledge on the influence of design and configuration of artificial reefs on specific taxa or functional groups could then be used to minimize the abundance of certain species and any other associated effects.

Another effect that may be wanted or unwanted depending on species and ecological condition is the potential change in dispersal patterns of fish and crustaceans (Page et al. 2006; Glasby et al. 2007; Vaselli et al. 2008). Reef-dwelling species can be limited in their distribution by there being too great a distance between hard-bottom areas. Large clusters of wind- and wave-energy installations could fill some of these gaps with hard substrata, and hence change the dispersal patterns and the biogeographic distribution of species within a region. Changed dispersal patterns have already been demonstrated for sessile organisms, including non-indigenous species, at windfarms in the North Sea and the Baltic Sea and in areas with many coastal structures and petroleum platforms (see Wilhelmsson et al. 2010, for references). A detailed survey of the turbines at the Belgian C-Power Project windfarm revealed that several non-indigenous species were common in the intertidal fouling community; the oyster *Crassostrea gigas*, the barnacles *Elminius modestus* and *Megabalanus coccopoma*, the amphipod *Jassa marmorata*, the crab *Hemigrapsus sanguineus* and the midge *Telmatogeton japonicus* (Degraer et al. 2011). Further, the characteristic high intertidal splash zone on offshore wind turbines is often colonized by a conspicuous *Telmatogeton* zone (Leonhard and Pedersen 2006; Brodin and Andersson 2009). As development of offshore wind power and wave energy progresses according to plans and already issued licenses, these effects may be important for certain species or regions.

OWFs can form attractive fishing grounds (Fayram and de Risi 2007). Fishing would in most cases have to be limited to setnets, cages and hook-and-line techniques, and recreational fisheries could in particular benefit. In the test park for wave power off Lysekil in Sweden (Langhamer and Wilhelmsson 2009), commercial lobster and crab fishing is reportedly already taking place. It is interesting to note that at least for shore crabs, the added production through artificial habitat deployment might balance or outweigh what is being harvested (Sheehan et al. 2008). In most cases, however, increases in overall fish and crustacean biomass (added production) through the provision of habitat are difficult to prove, and there is generally a risk of aggravating overfishing by simply concentrating fishery resources. Recreational diving organizations are, though, showing interest in windfarms as dive sites, the value of ARs created for recreational diving having been highlighted by Brock (1994), who showed that such structures can provide greater revenue as dive sites than when used for commercial fishing. Indeed, Wilhelmsson et al. (1998) estimated the minimum gross annual income generated from dives on two shipwrecks in the Red Sea at US $ 368,000.

Concluding Remarks

OWFs may consist of hundreds (wind power) to thousands (wave power) of units, requiring substantial areas (up to hundreds of km^2) where fishing would inevitably be limited or prohibited. Although there will always be a risk of fishing effort being redirected to other areas, such areas, with their often numerous AR patches, may positively influence local fish and decapod stocks. It is in this context worth noting that the benefits of management strategies combining MPAs and AR deployment, e.g. to recover depleted fish stocks, are increasingly being acknowledged. If enforcement of an MPA is ineffective, ARs can further inhibit bottom trawling, and for species where the quality of habitat is enhanced through ARs, this might further enhance the effects of an MPA (Pitcher et al. 2002; Claudet and Pelletier 2004). Restrictions on fishing can alternatively protect the fish resources associated with ARs (Pitcher et al. 2002).

Considering the scale of offshore renewable energy development, the AR deployment and trawl exclusion may be of regional importance for habitat-limited and/or heavily fished species, provided the affected species spend a notable part of their life cycle within the area and that there is no significant negative impact on reproductive behaviour or feeding efficiency.

In terms of the effects of trawling exclusion and limitation of other fisheries, the evidence from operational OWFs is currently weak. Theoretically though, the biggest OWFs do have the potential to contain greater densities of commercial fish, to enhance local species diversity and to generate spill-over effects to adjacent areas.

Species-specific responses to ARs, FADs and MPAs vary, and the available data largely limit predictions to scientific speculation. More robust assessments will require well-designed field studies in differently configured OWFs. Intensified research on habitat use, habitat requirements and behavioural attributes of individual species, and in the environments around each OWF is necessary. The results would also better equip the toolbox for manipulating the abundance of certain taxa through structural design and configuration of an OWF, where desired. Although more complete and robust data on the impacts of OWFs on the areas as a whole, including the cumulative effects of a large number of wind turbines and wave-energy devices, can only be gathered at large farms, research on smaller clusters of devices or even individual devices can provide guidance. Many potential impacts can with appropriate caution be considered as the summary effects of individual energy devices. Provided sample size and design (i.e. the number and distribution of devices) fulfil minimum requirements for appropriate statistical analyses, measurements of distance-related impacts of single devices on fish and crustaceans can, therefore, still generate valuable data,

Acknowledgements We thank IUCN, Vattenfall AB, Sida and E.ON for supporting the work within which knowledge for parts of this chapter was gathered.

References

Ambrose RF, Anderson TW (1990) Influence of an artificial reef on the surrounding infaunal community. Mar Biol 107:41–52

Ambrose RF, Swarbrick SL (1989) Comparison of fish assemblages on artificial and natural reefs off the coast of southern California. Bull Mar Sci 44:718–733

Baine M (2001) Artificial reefs: a review of their design, application, management and performance. Ocean Coast Manage 44:241–259

Bergström L, Sundqvist F, Bergström U (2012) Effekter av en havsbaserad vindkraftpark på fördelningen av bottennära fisk. Swedish Environmental Protection Authority, Report 6485, January 2012. ISBN 978-91-620-6485-3 (in Swedish)

Beukers-Stewart BD, Vause BJ, Mosley MWJ, Rossetti HL, Brand AR (2005) Benefits of closed area protection for a population of scallops. Mar Ecol Prog Ser 298:189–204

Bohnsack JA (1989) Are high densities of fishes on artificial reefs the result of habitat limitation or behavioural preference? Bull Mar Sci 44:934–941

Bohnsack JA, Harper DE, McClellan DB, Hulsbeck M (1994) Effects of reef size on colonization and assemblage structure of fishes at artificial reefs off southeastern Florida, USA. Bull Mar Sci 55:796–823

Bohnsack JA, Johnson DL, Ambrose RF (1991) Ecology of artificial reef habitats and fishes. In: Seaman W, Sprague LM (eds) Artificial habitats for marine and freshwater fisheries. Academic Press, San Diego, pp 61–84

Brock RE (1994) Beyond fisheries enhancement: artificial reefs and ecotourism. Bull Mar Sci 55:1181–1188

Brodin Y, Andersson MH (2009) The marine splash midge *Telmatogeton japonicus* (Diptera; Chironomidae)—extreme and alien? Biol Invas 11:1311–1317

Butler MJ, Herrnkind WF (1997) A test of recruitment limitation and the potential for artificial enhancement of spiny lobster (*Panulirus argus*) populations in Florida. Can J Fish Aquat Sci 54:452–463

Castro JJ, Santiago JA, Santana-Ortega AT (2002) A general theory on fish aggregation to floating objects: an alternative to the meeting point hypothesis. Rev Fish Biol Fish 11:255–277

Christie MR, Tissot BN, Albins MA, Beets JP, Jia Y, Ortiz DM, Thompson SE et al (2010) Larval connectivity in an effective network of marine protected areas. PLoS ONE 5:e15715. doi:10.1371/journal.pone.0015715

Clark S, Edwards AJ (1995) Coral transplantation as an aid to reef rehabilitation: evaluation of a case study in the Maldive Islands. Coral Reefs 14:201–213

Claudet J, Osenberg CWL, Benedetti-Cecchi P, Domenici JA, Garcia-Charton A, Perez-Ruzafa F, Badalamenti J et al (2008) Marine reserves: size and age do matter. Ecol Lett 11:481–489

Claudet J, Pelletier D (2004) Marine protected areas and artifical reefs: review of the interactions between management and science. Aquat Living Resour 17:129–138

Côté IM, Mosqueira I, Reynolds JD (2001) Effects of marine reserve characteristics on the protection of fish populations: a meta-analysis. J Fish Biol 59(Suppl. A):178–189

Couperus B, Winter E, van Keeken O, van Koten T, Tribuhl S, Burggraaf D (2010) Use of high resolution sonar for near-turbine fish observations (DIDSON). IMARES Wageningen UR, Report C138/10, 29 pp

Dagorn L, Fréon P (1999) Tropical tuna associated with floating objects: a simulation study of the meeting point hypothesis. Can J Fish Aquat Sci 56:984–993

Danovaro R, Gambi C, Mazzola A, Mirto S (2002) Influence of artificial reefs on the surrounding infauna: analysis of meiofauna. ICES J Mar Sci 59:S356–362

Degraer S, Brabant R, Rumes B (2011) Offshore wind farms in the Belgian part of the North Sea: selected findings from the baseline and targeted monitoring. Royal Belgian Institute of Natural Sciences, Management Unit of the North Sea Mathematical Models, Marine Ecosystem Management Unit, 157 pp + Annex

Deveen JF (1978) Selective tidal transport in the migration of North Sea plaice (*Pleuronectes platessa*) and other flatfish species. Neth J Sea Res 12:115–147

Einbinder S, Perelberg A, Ben-Shaprut O, Foucart MH, Shashar N (2006) Effects of artificial reefs on fish grazing in their vicinity: evidence from algae presentation experiments. Mar Environ Res 61:110–119

EWEA (2009) Oceans of opportunity: Harnessing Europe`s largest domesting energy resource. In: Azau S, Rose C (eds). http://www.ewea.org/fileadmin/ewea_documents/documents/publications/reports/Offshore_Report_2009.pdf

Fabi G, Fiorentini L (1997) Molluscan aquaculture on reefs. In: Jensen AC (ed) European artificial reef research, Proceedings of the 1st EARRN Conference, Ancona, Italy, March 1996. Southampton Oceanography Centre, Southampton, pp 123–140

Fayram AH, de Risi A (2007) The potential compatibility of offshore wind power and fisheries: an example using bluefin tuna in the Adriatic Sea. Ocean Coast Manag 50:597–605

Frederiksen M, Edwards M, Richardson AJ, Halliday NC, Wanless S (2006) From plankton to top predators: bottom-up control of a marine food web across four trophic levels. J Anim Ecol 75:1259–1268

Gao QF, Shin PKS, Xu WZ, Cheung SG (2008) Amelioration of marine farming impact on the benthic environment using artificial reefs as biofilters. Mar Pollut Bull 57:652–661

Glasby TM, Connell SD, Holloway MG, Hewitt CL (2007) Nonindigenous biota on artificial structures: could habitat creation facilitate biological invasions? Mar Biol 151:887–895

Godoy EAS, Coutinho R (2002) Can artificial beds of plastic mimics compensate for seasonal absence of natural beds of *Sargassum furcatum*? ICES J Mar Sci 59:S111–S115

Grove RS, Sonu CJ, Nakamura M (1991) Design and engineering of manufactured habitats for fisheries enhancement. In: Seaman W, Sprague LM (eds) Artificial habitats for marine and freshwater fisheries. Academic Press, San Diego, pp 109–149

Guénette S, Pitcher TJ, Walters CJ (2000) The potential of marine reserves for the management of northern cod in Newfoundland. Bull Mar Sci 66:831–852

Guichard F, Bourget E, Robert JL (2001) Scaling the influence of topographic heterogeneity on intertidal benthic communities: alternate trajectories mediated by hydrodynamics and shading. Mar Ecol Prog Ser 217:27–41

Hallier JP, Gaertner D (2008) Drifting fish aggregation devices could act as an ecological trap for tropical tuna species. Mar Ecol Prog Ser 353:255–264

Halpern BS (2003) The impact of marine reserves: do reserves work and does reserve size matter? Ecol Appl 13:117–137

Helvey M (2002) Are Southern California oil and gas platforms essential fish habitat? ICES J Mar Sci 59:S266–271

Inger R, Attrill MJ, Bearhop S, Broderick AC, Grecian WJ, Hodgson DJ, Mills C et al (2009) Marine renewable energy: potential benefits to biodiversity? An urgent call for research. J Appl Ecol 46:1145–1153

IUCN (1988) Proceedings of the 17th session of the General Assembly of IUCN and the 17th Technical Meeting, San José, Costa Rica, 1–10 February 1988. Gland, Switzerland, 322 pp

Jensen A (2002) Artificial reefs of Europe: perspective and future. ICES J Mar Sci 59:S3–13

Jensen AC, Collins KJ, Free EK, Bannister RCA (1994) Lobster (*Homarus gammarus*) movement on an artificial reef: the potential use of artificial reefs for stock enhancement. Crustaceana 67:198–211

Jessee WN, Carpenter AL, Carter JW (1985) Distribution patterns and density estimates of fishes on a Southern California artificial reef with comparisons to natural kelp–reef habitats. Bull Mar Sci 37:214–226

Kaiser MJ, Clark KR, Hinz H, Austen MCV, Somerfield PJ, Karakassis I (2006) Global analysis and recovery of benthic biota to fishing. Mar Ecol Prog Ser 311:1–14

Kurz RC (1995) Predator–prey interactions between gray triggerfish (*Balistes capriscus* Gmelin) and a guild of sand dollars around artificial reefs in the North-eastern Gulf of Mexico. Bull Mar Sci 56:150–160

Langhamer O, Wilhelmsson D (2009) Colonisation of fish and crabs of wave energy foundations and the effects of manufactured holes: a field experiment. Mar Environ Res 68:151–157

Langhamer O, Wilhelmsson D, Engström J (2009) Artificial reef effect and fouling impacts on offshore wave power foundations and buoys: a pilot study. Estuar Coast Shelf Sci 82:426–432

Leonhard SB, Pedersen J (2006) Benthic communities at Horns Rev before, during and after construction of Horns Rev offshore wind farm. Vattenfall. Final Report/Annual Report 2005, 134 pp

Leonhard S, Stenberg C, Støttrup J (2011) Effect of the Horns Rev 1 offshore wind farm on fish communities follow-up seven years after construction. DTU Aqua, Orbicon, DHI, NaturFocus. DTU Aqua Report, 246–2011, 99 pp

Lindeboom HJ, Kouwenhoven HJ, Bergman MJN, Bouma S, Brasseur S, Daan R, Fijn RC et al (2011) Short-term ecological effects of an offshore wind farm in the Dutch coastal zone; a compilation. Environ Res Lett 6:e035101, 13 pp. doi:10.1088/1748-9326/6/3/035101

Linley EAS, Wilding TA, Black K, Hawkins AJS, Mangi S (2007) Review of the effects of offshore windfarm structures and their potential for enhancement and mitigation. Report from PML Applications Ltd to the Department of Trade and Industry, Contract RFCA/005/0029P

Luckhurst BE, Luckhurst K (1978) Analysis of influence of substrate variables on coral reef fish communities. Mar Biol 49:317–323

Maar M, Bolding K, Petersen JK, Hansen J, Timmerman K (2009) Local effects of blue mussels around turbine foundation in an ecosystem model of Nysted offshore wind farm Denmark. J Sea Res 63:159–174

Milon JW (1989) Artificial marine habitat characteristics and participation behavior by sport anglers and divers. Bull Mar Sci 44:853–862

Moreno G, Dagorn L, Sancho G (2007) Using local ecological knowledge (LEK) to provide insight on the tuna purse seine fleets of the Indian Ocean useful for management. Aquat Living Resour 20:367–376

Nakamura M (1985) Evolution of artificial fishing reef concepts in Japan. Bull Mar Sci 37:271–278

OSPAR 2006 Guidance on developing an ecologically coherent network of OSPAR marine protected areas. Biodiversity series. OSPAR Agreement 2006-3. www.ospar.org

Overholtzer-Mcleod KL (2006) Consequences of patch reef spacing for density-dependent mortality of coral reef fishes. Ecology 87:1017–1026

Page H, Dugan MJE, Culver CS, Hoesterey JC (2006) Exotic invertebrate species on offshore oil platforms. Mar Ecol Prog Ser 325:101–107

Petersen JK, Malm T (2006) Offshore windmill farms: threats to or possibilities for the marine environment. Ambio 35:75–80

Piet GJ, Rijnsdorp AD (1998) Changes in the demersal fish assemblage in the south-eastern North Sea following the establishment of a protected area ("plaicebox"). ICES J Mar Sci 55:420–429

Pitcher TJ, Buchary EA, Hutton T (2002) Forecasting the benefits of no-take human-made reefs using spatial ecosystem simulation. ICES J Mar Sci 59:17–26

Polovina JJ (1991) Fisheries applications and biological impacts of artificial habitats. In: Seaman W, Sprague LM (eds) Artificial habitats for marine and freshwater fisheries. Academic Press, San Diego, CA, pp 154–174

Powers SP, Grabowski JH, Peterson CH, Lindberg WJ (2003) Estimating enhancement of fish production by offshore artificial reefs: uncertainty exhibited by divergent scenarios. Mar Ecol Prog Ser 264:265–277

Relini M, Relini OL, Relini G (1994) An offshore buoy as a FAD in the Mediterranean. Bull Mar Sci 55:1099–1105

Reubens JTS, Degraer SR, Vincx M (2011a) Aggregation and feeding behaviour of pouting (*Trisopterus luscus*) at wind turbines in the Belgian part of the North Sea. Fish Res 108:223–227

Reubens J, Degraer S, Vincx M (2011b) Spatial and temporal movements of cod (*Gadus morhua*) in a wind farm in the Belgian part of the North Sea using acoustic telemetry, a VPS study. In: Degraer S, Brabant R, Rumes B (eds) Offshore wind farms in the Belgian part of the North Sea: selected findings from the baseline and targeted monitoring. Royal Belgian Institute of Natural Sciences, Management Unit of the North Sea Mathematical Models, Marine Ecosystem Management Unit, pp 39–46

Risk MJ (1972) Fish diversity on a coral reef in the Virgin Islands. Atoll Res Bull 153:1–6

Roberts CM, Hawkins JP, Gell FR (2005) The role of marine reserves in achieving sustainable fisheries. Philos Trans Roy Soc B 360:123–132
Rogers SI (1997) A review of closed areas in the United Kingdom exclusive economic zone. Cefas Science Series Technical Report, 106. Cefas, Lowestoft, 20 pp
Russ GR, Alcala AC (2011) Enhanced biodiversity beyond marine reserve boundaries: the cup spillith over. Ecol Appl 21:241–250
Sánchez-Jerez P, Gillanders BM, Rodríguez-Ruiz S, Ramos-Esplá AA (2002) Effect of an artificial reef in *Posidonia* meadows on fish assemblage and diet of *Diplodus annularis*. ICES J Mar Sci 59:S59–68
Sanchirico JN, Malvadkar U, Hastings A, Wilen JE (2006) When are no-take zones an economically optimal fishery management strategy? Ecol Appl 16:1643–1659
Seaman W (2004) Artificial habitats and the restoration of degraded marine ecosystems and fisheries. In: Proceedings of the 39th European Marine Biology Symposium, Genoa. Springer, Dordrecht, pp 143–155
Seaman W, Sprague LM (1991) Artificial habitat practices in aquatic systems. In: Seaman W, Sprague LM (eds) Artificial habitats for marine and freshwater fisheries. Academic Press, San Diego, pp 1–27. ISBN 0-12-634345-4
Sheehan EV, Thompson RC, Coleman RA, Attrill MJ (2008) Positive feedback fishery: population consequences of "crab-tiling" on the green crab *Carcinus maenas*. J Sea Res 60:303–309
Soria M, Dagorn L, Potin G, Fréon P (2009) First field-based experiment supporting the meeting point hypothesis for schooling in pelagic fish. Anim Behav 78:1441–1446
Spalding M, Wood L, Fitzgerald C, Gjerde K (2010) The 10 % target: where do we stand? In: Toropova C, Meliane I, Laffoley D, Matthews E, Spalding M (eds) Global ocean protection: present status and future possibilities. IUCN, Gland, Switzerland, pp 25–40
Stephan CD, Lindquist DG (1989) A comparative analysis of the fish assemblages associated with old and new shipwrecks and fish aggregation devices in Onslow Bay, North California. Bull Mar Sci 44:698–717
Thrush SF, Dayton PK (2002) Disturbance to marine benthic habitats by trawling and dredging: implications for marine biodiversity. Ann Rev Ecol Syst 33:449–473
Vaselli S, Bulleri F, Benedetti-Cecchi L (2008) Hard coastal-defence structures as habitats for native and exotic rocky-bottom species. Mar Environ Res 66:395–403
Walmsley SF, White AT (2003) Influence of social, management and enforcement factors on the long-term ecological effects of marine sanctuaries. Env Conserv 30:388–407
Whitmarsh D, Santos MN, Ramos J, Monteiro CC (2008) Marine habitat modification through artificial reefs off the Algarve (southern Portugal): an economic analysis of the fisheries and the prospects for management. Ocean Coast Manag 51:463–468
Wilhemsson D, Langhamer O (2010) The potential for wave energy devices to provide artificial habitats and protect areas from fishing. IUCN, Vattenfall AB, 87 pp
Wilhelmsson D, Malm T (2008) Fouling assemblages on offshore wind power plants and adjacent substrata. Estuar Coast Shelf Sci 79:459–466
Wilhelmsson D, Malm T, Öhman M (2006) The influence of offshore wind power on demersal fish. ICES J Mar Sci 63:775–784
Wilhelmsson D, Malm T, Thompson R, Tchou J, Sarantakos G, McCormick N, Luitjens S et al (2010) Greening blue energy: identifying and managing the biodiversity risks and opportunities of offshore renewable energy. IUCN, Gland, Switzerland, 102 pp. ISBN 978-2-8317-1241
Wilhelmsson D, Öhman MC, Ståhl H, Shlesinger Y (1998) Artificial reefs and dive tourism in Eilat, Israel. Ambio 27:764–766
Wilson JC, Elliott M (2009) The habitat-creation potential of offshore wind farms. Wind Energy 12:203–212

Marine Renewable Energy, Electromagnetic (EM) Fields and EM-Sensitive Animals

6

Andrew B. Gill, Ian Gloyne-Philips, Joel Kimber and Peter Sigray

Abstract

In the marine environment there are natural magnetic and electric fields associated with both physical and biological sources, and there are anthropogenic electromagnetic fields (EMFs) that permeate it. Many marine animals can detect electric and magnetic fields and utilize them in such important life processes as movement, orientation and foraging. Here, these EMFs are explored and discussed in terms of how they arise, their properties (particularly those that are measurable) and the animals that have the ability to detect them. Then the evidence base for whether anthropogenic EMFs can affect sensitive receptor animals is explored. As marine renewable energy developments (MREDs) expand rapidly worldwide, with multiple devices and networks of subsea cables that emit EMFs into the marine environment, it is necessary to focus on their interaction with marine animals. The MRED industry has to take EMFs into account, so the industry perspective is also covered. Finally, suggestions are made on how research on EMFs associated with MREDs (and other sources) and its interaction with marine animals should advance in future.

Keywords

B fields · E fields · Electrosensitive · EMF · Magnetosensitive

Overview and Terminology

Humans are generally unaware that they live within an electromagnetic world. The concept that we are surrounded by charged particles may seem ethereal but is more real than generally acknowledged. We are familiar with an occasional lightning storm, but we are also bombarded continually with electromagnetic emissions from the sun and encompassed by the Earth's own geomagnetic field (and other sources, such as granite geology). At a local level, humans are immersed among anthropogenic electromagnetic emissions that emanate from the plethora of electrical appliances and technologies that have been developed to become part of everyday life.

Animals with which humankind shares the environment are also exposed to electromagnetic fields (EMFs) both natural and anthropogenic in origin. Several animals are known to be able to detect EMFs (or more specifically the component electric and/or magnetic field) and to use them for activities that are vitally important in terms of resource gain and movement around their environment. This is particularly true of marine animals, many of which undertake large-scale

A. B. Gill (✉)
Environmental Science and Technology Department, School of Applied Sciences, Cranfield University, Cranfield, Bedfordshire MK43 0AL, UK
e-mail: a.b.gill@cranfield.ac.uk

I. Gloyne-Philips · J. Kimber
Centre for Marine and Coastal Studies Ltd (CMACS Ltd), 80 Eastham Village Road, East Ham, Wirral CH 62 0AW, UK

P. Sigray
Swedish Defence Research Agency—FOI, Stockholm 164 90, Sweden

M. A. Shields, A. I. L. Payne (eds), *Marine Renewable Energy Technology and Environmental Interactions,* Humanity and the Sea,
DOI 10.1007/978-94-017-8002-5_6,

movements that apparently follow the orientation of the Earth's geomagnetic field (Kirschvink 1997). Moreover, some animals possess specialist electroreceptive organs that can detect weak bioelectric fields emitted by their prey and conspecifics.

Although knowledge of how marine animals use magnetic and electric fields is increasing, there is still scant understanding of how animals interact with anthropogenic sources of EMF. The purpose here, therefore, is to provide an overview of what is currently known about EMFs in the marine environment and to evaluate how electromagnetically sensitive receptor animals interact with the EMFs associated with marine renewable energy developments (MREDs). The latter are being developed to transform renewable sources of energy into electricity and are therefore a new and rapidly expanding feature of our coastal and marine landscape; the time to ask appropriate questions for planning and development is now. It should also be recognized that there are other sources of anthropogenic EMF in the seas and although the focus of this chapter is on MREDs, one needs to be aware too of these other sources. To date, however, these other sources, which include power and telecommunications cables and some pipeline structures, have not been adequately considered in the context of their interaction with the marine environment.

We start by highlighting what is currently understood about the natural electromagnetic marine environment and the anthropogenic sources of EMF (including MREDs). This is followed by an evaluation of animals regarded as sensitive receptors to EMF (and therefore potentially affected), natural or artificial. The sources of EMF are first described and focus then turns to marine renewable energy (in all forms except for shore-based devices) for two reasons: first, a MRED is for the generation of electricity, so the subsea cables that transport electricity between devices and to the shore emit EMFs (as do some of the devices); second, the rapid deployment of MREDs and extensive planning worldwide is creating an unprecedented change to coastal and offshore environments in terms of EMFs because of the high density of cables associated with the seabed that network the arrays of devices. Further, generating electricity from marine renewable resources has spawned significant plans for subsea cable networks for importing and exporting electricity through the seas.

The current level of understanding EMFs has influenced the guidance concerning environmental factors related to consenting (or permitting) and deploying MREDs. We therefore need to look to the future by taking an integrated and realistic approach to how one might improve understanding of this poorly known topic, though set within the context that renewable energy development should not be hindered by uncertainty in the knowledge base. The approach has to be collaborative, with science and industry moving forward together using scientific evidence to assist decision-makers and stakeholders to promote renewable energy deployments within an adaptive framework that can respond to improved knowledge from focused research and targeted environmental monitoring.

The topic area is obviously still lacking much detail, so several issues are surrounded with great uncertainty, perhaps causing a sense of frustration in those looking for answers; this needs to be taken into account. The aim, therefore, is to provide clear understanding and to suggest how one can best advance knowledge in effectively reducing some of the uncertainty around the topic, thereby providing the correct focus and a route map for the future.

The electromagnetic world has some specific terminology, much of which is unfamiliar to many, so as a start, terms are listed here and described in an attempt to allow the reader to understand the subtle differences between some terms.

- B field—the magnetic flux density (more commonly referred to as the magnetic field);
- cable rating—the specifications attributed to a cable in terms of cross-sectional area, maximum voltage (volts, or V), current carrying capacity (amps or amperes, or A) and maximum power (watts, or W);
- cgs—centimetre-gramme-second units (or Gaussian units), as opposed to SI units;
- current (A)—the flow of electrical charges within a cable (or other media);
- E field/iE field—electric field/induced electric field;
- electric dipole field—two separate electric charges give rise to an electric dipole field;
- electric gradient/electric potential gradient—the rate of change in electric charge between two points in terms of their distance apart, which is equal to the electric field ($V\ m^{-1}$);
- electroreceptor—a sensory organ found within some animals that allow them to detect DC and AC (see below) electric fields in the environment;
- EM—electromagnetic;
- EMF—electromagnetic field;
- gauss (G)—a unit of magnetic induction (the cgs unit of measurement of a magnetic field), 1 G equalling 10^{-4} tesla (T);
- geomagnetic field—Earth's magnetic field;
- hertz (Hz)—a unit of frequency, where the number of hertz equals the number of cycles per second;
- HVAC—high voltage alternating current;
- HVDC—high voltage direct current;
- magnetic dipole field—magnetic dipoles consist of north and south poles that give rise to a magnetic dipole field;
- magnetite—a ferromagnetic mineral; some animals have magnetite crystals within their bodies that are thought to be involved in magnetoreception and migration;

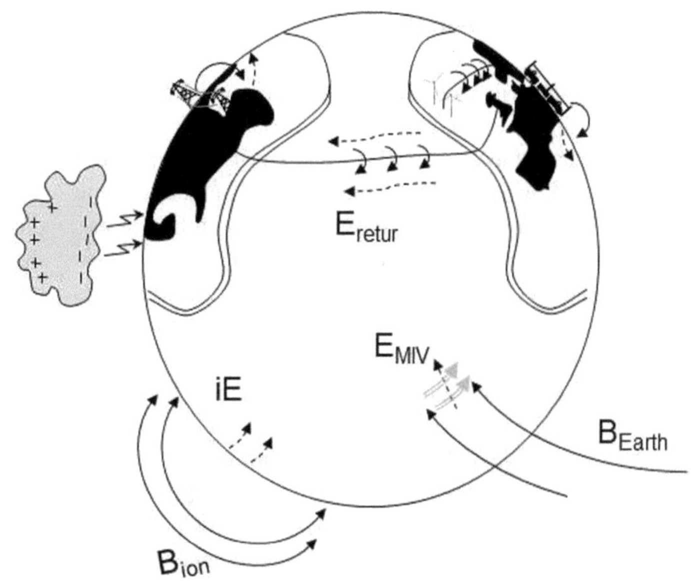

Fig. 6.1 Representation of natural and anthropogenic sources of electric (E) and magnetic (B) fields associated with the Earth. The B fields penetrate or are generated in the sea and the E fields (*dashed arrows*) are induced by the B fields (*solid arrows*). B_{Earth}—Earth's magnetic field; B_{ion}—magnetic fields from the ionosphere; E_{MIV}—motionally induced voltage electric field; E_{retur}—electric field from subsea cable return current fed through the sea

- migration—animal movement over distances along predictable routes;
- orientation—localized movement of an animal in relation to habitat features;
- SI—the international system of unit description;
- tesla (T)—a unit of magnetic induction or magnetic flux density (SI), where 1 T equals 10^4 G;
- three-phase core—three cables laid parallel (sometimes also twisted) surrounded by shielding, most commonly used for HVAC cables;
- voltage (V)—the electrical potential difference between two points (measured in V);
- V m^{-1} (volts per metre)—the unit for electric field measurement (SI).

Introduction to the Natural Electromagnetic World

The dominant natural source of magnetic field in the sea (as well as on land) is the Earth's geomagnetic field (Fig. 6.1, B_{Earth}). It has a variable strength that averages ~30,000 nanoTesla (nT) at the equator and 60,000 nT at the poles. The field is dipole-like (i.e. a pair of magnetic poles of equal magnitude but opposite polarity separated by a distance) and somewhat displaced relative to the geographic North Pole (hence the existence of a magnetic North Pole). It does not vary greatly in the short term, but does vary from century to century by about 6 %.

The sea itself is non-magnetic so has no effect on the prevailing geomagnetic field. This implies that the Earth's magnetic field propagates unaltered through the oceans. The second largest magnetic source in the environment is generated by solar wind, which is a stream of energetic particles ejected by the sun. The flux of the particles hits the upper atmosphere and creates ions, which form electric currents in the ionosphere that give rise to magnetic fields (Fig. 6.1, B_{ion}), which propagate to the Earth's surface and into the sea where the field is attenuated as predicted by the skin effect (see below). The magnetic field has a strength of some 1–10 nT at the Earth's surface on a solar quiet day. After a solar eruption, the solar wind is stronger and can give rise to magnetic storms where the magnetic fields at the Earth's surface can be several hundred nT (i.e. around two orders of magnitude less than the geomagnetic field).

The strongest generator of natural electric fields is atmospheric thunderstorms, whereby an updraft of air inside a cloud separates charges, resulting in a dramatic increase of the electric field between the top of the cloud and the ground. Eventually, when the electric field intensity is strong enough, strikes of lightning are generated over the land and sea.

A key difference between magnetic and electric fields is that magnetic fields penetrate the sea and the Earth's crust whereas electric fields are mainly bound to the sea and to water-filled sediments, so electric current flows in the marine environment (creating electric fields). The dominant natural electric source in the sea is generated by the phenomenon of motionally induced voltage (MIV) in the ocean, where the saline, and hence conducting, water moves through the Earth's magnetic field (Fig. 6.1, E_{MIV}). Assuming that the magnetic field is mainly vertical, as at higher latitudes, the electric field would be orientated in a horizontal direction. Variability in the electric field present is proportional to the velocity of the moving water, because the Earth's magnetic field at its lowest order can be regarded as constant in time. The dominant motion of the sea is governed by the tidal stream, so in the sea the electric field varies with tidal frequency, an effect that is weaker at the equator where the magnetic field is horizontal.

The Physics of EMF in the Sea

For a comprehensive description of electromagnetism the reader is referred to the rich variety of books dealing with electromagnetic theory (e.g. Paris and Hurd 1969; Stratton 2007). Here, however, a short introduction is given to some of the main concepts relevant to what follows in the chapter. In a physical description of electromagnetism, there are positive and negative electric charges that create electric fields and magnetic dipoles which give rise to magnetic fields. In a static case, when the sources are neither moving nor changing in time, the static electric and magnetic fields are not related and can exist independently of each other. The physics changes when the sources move or the field strength varies in time. Consequently, the electric and magnetic fields become intrinsically related and co-exist simultaneously as a combined electromagnetic field. Electromagnetic theory shows that the propagation of the field depends on the geometry and the electromagnetic properties of the environment as well as the source (Paris and Hurd 1969). An important aspect here is that sources can interact with each other by mediation of the electric and magnetic fields, so a primary source (e.g. a geomagnetic field or a cable) can exert forces on a secondary source (e.g. magnetite-based material within some migratory marine animals) when the latter is exposed to fields generated by the primary source.

An effect of electromagnetism that plays an important role in the sea is the attenuation of alternating magnetic fields in conductive media. An external magnetic field impinging on a conductive medium such as the sea will induce electric currents, known as eddy currents, which oppose the external magnetic field. The strength of the external magnetic field in the conductive medium decreases with penetration depth in the conductor. This phenomenon is called the skin effect, and the attenuation is heavily dependent on the frequency of the magnetic field. The skin depth is defined as the depth at which the field is attenuated to 0.37 of its initial value. A 50 Hz undulating magnetic field (generated by electric equipment such as cables) has a skin depth of ~35 m in Atlantic Ocean water, whereas a 1 MHz field has one of just 0.25 m. The skin effect is applicable for sources inside and outside the conductive media; for example, the magnetic field circulating an electricity-carrying subsea cable will decrease with increasing distance from the cable, and the attenuation will be stronger at higher frequencies.

A globally present background EMF phenomenon is the Schumann resonances that can be observed when one measures electric fields. Schumann resonances arise because the Earth and the ionosphere constitute an electromagnetic cavity, but because of its size, the EMF that can reside inside the Earth–ionosphere cavity will be at extremely low frequency (tens of Hz). The excitation source of the resonance is lightning, where the electric field is vertical and the magnetic field horizontal in the atmosphere. However, on entering the ocean, the electric field refracts at the sea-surface interface and becomes horizontal. The Schumann resonances are detected at specific frequencies where the three lowest are at 7.9, 14.3 and 20.8 Hz. The electric field of the resonances 50 m deep are ~10 nV m^{-1} and the magnetic field is in the picoTesla (pT) range, so it is difficult to detect because of the high background magnetic levels. Even if weak, the resonances show up in the spectrum of an electric field measurement and are often used as an indicator of properly working field equipment and it is not known if the resonances are utilized or sensed by aquatic animals.

An Introduction to the Anthropogenic Electromagnetic World

Existing Anthropogenic EMF Sources

As highlighted earlier, magnetic and electric fields occur naturally in the environment. Although they have altered throughout geological time, it is certain that animals have evolved within a complex magnetic and electrical environment and that several taxa have developed sensory systems that take advantage of their presence. In the past few hundred years, however, industrial and commercial interest in the sea has introduced anthropogenic sources of EMF that were not present in the environment before the pre-industrial era. Examples include subsea cables, anti-corrosion systems, and the most recent and arguably greatest potential development, marine renewable energy arrays and cable networks throughout the seas. Further, economic and political trends suggest that the number of these sources will continue to increase.

Land-Based EMF

Land-based sources generate electromagnetic fields that penetrate into the sea surface or via land–sea interfaces. Powerline infrastructure comprises the main land-based source, with magnetic fields of the power grid appearing at known frequencies, e.g. 50 and 60 Hz (and harmonic frequencies). Power cables are not always visible; many are buried, especially in urban areas. Their electromagnetic influence is similar, however, to that of overhead power lines (although the cable design and configuration leads to emissions at lower intensity). Nevertheless, if situated near the sea or even along a shore or beach they will elevate the electromagnetic fields in the nearshore environment, with the affected area depending on a number of factors, e.g. electric current and cable characteristics and armour. Environmental properties such as water conductivity and sediment composition have less influence on levels of EMF. A crude estimate is that a power cable carrying an electric AC current of 500 A will give rise to a magnetic field of about 1 μT at a distance of 1 m from the cable. The extent of the power-grid systems on land is often designed to feed return currents through the ground, meaning that electric current flows through both ground and sea, giving rise to electric fields.

EMF from land-based sources also includes the traction network of the railway system across the world, which runs on different frequencies relative to the national power grid (commonly 16⅔, 25 and 50 Hz), as well as DC. The electric powering of trains is generally fed from overhead current lines and redistributed to the power source through the rail tracks, with the implication that relatively large electric currents flow in the ground, and if in close proximity, will also enter the sea. The fields are emitted as magnetic fields that are refracted into the sea either from the surface or from the seabed. In most national railway systems, the rails are continuous and welded, which leads to a continuous flow of electric current. However, some local railway lines have tracks with non-welded joints, which might give rise to varying resistance and result in transient currents when passing the joints. These transients are observable as broadband elevation of background levels in the frequency spectrum.

All high-frequency sources such as radio and radars do not penetrate into the sea because they are effectively attenuated by the skin effect. The influence, if any, of land-based EMF sources will be near the shore and in the intertidal zone, decreasing in the offshore environment.

Sea-Based EMF

There are a number of sources that generate EMF in the sea. Offshore sources are mainly associated with cables that emit fields directly into the sea, but ships, boats and busy navigation areas are associated with elevated levels of EMF. Two sources that generate EMF in ports and marinas are cables and electric machinery, which can be situated on land or in the sea. Moreover, both ships and boats emit EMF. Part of the field is generated by the electrical equipment on board the vessels, but the dominant sources are the electrochemical action between dissimilar metals and the anti-corrosion system. Corrosion currents from passive or active cathodic protection systems form either electric or magnetic fields around the hull. The electric current has both a static (DC) and modulating (AC) component. Static magnetic fields surround a ship as a result of the ferro-magnetic steel in its main construction. In addition, the steel hull interacts with the Earth's magnetic field, giving rise to eddy currents flowing in the hulls that produce an induced magnetic field with strength dependent on the location and heading of the vessel. In ports and marinas, metallic objects such as tubes, rails and groundings, which are in contact with both seawater and land, will result in corrosion currents flowing in the sea. Further, metallic objects that consist of different metals in contact with each other and the sea will give rise to corrosion currents and hence a localized electric field. There are many metallic objects in ports and marinas as well as along shorelines.

Where there are ships there will be locally generated EMF in the sea. This implies that in congested areas such as shipping lanes, there will be anthropogenic electric and magnetic fields present. The fields will move with the ship, potentially covering large areas of the sea relatively consistently for busy shipping lanes.

Subsea Cables

In the sea, the primary sources of anthropogenic EMF are submarine cables located on or in the seabed, e.g. three-phase AC, high-voltage AC and DC, low-voltage AC and DC and telecommunication cables. Regardless of power rating or material, the current along a cable generates magnetic fields that circle the conductor (Stratton 2007; Slater et al. 2010). For DC cables, the sediment properties do not influence the magnetic field. This is not the case for AC cables where the asymmetry and rotational characteristics of the magnetic field will give rise to eddy currents in the sea and in the sediment, provided the sediment contains seawater. Moreover, the water–sediment asymmetry will distort the magnetic field. Where *in situ* measurements are not available, the B field and the induced E field can be approximated through analytical or numerical modelling (see Slater et al. 2010; Gill et al. 2012a).

A second important factor associated with subsea cables is the way the return current is handled (Slater et al. 2010). Even in high-voltage DC cables used for transmitting power over long distances, the return current has in some cases been fed through the sea (Fig. 6.1, E_{retur}). This has the effect that electric currents will flow in the sea, covering large areas and long distances. An example of this is the high voltage direct current cable (HVDC) FenoScan that runs between Forsmark in Sweden and Rauma in Finland (Öhman et al. 2007). The return current through the sea is 1,600 A at full power, even if the voltage difference between the two endpoints is low. In this case, the electric current traverses a distance of 180 km through the Bothnian Sea, giving rise to both a dipole-like electric current and an electric field distribution (Fig. 6.1, E_{retur}). About halfway between the two endpoints, the electric current distribution reaches its maximum width.

Engineers have recognized that using the sea as the return is not a preferred design if based on environmental and efficiency issues, and there are now several DC-cabling techniques available. Therefore, to preclude currents returning through the sea, different types of bipolar technique are employed, e.g. the main and the return current being fed in two separate cables or conductors (Slater et al. 2010). The SwePol link cable between Sweden and Poland has two separate cables laid in parallel on the seabed, a configuration that solves the problem with currents in the sea but generates relatively strong magnetic fields that reduce with decreasing distance between the two cables because of the cancelling effect of the two opposing currents. Another technique employs two electric conductors in a single cable, so the distance between the currents is minimal. This technique lowers the emitted magnetic fields and is therefore preferable when considering the environmental aspect of EMF emissions. Such cables have been used in several power systems, for example in the North Sea.

Long-distance telecommunication cables utilize the same technique as HVDC cables but with a current of ~1 A. The current is used for powering electronic devices mounted along the cable with the purpose of reamplifying the optical signal. The return current is fed through the sea, giving rise to an electric field (Fig. 6.1, E_{retur}). HVDC cables are used when the distance to the power grid is relative long. For a windfarm situated near the coast (i.e. <30 km from shore), three-phase AC transmission cables have generally been employed to date. It is well known that the currents in these systems sum to zero, so the magnetic field reaches a low level, typically in the μT to pT range several metres distant from the cable. However, because of the concentric configuration of the conductors, there is a region close to the cables where the intensity of the magnetic field is not negligible.

In archipelagos and close to beaches, the density of low-voltage cables (e.g. 220 or 110 V) is relatively high. These are often intentionally or unintentionally grounded to earth, resulting in return currents flowing through the sea, so near-shore, the electric field from the power grid is omnipresent and the electric field is typically in the $\mu V\ m^{-1}$ range (Soderberg 1969).

Although already discussed in connection with grounding on land, it is worth stressing here the existence of sea-based grounding systems. These are used, for example, in HVDC, low-voltage and telecommunication systems as well as for grounding infrastructure. Specially designed and deployed electrode systems are used in direct contact with the sea (e.g. copper or platinum), so these groundings are confined to specific areas where the electric currents converge and the electric fields are elevated.

Estimating the magnetic fields generated by cables is difficult because there are many cable types, configurations and materials. The electric current will directly scale the magnetic field, so stronger currents will induce stronger magnetic fields. There are a number of different cables commercially available, consisting of one, two or three conductors. A rule of thumb valid for both AC and DC cables is that more conductors give rise to weaker fields because the magnetic field generated by the individual electric currents cancels out some of the emitted field. Other factors that influence the generated magnetic fields are the magnetic properties of the armouring of the cable and the helicity (twisting) of the conductors in a cable; the extent of the magnetic field depends on both these factors. A very approximate estimate is that a single DC cable with one conductor carrying 1,000 A generates a field of 150 μT at 1 m distance, whereas a three-phase cable carrying 100 A generates 1 μT at 1 m. However, such general emitted levels need to be referred to with caution because it is much more appropriate to perform a specific assessment for each individual cable with its individual properties and characteristics. Different cases of cables are considered in detail by Öhman et al. (2007).

Marine Renewable Energy Sources

The principle energy product harnessed from marine resources is electricity. In general, all the devices used in offshore or coastal waters, whether wind, wave, tidal or other technology, convert mechanical (wind, wave, tidal), thermal (ocean thermal energy conversion) or chemical gradient (osmotic pressure) energy from the marine environment into electrical power. The devices used then transmit this electrical power through a cable system/network, for which the design, specifications and type will have implications for the EMF associated with each form of generation. Therefore, there are technology-specific and common EMF sources (such as cables) that need to be considered by device-specific manufacturers and developers. Further details on each technology are available in Gill (2011).

Wind

At present, the dominant MRED technology harnesses wind power. Its dominance is reflected in the extensive construction and future plans for deployment on land and at sea throughout the world. Plans are also being made to increase the size of the wind-turbine arrays and the power output of each device. The consequence with respect to EMF is that more will be emitted through a combination of greater areal extent (i.e. their environmental footprint) and greater emissions from more power output.

The electricity is generated in the turbine nacelle at the top of a tower and transmitted down the inside of the tower to an exit J-tube (Fig. 6.2), where the cable then runs down the outside of the tower into the water column and down to the seabed. The cable is then normally buried in 1–1.5 m of sediment, but where the substratum is hard, concrete mattresses, rock protection, or other bespoke forms of protection are deployed. The principal reason for burying a subsea

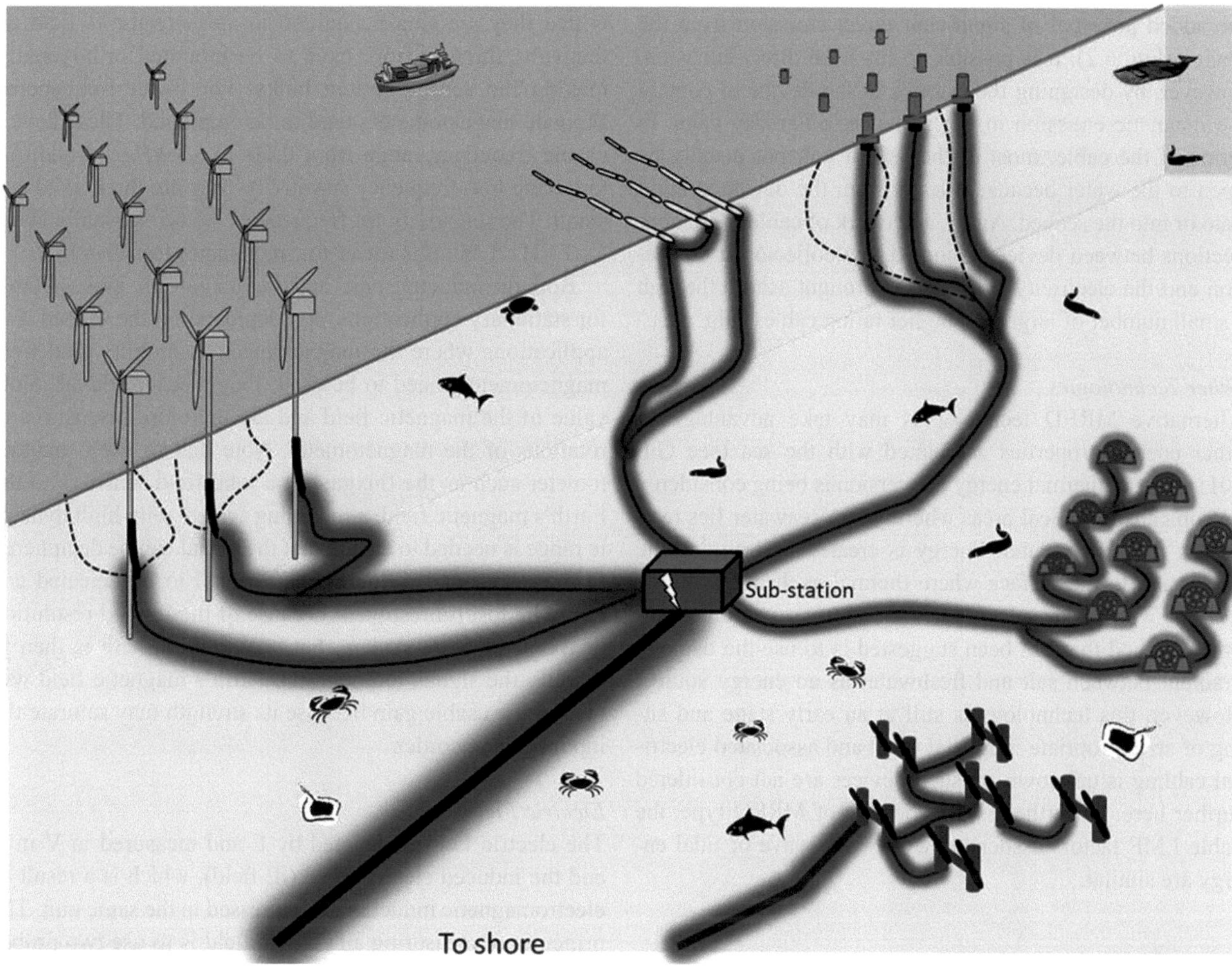

Fig. 6.2 Composite representation of MRED devices and cable connections and networks. Cable positions are indicative but there are many different configurations, some routed through an offshore substation (as shown), others straight to an onshore substation. The *dark zone* along the cables highlights the EMF emission, which is greater in extent for cables carrying greater power. Not to scale

cable is to protect it, perhaps from contact with vessel anchors and fishing gear or abrasion from the seabed. From an EM perspective, there is no dampening of the B field emission, as illustrated by the shaded zone along each of the cables in Fig. 6.2. The EMF will peak at the skin of the cable, so its burial will ensure a physical barrier between the cable and epibenthic animals (but not infauna). However, burial will not dampen the EMF to levels below known lower limits of detection by EM-sensitive animals. Other than this physical barrier, there are no known reasons to use burial as a form of mitigation for the environmental effects associated with EMF.

The cable from the J-tube to the seabed joins with other cables to form a network that ends either at a collector turbine or a substation (Fig. 6.2). Where cables come together, the EMF becomes more complex because of the different orientations and geometries of the emissions. The expectation would be that some emissions cancel each other out, whereas others will be additive and increase the EMF. This is a very poorly understood aspect of the EMF related to subsea cable networks. From the substation, the electricity is transmitted ashore through one or more cable of larger specification, coming ashore either in the tidal zone or up-river.

Wave

Conversion of wave energy to electricity requires a device to be able to react to the wave regime. In general, the device will possess a cable that runs from the device at the surface (e.g. sea-surface motion type) or near the sea surface (e.g. wave oscillating device) down through the water column to the seabed. The system then follows a similar pattern to windfarm subsea cable networks (Fig. 6.2).

Tidal

Energy-producing tidal devices are constructed in the mid/lower water column and/or near the seabed (Fig. 6.2). Some such devices contain large permanent magnets so there is the added potential of significant direct emission from the device (Fig. 6.2). It is possible to limit the direct emission, however, by designing the exterior of the device to contain the magnetic emission in, for instance, a Faraday cage. In terms of the cable, most of the length will not usually be open to the water because it leads from the device straight onto or into the seabed. Again, a network of cables with connections between devices is joined to a collector or substation and the electricity generated is brought ashore through a small number of larger but higher rating cables (Fig. 6.2).

Other Technologies
Alternative MRED technologies may take advantage of other energy properties associated with the sea (see Gill 2011). Ocean thermal energy conversion is being considered in tropical/subtropical areas where warm seawater lies over colder, deeper seawater. Energy is created by bringing the cold water to the surface where thermal exchange with the warmer water takes place, creating an energy source. Another method that has been suggested is to use the osmotic gradient between salt and freshwater as an energy source. However, this technology is still at an early stage and siting of an appropriate industrial plant and associated electrical cabling is unknown, so such devices are not considered further here. Nevertheless, irrespective of MRED type, the cable EMF factors associated with wind, wave or tidal energy are similar.

EMF Properties and Quantification

It is apparent that MREDs emit EMF into an environment that possesses natural EMFs, so as with any anthropogenic emission into the environment, it is necessary to understand the levels, characteristics and extent of the emission. In order to measure the EMF present in the sea, one needs to consider both the electric and the magnetic field as separate entities with different properties requiring quantification.

Attributes

Magnetic Fields
There are two magnetic fields, denoted B (the magnetic flux density) and H (the magnetic field intensity), related by a multiplicative constant. In non-magnetic materials such as the sea, it is normal to discuss the magnetic field in terms of B alone, as measured in tesla (T) in SI units and in gauss (G) in cgs units. There are many commercial sensors available that can quantify magnetic fields; the most commonly used are inductive sensors, which consist of a conducting coil wound around a magnetic material. These sensors have a sensitivity of <1 pT $\sqrt{Hz}^{-1}$ for frequencies >1 Hz. A disadvantage is that they are single axial, so to measure the B field effectively, three sensors have to be mounted orthogonally, making the sensor system bulky. For lower frequencies, fluxgate magnetometers tend to be employed. These have a usable frequency range from 0 Hz to ~3 kHz. In addition to having low-frequency sensitivity, they are three-axial and small. The sensitivity for frequencies >1 Hz is usually about 5 pT $\sqrt{Hz}^{-1}$, but it is lower for frequencies less than that.

Both the inductive coils and the fluxgate are most suitable for stationary applications, e.g. deployed on the seabed. For applications where the magnetometer is mobile, total field magnetometers need to be used; they measure the absolute value of the magnetic field and are therefore insensitive to rotations of the magnetometer. Note that to use a magnetometer such as the fluxgate or a total field sensitive to the Earth's magnetic field, a recording system with high dynamic range is needed to ensure that the signal can be deciphered from background influence. The signal to be detected can easily be lost (unresolved) because of the limited resolution of the acquisition system, but one solution here is then to amplify the signal, although the Earth's magnetic field will restrict the usable gain because its strength may saturate the input of the recorder.

Electric Fields
The electric field is denoted by E and measured in V m^{-1}, and the induced electric field (iE field), which is a result of electromagnetic induction, is expressed in the same unit. The principle of measuring an electric field is to use two probes that are in electrical contact with the sea. The measured voltage and the distance between the two probes are used to establish the electric field in V m^{-1} (known as the electrical voltage gradient). Consequently, the distance between the probes will scale linearly with the obtained voltage. To establish the electric field in 3-dimension requires at least four probes orthogonal to each other, with one central probe placed at the origin, but often the probes are used in pairs, making a total of six probes per three-axial sensor system.

Relative to magnetometers, there are far fewer commercial products available. An electric field sensor system consists of two main components, probes and low-impedance amplifiers. The probes need both low contact resistance and large wet-surface area to keep the intrinsic electrical noise low. In marine applications, two main groups of probe are used; non-polarizable and polarizable electrodes, the most common being the non-polarizable silver–silver chloride electrode (Ag–AgCl). This electrode has low intrinsic noise characteristics and the advantage of being sensitive to frequencies as low as 0 Hz. However, it does have several drawbacks; it breaks easily, is sensitive to hydrodynamically induced motion, is sensitive to the surrounding environment, and an electrode pair often gives rise to large voltage bias (in the mV range). Further, for long-term applications, the electrode pair measurements have a tendency to drift as a consequence of changes in material properties, salinity and temperature differences at the electrodes and biological fouling. However, these impediments can be overcome partly by using a semi-transparent protection between the electrode and the seawater as well as a specially designed amplifier. An electrode that belongs to the second group is the carbon fibre electrode, which has a useable frequency range from 0.01 Hz up to several kHz. It is a robust electrode and can be handled without special care, although it also has to be protected from hydrodynamically induced motion. Its disadvantages are that it costs much more and it cannot be used at very low frequencies.

To measure weak electric fields in the sea requires low-noise amplifiers with low-impedance inputs. It is vital that amplifiers be isolated from external ground to avoid Earth currents, and special care needs to be taken when the data acquisition system is connected to the power grid. There are two types of amplifier; chopped and linear. The first type is used with Ag–AgCl electrodes, and because of the chopping technique of the amplifier input, the leakage currents at the input evens out, resulting in a prolonged life of the electrode. It also has the advantage of low noise levels at low frequencies. The second type is based on an ordinary linear amplifier technology, except that the input impedance is low and noise much reduced for low frequencies. The typical noise floor at 1 Hz for the two types of amplifier is ~5 nV $\sqrt{Hz}^{-1}$.

The same methodology is used when measuring electric and magnetic fields. Measurements tend to be performed with the sensor in a stationary position, because of the directional sensitivity of the sensors. Sensor rigs elevating the sensors from the seabed might be used to measure within the water column, with the risk that the hydrodynamically induced motion of the rig may influence the resolution of the data collected. To establish the fields as a function of distance to source, measurements can be taken in a stationary position and the sensors then moved to repeat the measurement at several locations. An alternative is to make use of two sensors that sample simultaneously, the first in a stationary position and the second being moved around on the seabed. This system has the advantage that the measured data from the moved sensor can be referenced against the stationary sensor; this configuration is pertinent when the field is temporally varying on short time-scales. Measurements in the water column are difficult to perform, however, because of the unavoidable motion of the sensor. For magnetic fields, though, total field magnetometers are favoured.

Electromagnetism and Animals

A relatively large number of marine animals is either known to be sensitive to electromagnetic fields in the marine environment or has the potential to detect them (see Peters et al. 2007, for a comprehensive list). Some are classed as electroreceptive and some as magnetosensitive. The former type detects directly emitted electric fields or electric fields induced from magnetic fields and the latter respond directly to emitted magnetic fields.

Electric Field Detection

The most common electroreceptive marine animals are chondrichthyans, i.e. elasmobranchs (sharks, skates and rays) and holocephalans (chimaeras); all have specialized electroreceptive organs, the Ampullae of Lorenzini, which are well studied and described (Tricas and Sisneros 2004). Their electroreceptive system is very sensitive, allowing the round stingray, *Urobatis helleri*, for example, to detect electrical voltage gradients (i.e. electric fields) as low as 5–20 nV m^{-1} (Kalmijn 1982; Tricas and New 1998). The electrosense is used to detect the DC and AC bioelectric fields emitted by prey, conspecifics and potential predators, and is also thought to aid orientation and navigation (see below). There is evidence available too to suggest that some elasmobranchs may be repulsed during encounters with E fields in excess of several hundreds of mV; for example, Kimber (2008) and Yano et al. (2000) demonstrated repulsion of some species of elasmobranch at 400 and 1,000 μV m^{-1}, respectively.

Other electrosensitive marine fish (or fish-like taxa) with similar specialized electroreceptors are the Agnatha (jawless fish; e.g. lampreys), Acipenseriformes (sturgeons and paddlefish) and Coelacanthiformes (coelacanths). Teleost (bony) fish such as salmon, tuna, plaice and cod have been postulated as being electrically sensitive, but they do not possess specialized electroreceptors (at least, none have been found to date) and are thought to be able to detect induced voltage gradients associated with water movement through magnetic fields, such as tidal movements (Metcalfe et al. 1993). It is likely that such fish respond to the E fields associated with peak tidal movements (Pals et al. 1982). However, as the actual sensory mechanism is not yet properly understood, some reviews have cast doubt on these abilities (Bullock 1986), although it does appear that some fish species may be repulsed by strong E fields (6–15 V m^{-1}, or even more; Uhlmann 1975; Poléo et al. 2001). Even the electrogenic stargazers (Uranoscopidae) do not appear to utilize electroreception (Bradford 1986; Alves-Gomez 2001). The Anguillidae (migratory eels, with adults living in freshwater but juveniles and breeding grounds in the sea) are an exception, however, having been demonstrated as being sensitive to weak electric AC and DC electric fields (Enger et al. 1976; Berge 1979).

In terms of other taxa, there are few examples of animals that have electrosensory apparatus, although this statement is based on a very small set of studies. A recent demonstration of electroreception of AC fields in a dolphin (Czech-Damal et al. 2011) suggests that the widely held belief that cetaceans are not sensitive to E fields may be incorrect, a sensitivity threshold of just 460 μV m^{-1} having been recorded, approximately three orders of magnitude greater than that of elasmobranchs.

Magnetic Field Detection

Magnetically sensitive marine animals can be categorized into two groups based upon their mode of magnetic field detection: those that utilize induced E field detection, and those capable of direct B field detection. The first group relates to species that are electroreceptive (see above), i.e. species that sense the presence of a magnetic field indirectly by detection of the electrical field that is induced (iE field) by the movement of water through a magnetic field or by their own movement through that magnetic field. In a natural scenario, electric field induction usually results from an animal positioning itself in a tidal current, and animals may actually time certain activities (e.g. foraging) by detecting the diurnal cues resulting from varying tidal flows. The second group is believed to use either magnetic particles (magnetite) within their own tissues (Kirschvink 1997) or photoreceptor molecules (cryptochromes) within their eyes (Solov'yov et al. 2010) to detect magnetic fields. Although the precise sensory mechanism remains unknown, it is generally acknowledged that such animals can utilize magnetic cues (such as the geomagnetic field) to orientate themselves in their environment during migration (geonavigation; for reviews, see Walker et al. 1992; Kirschvink 1997). Marine animals considered to possess the capability to detect magnetic fields in the sea include cetaceans (whales, dolphins and porpoises; see Kirschvink et al. 1986), chelonians (turtles; see Lohman and Lohman 1996), certain teleosts (e.g. flatfish, salmonids and eels; see Souza et al. 1988; Metcalfe et al. 1993), crustaceans (lobsters, crabs, prawns and shrimps; see Everitt 2008; Ugolini and Pezzani 1995) and molluscs (snails, bivalves and cephalopods; see Willows 1999).

There have been suggestions that pinnipeds (seals, sea lions and walruses) and sirenians (manatees and dugongs) are capable of geomagnetic navigation, but despite some species undertaking long-range, accurate migrations (e.g. harp seals migrate ~5,000 km), neither magnetite nor cryptochromes have been found in either group (Riedman 1990; Sheppard et al. 2006). It is currently suggested that the migrations of these marine mammals are based on olfactory or mechanosensory cues.

On occasion, queries arise about flying animals and EMF. The potential issue is most related to wind turbines and whether any EMF emitted will be detectable by airborne animals. As far as we are aware this has not yet been studied, but if it were it would be important to consider it in comparison with other, more certain responses (e.g. aerial avoidance, collision) and the consequent effect on migration, foraging and mortality rate of these animals. Although birds do orientate to geomagnetic cues (linked to migration and orientation), there is no apparent evidence that they use EMF at the scale associated with an MRED EMF. Moreover, there is no known mechanism by which seabirds when diving can detect the EMF emanating from MREDs and subsea cables under water. Hence, with the lack of any evidence to the contrary, we merely assume here that EMFs are not likely to be encountered or detected during aerial migration or seabird foraging, so are not considered further.

A demonstrable link between bats and EMF relates to the use of radar to reduce mortality at onshore wind turbines. Radar emits an EMF, which appears to cause bats to avoid wind turbines (Nicholls and Racey 2009), thereby acting as an effective deterrent and mitigation. That study, however, was not dealing with the EMF being detected by bats but rather its use at radar frequencies to reduce collision mortality for bats foraging on insects attracted to the turbines.

Interactions Between Marine Animals and EMF

The Evidence Base

The topic of anthropogenic EMF and its interaction with marine animals (whether positive or negative) is one of the least understood and most complex of all the environmental questions related to MREDs. As highlighted already, understanding of how marine animals experience and use either natural magnetic or electric fields is poor, but the knowledge relating the same animals to anthropogenic sources (e.g. subsea cables) and their properties within the environment is worse. As is evident from the sections below, there are published cases of animals apparently responding to EMF sources in the marine environment, the most convincing being those that have attempted to address a specific research question or hypothesis focused on improving the scientific evidence base covering marine animals and their interaction with EMF.

Determining Effect vs. Impact

As there is limited understanding of EMFs and how marine animals react to them, it is appropriate to consider how that lack of knowledge should be addressed, set perhaps within the context that there are a number of other factors that can interact and potentially affect the same animals. Another crucial aspect is that there may be evidence for an effect (e.g.

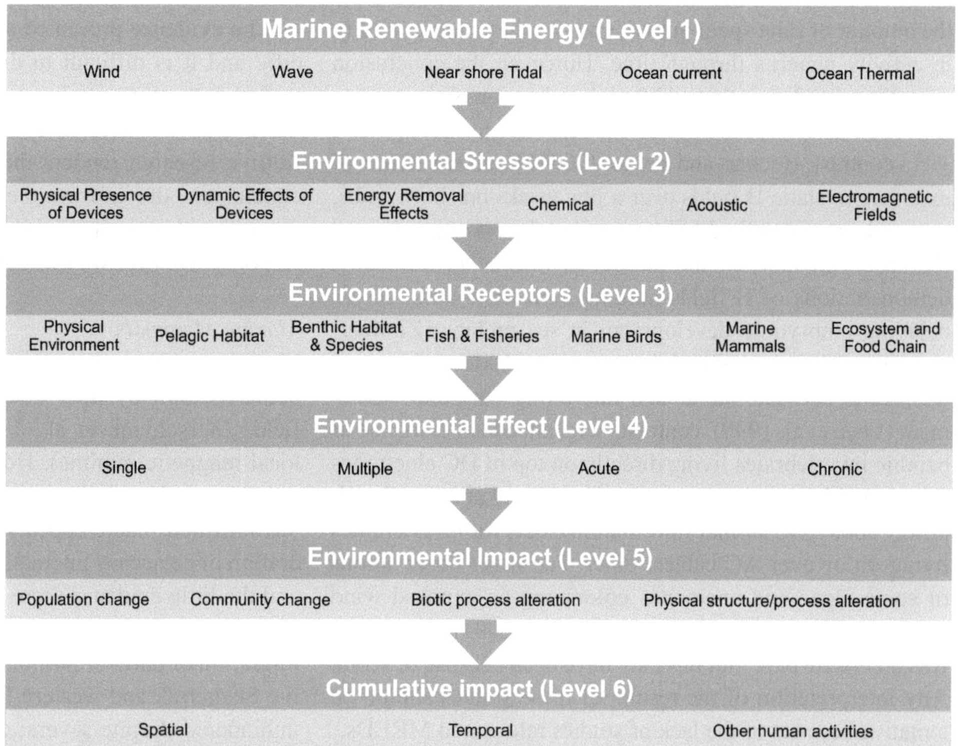

Fig. 6.3 Framework for the consideration of environmental effects of marine renewable energy encompassing different scales (after Boehlert and Gill 2010). Each technology will have associated stressors that affect different receptors. The effects vary across scales and receptors; if the effects are sufficient to have impacts, those impacts can apply across different levels (1–6, top to bottom) from population through biological and physical processes. Cumulative impacts need to be considered as an additional dimension to the impacts

diverting the response of eels to subsea cables), but whether this constitutes a biologically significant impact is key. Figure 6.3 shows a framework proposed by Boehlert and Gill (2010), which aimed to structure the research agenda around MREDs. It specifically defined levels in the framework and stressed that Level 4 represents the current state of knowledge for all environmental stressors considered, including EMFs. It will be extremely important when considering how the current evidence base is interpreted to recognize the difference between Levels 4 and 5 (i.e. the difference between effect and impact).

In order to interpret current status properly and perhaps more importantly to guide future research objectives, we support the proposition by Boehlert and Gill (2010) that research needs to move from its current Level 4 to Levels 5 and 6. In terms of EMF, following this suggestion will generate greater confidence in the stakeholders involved in future research to address those aspects necessary in assisting with environmental impact assessment (EIA), monitoring requirements and addressing the great uncertainty associated with the topic. We stress here too that the characteristics of EMF generated by AC and DC cables differ, so the responses by marine animals cannot be assumed to be similar for both types. However, owing to the paucity of information and the uncertainty surrounding precise differences in behavioural effects, both are here considered together. Most studies deal with single-source EMF rather than cable-type emissions, but they are included as the best evidence currently available.

Magnetic Fields

Several marine organisms are considered to be magnetosensitive or responsive to magnetic fields, so potentially to be able to detect and respond to anthropogenic B fields. Below, we consider the full range of organisms for which to date there has been study of their response to magnetic fields.

Marine Taxa (Bacteria and Algae)
Compass orientation, demonstrated by migration in magnetic fields as weak as 50 μT, is evident among bacteria (Kirschvink 1980) and algae (Lins de Barros et al. 1982). However, no effects of HVDC subsea cable B fields on their distribution or physiology have been recorded (Poléo and Harboe 1996).

Invertebrates
Despite a number of marine invertebrates being magnetosensitive, there is little and indeed contradictory evidence of interactions with anthropogenic sources of magnetic fields. The brown shrimp (*Crangon crangon*) has been recorded as being attracted to AC B fields of the magnitude expected around windfarms (ICES 2003), and shore crabs (*Carcinus maenas*) have been demonstrated to be less aggressive in the presence of an AC B field generated to match the magnitude of windfarm cabling (Everitt 2008). Woodruff et al. (2012) provide some recent evidence of subtle changes in the behaviour of Dungeness crab (*Metacarcinus magister*), e.g. in

the amount of time spent buried, and changes and variability in activity patterns through time. However, the conclusion from those studies is that additional replication of the studies is needed to assess the behavioural response further.

In contrast, Bochert and Zettler (2004) found no effects of exposure to static B fields over a few weeks on shore crabs, nor on the round crab (*Rhithropanopeus harrissii*), an isopod (*Saduria entomon*) or the mussel *Mytilus edulis*. Equally, demonstrations of B fields ranging between 1 and 100 μT delaying embryonic development in sea urchins (Zimmerman et al. 1990) and of high frequency AC EMF causing cell damage to barnacle larvae and interfering with their settlement (Leya et al. 1999), contrasts with anecdotal evidence of benthic invertebrates living directly on top of DC electrodes (Nielsen 1986) with no apparent effects (Walker 2001; Swedpower 2003). No similar information exists for invertebrates living on or over AC cables, other than diver observations of some algae and anemones colonizing an exposed wind turbine J-tube (Marine Seen and CMACS 2004). The J-tube was otherwise bare, but this may have been because of scour. Any interpretation of the results of these studies should be tentative because of the lack of studies relevant to MREDs.

Fish

There is evidence that teleost fish possess magnetic receptors (see Kirschvink 1997, for a review), often supported by demonstrations of orientation behaviour. Similarly, the ability of chondrichthyans to detect magnetic fields by induction of electric fields (Kalmijn 1984) is supported by demonstrations of orientation behaviour towards magnetic fields in a number of species (e.g. Meyer et al. 2005). Whether B fields from subsea cables would affect these fish is unclear.

Bochert and Zettler (2004) found no significant effects of static B fields on flounder (*Platichthys flesus*) in laboratory tests. Recently, Woodruff et al. (2012) found inconclusive effects on behaviour in coho salmon (*Oncorhynchus kisutch*), but evidence of suppressed melatonin (a stress-related hormone) levels in juveniles. Swedpower (2003) found no measurable impact when subjecting salmon and trout to magnetic fields twice the magnitude of the geomagnetic field. Other laboratory studies by Woodruff et al. (2012) indicate that developmental processes during embryogenesis in rainbow trout (*O. mykiss*) may be affected by temporal exposures to EMF. Also, there were potential effects on growth and developmental stage in Atlantic halibut (*Hippoglossus hippoglossus*), but not in the closely related Pacific halibut (*H. stenolepis*). The European eel (*Anguilla anguilla*) deviates from its migration route in the presence of a 5 μT HVDC field for a short period and over a short distance (Westerberg 2000; Öhman et al. 2007). Atlantic salmon migration in and out of the Baltic Sea over a number of operating subsea HVDC cables seemed to continue unaffected (Walker 2001).

The evidence presented above is, of course, context specific and it is difficult to draw comparisons between controlled laboratory studies (with their own constraints) and those that are field-based. Moreover, the absence of definitive research renders the evidence highly uncertain, so conclusions drawn from the evidence to hand are currently tenuous, other perhaps than that there appears to be some response to EMFs by EM-sensitive fish.

Marine Mammals

Marine mammals have long been linked with the use of geonavigation by their detection of variation in magnetic fields (Kirschvink et al., 1986, correlated strandings with local magnetic minima). However, the ability has not been demonstrated experimentally, and how the sense operates is unconfirmed. There appears to be little documented consideration of cetaceans interacting with subsea cables, however, and the little evidence there is suggests that cetacean migration is not affected by subsea cable B fields. Moreover, the migration of harbour porpoises (*Phocoena phocoena*) across the Skagerrak and western Baltic Sea has been observed as unhindered despite several crossings over operating subsea HVDC cables (Walker 2001).

Electric Fields

Invertebrates

No marine invertebrates have been definitively demonstrated as being electrically sensitive (although it has been suggested that certain freshwater crayfish may possess an electric sense; Patullo and Macmillan 2007), but evidence is lacking (Steullet et al. 2007)).

Fish

In general, fish other than chondrichthyans are not thought to be noticeably sensitive to electricity, though teleosts may respond to strong electric fields of 6–15 V m^{-1} or more, at which levels the fish would be repelled from the source (Uhlmann 1975; Poléo et al. 2001). The electrosensitive sturgeon veer away or slow when approaching high voltage electricity lines (110 kV) passing over the water (Poddubny 1967). The European eel is sensitive to weak AC and DC fields (Berge 1979; Enger et al. 1976) and its life history embraces both marine and coastal waters, although based on limited evidence, the effect of subsea cable iE fields on eels would likely be similar to that elicited by B fields; minimal and temporary (Öhman et al. 2007). Walker (2001) also believed there would be no effects of HVDC on teleost fish, when investigating possible impacts of the Basslink HVDC between Australia and Tasmania.

By far the most likely group of marine animals to be affected by any iE fields are elasmobranchs, because of their acute sensitivity to electric fields (lower thresholds of 5–20 nV m^{-1}: Kalmijn 1982; Tricas and New 1998). They are repelled by strong anthropogenic electric fields (e.g. electric repellents), a fact that has raised concerns that cables inducing an electric field can act as barriers to movement (e.g. between feeding, mating and nursery areas).

Other than the use of very strong electric fields in shark repellents, avoidance behaviour for E fields at intensities within the range emitted by cables has only been documented twice: when small-spotted catsharks were presented with DC dipole electric fields of 1,000 μV m^{-1} (Gill and Taylor 2001), and when silky (*Carcharhinus falciformis*), whitetip reef (*Triaenodon obesus*) and zebra (*Stegostoma fasciatum*) sharks were presented with both DC and AC fields of 1,000 μV m^{-1} (Yano et al. 2000). Neither of these studies was designed to consider a range of field strengths, however, so it is difficult to be certain about an avoidance threshold. Nevertheless, other research has demonstrated unequivocal repeated, attraction behaviour to DC fields of ~60 μV m^{-1} (Kalmijn 1982; Kimber et al. 2011) and some avoidance has been observed at levels of 400–600 μV m^{-1} (Kimber 2008). Perhaps the threshold between dipole E field attraction and avoidance lies somewhere between ~400 and 1,000 μV m^{-1} for catsharks, although other species-specific thresholds are likely to exist.

The maximum iE fields induced by offshore windfarm standard, 132 kV, three-phase, AC cables have been demonstrated as being only slightly weaker than the smallest fields shown to elicit avoidance behaviour in elasmobranchs (CMACS 2003; Gill et al. 2005, 2012a). If it were assumed that behaviour of fish is similar for cable E fields and for fields emitted by a dipole, stronger fields would be expected to cause repulsion. The hypothetical consequence would then be that the cable would act as a barrier to movement and/or migration if the routes passed over them. Based upon the little information available, current thinking is that avoidance might take place within close proximity of higher rated AC cables and HVDC cables with currents creating larger B fields, which would in turn create larger iE fields (i.e. an avoidance zone).

There is considerable uncertainty as to whether laboratory-demonstrated repulsion from dipole sources of DC fields and to a lesser extent AC fields would translate into avoidance of cables in the real world. Some studies have focused on very low frequencies because, according to the best available physiological evidence, EM-receptive species are most sensitive in the range 0–20 Hz (Brown et al. 1974; New and Tricas 1998). The EMFs associated with subsea cables vary, but generally the AC systems run at 50–60 Hz depending on location. Nevertheless, based on the few studies summarized here, such frequencies appear detectable by sensitive fish receptors.

It is not clear whether any effects would be temporary or sustained. It is, however, apparent that a species capable of moving off the seabed into the water column should be able to cross cables, reducing its encounter with the higher intensity emissions. However, whether predominantly benthic species such as skates and rays would do so is uncertain and more research in this particular area would be valuable.

Electroreceptive species are responsive to E fields below those that elicit repulsive reactions and utilize them for a number of behaviours: prey, predator and mate detection and navigation (Tricas and Sisneros 2004). Hence, there is a question over whether such species can be attracted to and perhaps confused by anthropogenic E field sources that lie within similar ranges to natural bioelectric fields. All living animals emit weak E fields of three types: those associated with high-frequency AC caused by muscle action (including heart, gill and motor function muscles); DC associated with the difference in potential arising from membranous and epithelial proximity to water in body cavities (mouth, respiratory and anal); and low frequency AC caused by the alternating expansions and contractions of body cavities modulating the DC. Again, the evidence base is poor, but it appears that the extent and strength of these E fields varies significantly among taxa and that in general, bioelectric fields increase in intensity with increasing body size of the species (Kalmijn 1972; Haine et al. 2001). Measurement of these bioelectric fields in seawater is difficult and success varies between the few studies that have attempted it, but in general they seem to range between 1 μV (small molluscs) and 500 μV (small fish). Larger marine animals (i.e. large cephalopods and fish, and marine mammals) most likely emit bioelectric fields of even greater intensity, based on the evidence from biomedical research that a larger body volume scales up the bioelectric field (Malmivuo and Plonsey 1995). To our knowledge, though, research has not yet determined whether animal bioelectric fields within the same species increase in intensity with greater body size, although it would appear reasonable to assume that they do.

Marra (1989) recorded four power transmission failures in a transatlantic fibre-optic cable in the mid-1980s. Upon raising the cable for repairs, bite marks and embedded teeth were found at the damaged sections. Further investigation revealed the damage to be attributable to shark bites in all four instances. Attraction to iE fields induced around the cable (perhaps regarding them as related to prey) was considered the most likely reason for the shark response. Using biologically based reasoning, the cables were reinforced and shielded along sections of the cable that lay within the depth range that the sharks were known to inhabit, and there were no further problems with the cables. What happened to the sharks that bit the cables is unknown.

Laboratory behavioural studies have demonstrated that both AC and DC artificial electric fields stimulate similar feeding responses in elasmobranchs (Kalmijn 1982; Tricas and Sisneros 2004; Kimber et al. 2011). Recent work using small-spotted catsharks as a model benthic elasmobranch has demonstrated that despite their ability to distinguish certain artificial E fields (strong vs. weak; DC vs. AC), the sharks seemed either unable to distinguish or showed no preference between anthropogenic (dipole) and natural (live crab) DC E fields of similar strength (Kimber et al. 2011). If it is assumed that these species respond in a similar way to cable EMF, then this raises the question of whether the predators might waste time and energy "hunting" electric fields, such as those associated with subsea cables, in their search for the bioelectric fields associated with their prey.

A recent experiment to improve understanding of how elasmobranchs (as the most EM-sensitive species known) interact with subsea cables involved large, netted enclosures (known as mesocosms) over a section of subsea cable within an area of seabed similar to that favoured for windfarm development. This technically challenging approach was chosen to study the response of elasmobranch species to controlled EMF emissions to be assessed within a semi-natural setting at a scale appropriate to the MRED sector (Gill et al. 2009). The study was aimed specifically at answering the question of whether AC EMF emissions from subsea, electricity cables (50 Hz) of the type emitted by the offshore renewable energy industry could be detected by fish. The research provided the first evidence of electro-sensitive fish responding to EMF and found that small-spotted catsharks were more likely to be found within the zone of EMF emissions when the cable was energized, and that some thornback rays (*Raja clavata*) showed increased movement around the cable when it was switched on. Responses were unpredictable and did not always occur, appearing to be species- and individual-dependent. Whether these results imply any ecological impacts cannot be determined yet. What is clear, however, is that it is important to follow up these studies with targeted research that addresses aspects such as emergent properties that may arise from the responses recorded (i.e. those that could translate to population level effects) and consideration of how the fish might respond to different EMF emissions and whether they can habituate to the presence of an EMF.

Industry Perspective

The marine renewable energy industry has seen enormous expansion over a very short time, particularly in northern Europe. In the United Kingdom (but also elsewhere), there has been a drive to increase the proportion of energy generated via renewable sources as the industry responds to business opportunities created by national policy decisions and supporting energy tariffs. Consequently, there has been a steep learning curve as developers new to the offshore environment negotiate planning and development systems and processes, which in turn have been adapted rapidly by national environmental regulators and researchers and consultants, who attempt to answer the questions that arise.

In relation to the set of environmental concerns associated with MREDs, including EMF, the industry has been proactive in recognizing them at planning stages, such as during an EIA and supporting activities, but both industry and regulators have struggled to deal with the manifold gaps in knowledge. Such gaps present a significant challenge that environmental monitoring is currently being tasked with addressing. In the case of EMFs, there is a lack of clarity and high uncertainty relating to what should be monitored and which methodology and scale of monitoring is appropriate. For example, monitoring at offshore windfarms of the EMF effect on species abundance and distribution has tended to employ broad-scale, semi-quantitative sampling methods such as 2-m beam trawls. Such methodology is not well suited to sample for the key species of interest, such as elasmobranchs, because their large size allows them to swim fast enough to avoid capture (Wardle 1993), and trawling within close proximity to windfarm structures, including cables, is anyway restricted.

There now appears to be a risk, at least in England and Wales, that initial progress made into understanding the importance of anthropogenic sources of EMF in the marine environment, notably under the Collaborative Offshore Wind Research into the Environment (COWRIE) programme, could stall. The reality is that whereas the ability of certain species to detect iE fields associated with the most commonly deployed submarine power cabling has been demonstrated (Gill et al. 2009), itself a significant advance on understanding at the start of the expansion of offshore renewable energy, whether the findings can assist in determining environmental impact remains unclear. Although it appears from existing studies of deployed cabling that the barrier effects to fish movement have either not arisen or have caused what is regarded as a minor temporary response by animals, there is no evidence base from which to extrapolate this observation to larger developments with cabling of higher power or to sites with specific sensitivities such as important migration routes. Further, subtle effects such as attraction to areas of cabling and possible confusion of anthropogenic EMF for prey bioelectric fields might yet represent important ecological effects, which no monitoring undertaken to date has had the power to detect.

In the absence of clear evidence, developers can become frustrated that EMF remains a cause of concern when they perceive that significant resources have been allocated to related monitoring and research (in reality, much of the monitoring is more widely targeted at identifying only crude changes in target communities or habitats, and research budgets have been relatively modest). Other interest groups, including the fishing community, may have perhaps focused on the possible impacts of EMF in relation to more modest scale developments, leading to a situation where there is pressure on regulators to provide balance. However, without the necessary evidence for policy-making and with strong political pressure to progress MREDs, there may be a tendency to slacken the focus on more challenging issues such as EMF.

Although concepts such as birds striking wind turbines, migratory fish and diving birds colliding with wave and tidal turbines, or marine mammals and fish being injured or displaced by underwater noise from construction can be relatively easier to comprehend, interactions between disparate marine groups and magnetic/electric fields are more difficult to categorize and understand from a human perspective. This is not, however, a reason to ignore the situation. In fact, the opposite is true, it is important to understand properly whether or not EMF needs to be included in environmental considerations and therefore in the consenting (or permitting) and EIA processes.

Environmental Regulations

Research into possible interactions between marine fauna and anthropogenic EMFs is still in its infancy and is associated with great uncertainty. National regulators have not yet set any specific legislative requirements on subtidal EMF generation from a marine ecological perspective. This very much contrasts, for example, with limits imposed on EMF propagation at the sea surface in lieu of potential effects on ship navigation or in terrestrial situations in relation to human health.

Nonetheless, with fields of the magnitude anticipated from submarine power cabling demonstrated to lie within the sensitivity ranges of a variety of marine animals, and considering the burgeoning marine renewable energy industry and the related expansion in offshore grid connections, there is a growing need to understand better the effects and potential impacts (Gill 2005; Gill and Kimber 2005; Öhman et al. 2007; Sutherland et al. 2008). This is particularly pressing for many electromagnetically sensitive species that are also commercially exploited, e.g. eels, salmon, thornback rays, lobsters and crabs, with some anyway having suffered notable population declines in recent years (skates and rays—Baum et al. 2003; Myers and Worm 2003; European eels—Freyhof and Kottelat 2003). Although the EU's Marine Strategy Framework Directive (MSFD) has identified the introduction of energy (including EMF) at levels that do not adversely impact the marine environment as one of the descriptors of "Good Environmental Status" (GES), and EMF is included in some monitoring guidelines for offshore wind developments, the lack of certainty highlighted above can lead to de-emphasizing such an issue. In turn, this would create difficulties in securing the funding to research EMF because of a lack of knowledge, compounding the uncertainty and potentially marginalizing the issue. It is therefore vital that the knowledge gap be addressed so that the importance of EMF can be estimated, possible limits determined, and a balanced perspective applied to environmental guidance and regulation.

Current Practice in Terms of Monitoring and Mitigation

To date, most assessments of EMF generation in the marine environment have looked at the 50 Hz AC cables used extensively among relatively small, inshore windfarms. The assessments have drawn largely upon industry research supported by the COWRIE group. As the demand on transmission networks and associated supporting infrastructure increases, and as offshore windfarms become larger and are installed farther offshore, HVDC cables are being proposed. For example, although three-core 33–132 kV AC cables are currently deployed, 220 kV three-core and 275 kV single-core (either in trefoil or separated) designs are being developed. Also, it is 150–450 kV DC cables that are deployed currently, but 500 and 600 kV designs are being developed, and these will generate EMFs with characteristics different from those of AC cables. Assessing EMF generation in the marine environment therefore needs to be an ongoing process simply to keep pace with rapidly changing technology.

During EIAs of offshore renewable developments, the main considerations relating to potential impacts of anthropogenic B fields on marine fauna are impairment of navigation and physiological effects. The concerns relating to E fields are repulsion of animals (possibly causing a barrier effect to movements), attraction or confusion with bioelectric fields and the effects on animal energetics and physiology. Theoretically, should there be such effects (and depending on their severity), the potential to impinge upon an individual animal's ecological fitness may arise by reducing growth or reproduction (either directly if physiological, or indirectly if there is impairment of food resource location and acquisition). The challenge presented is in first understanding whether such effects are on individuals, and more demanding still, if they represent significant impacts on populations and ecosystems. Second, if these effects translate to impacts, then appropriate data collection will be required via specific guidance for environmental monitoring.

It is important to remember that the assessment of environmental impact involves balancing industrial requirements and costs with environmental concerns. For instance, sea electrodes are an inexpensive method of distributing electricity. However, they have been associated with deleterious environmental effects because of the strong electric fields and the generation of pollution products via electrolysis. Hence, their use is strongly advised against and generally avoided if possible by developers. In contrast, bundling bipole HVDC cables together confers environmental benefit via cancellation of opposing current flows, but the method is not financially or technologically feasible in depths > 40 m (approximately) at present.

Currently, consideration of EMFs during the EIA process for MREDs consists of literature review and desk study. Modelling, with many assumptions, is often undertaken by electrical engineers to estimate the strength and extent of EMF generation, and marine biologists then attempt to assess the potential for marine fauna to be affected using the limited literature and research available. This is not regarded as particularly satisfactory because it means that uncertainty remains great and understanding is not advanced, which subsequently does not assist with future decisions and developments. Both the EIA process and the supporting modelling could benefit from targeted research on EMFs and animal interactions and closer collaboration between modellers and biologists/ecologists, to incorporate the reality of the interaction and the complexity of EMFs in the marine environment (covered also in the section below).

At present, mitigation proposed in relation to EMFs is generally cable burial (which as pointed out above is not fully effective) and where uncertainty is acknowledged monitoring is generally proposed. Monitoring is often limited in scope, focusing on study of the differences in broad-scale distribution of EM-sensitive species (mainly elasmobranchs) between pre-construction and operational phases. For example, at Burbo Bank offshore windfarm in the eastern Irish Sea, 4-m commercial fish trawls were undertaken at a number of survey locations within and around the array area over a number of years, with the aim of determining whether elasmobranchs such as small-spotted catsharks, nursehounds (*Scyliorhinus stellaris*), starry smoothhounds (*Mustelus asterias*) and thornback rays were being deterred from approaching or entering the windfarm area. Although the results demonstrated that fish still frequented the windfarm during operation (Seascape 2010), the methodology was limited by the inability of the vessel to trawl in close proximity to turbines and directly over buried cables (because of the risk of collision and snagging, respectively). Moreover, with such a generic approach, it is not possible to ascertain whether EMF was the cause of any differences in distribution.

A more recent and ongoing survey at the Gwynt y Môr offshore windfarm in North Wales aims to tackle some of the problems by employing a local tanglenet fisher (who targets large skates and rays) to set nets closer to planned cable routes (CMACS 2011). Restrictions are still imposed by the developers to prevent fishing within 50 m of cable routes, but the nets can be set between cable arrays in an effort to provide a more representative and targeted method of fish sampling. However, the issue of determining the cause of distribution differences still persists.

Owing to the limited knowledge and generic monitoring undertaken, tentative and precautionary assessments are used to provide conclusions and advice to developers and governing bodies. The following general advice is currently proposed:

i. cable burial is largely ineffective in reducing EMF (see above), but it does provide a physical barrier preventing many animals (excluding infauna) from encountering the strongest fields;
ii. if possible, bipole cables should be bundled rather than being deployed separately, to reduce EMF generation via cancellation of fields;
iii. sea electrodes should be avoided because of the deleterious environmental effects associated with strong EMF and pollution;
iv. most potential effects are currently assumed to be minor and limited to within tens of metres, but future assessment needs to consider higher-rated cables and transmission currents;
v. benthic elasmobranchs are highlighted as potentially the most vulnerable taxa because of their acute sensitivity, their use of EMF for wide-ranging behaviours and the threatened population status of many species.

Potentially significant impacts such as barriers to migration cannot be completely ruled out (especially in the future for the more-powerful cables being developed), although conclusive determination of whether there will be any effect, let alone an ecologically significant one, is seriously hampered by insufficient knowledge and uncertainty.

The Future

The expected increased commercialization of the sea and its resources, where new energy developments and power cables will be established along with current offshore oil and gas installations fed power from land, means that it is vital to increase the knowledge base on which environmental assessments relating to EMF are undertaken, so increasing confidence in the conclusions and advice, i.e. significantly reducing the uncertainty.

Measurements of EMF generated by subsea cables *in situ* (see above) would significantly improve understanding of the strengths, geometries and potential interactions of the fields, rather than relying on modelled data with many concomitant assumptions. Further behavioural studies of EM-sensitive marine animals would improve understanding of how such animals might respond to subsea-cable EMF, and we advocate a set of studies ranging from work in the laboratory to work in the field. Laboratory experiments can be set up to determine whether different species and life stages are sensitive to the types and strengths of EMF generated by offshore MREDs (as demonstrated by Woodruff et al. 2012). Standard behavioural methods can be applied to studies that assess attraction or repulsion to different EMFs and whether they are able to differentiate between varying field types or intensities (e.g. Kimber et al. 2011).

At an MRED site, localized behavioural observations of target species (be they elasmobranchs, salmonids or eels) could provide valuable knowledge of how EM-sensitive animals react in close proximity to the cables that generate EMF in their natural environment, which can be compared then with observations around inactive cables. There are several *in situ* techniques with the potential to assist in determining animal response when encountering cables, for example baited camera traps and drop-down or towed cameras, although local conditions will play a role in determining the most appropriate technology to apply. At larger scales, tagging species of interest with tracking devices and/or loggers could provide real time, fine-scale analysis of movements or spatial distribution within the footprint of MREDs.

Finally, a hybrid of laboratory and field approaches, such as the COWRIE mesocosm study (see Gill et al. 2009) can be undertaken to address specific research questions relating to the EMF at a scale appropriate to the MRED industry. The mesocosm approach takes a controlled experiment within large enclosures deployed over subsea cables to conduct replicable and statistically robust analysis of animal behaviour in relation to specific stimuli in a semi-natural setting. The approach can also help answer a key question: whether EMF is the reason for the responses observed or how other stressors are acting in combination with it (Gill et al. 2012b). In our opinion, a combination of laboratory, field and mesocosm studies would provide a valuable and detailed knowledge base and be a big improvement on current, more broad-scale, less-specific methodology.

In addition to studying the potential impacts of EMFs on individual animals and different species, it is also important to determine whether there might be an ecological impact at a population or an ecosystem level for which supplementary data would be required on whether key ecological functions, such as breeding or feeding success, were being affected. Quantifying such ecological effects will require analysis of the potential for altered growth, health, reproductive success or survival of individual animals. If these attributes are indeed negatively affected, then population-level studies will be required to assess population distributions and demographics. Many EM-sensitive species rely on their sensory ability for prey detection, predator avoidance, searching for mates and orientation and navigation, so research needs to focus on these aspects. It is important when assessing effects at a population level that the scale of the interaction between MREDs and the species is taken into account. The home ranges, functional habitat availability and existing distribution of the animals with respect to the areal extent of the cable network within the MRED (or in multiple MRED sites within an area) should be incorporated into any such analysis to ensure that the real impacts are being considered. A population-level analysis also needs to take into account the major existing stressors of overfishing and habitat degradation that have already caused serious population declines of many EM-sensitive species (Baum et al. 2003), declines exacerbated by the slow life history traits of a number of the species (Frisk et al. 2005).

A further requirement is to consider MRED-associated EMFs in relation to other sources. On the seabed, pipelines, other electricity and telecommunication cables need to be assessed because some may emit stronger EMFs than MRED emissions. Shipping and shoreline facilities, such as ports and marinas, may also need to be taken into account. When cumulative impacts are being assessed, it will be important to include the existing sources of EMF and how much of the functional habitat of the EM-sensitive animals they occupy, in addition to assessing multiple MRED cables, cable arrays and networks cumulatively.

Improving knowledge and monitoring, as suggested above, will contribute to clarification for stakeholders of the interaction of marine animals with EMFs, an interaction which is often much misunderstood, and will help in producing more focused and clear assessments and recommendations about future developments. Rather than simply rehashing the limited, earlier work already in the literature, and funding somewhat inadequate monitoring, developers and consultants may eventually be able to discharge related consents more confidently, warranting the initial, potentially substantial costs associated with such research and monitoring.

Acknowledgements We thank the many people who have been involved in the research and discussions around this topic, Mark Shields for his invitation to write this chapter and for helping get it into its final form, Andy Payne for his editorial steer and the reviewers for providing insightful comments.

References

Alves-Gomes JA (2001) The evolution of electroreception and bioelectrogenesis in teleost fish: a phylogenetic perspective. J Fish Biol 58:1489–1511

Baum JK, Myers RA, Kehler DG, Worm B, Harley SJ, Doherty PA (2003) Collapse and conservation of shark populations in the Northwest Atlantic. Science 299:389–392

Berge JA (1979) The perception of weak electric AC currents by the European eel, *Anguilla anguilla*. Comp Biochem Physiol 62A:915–919

Bochert R, Zettler ML (2004) Long-term exposure of several marine benthic animals to static magnetic fields. Bioelectromagnetics 25:498–502

Boehlert GW, Gill AB (2010) Environmental and ecological effects of ocean renewable energy development-current synthesis. Oceanog 23:68–81

Bradford MR (1986) African knifefishes: the Xenomystines. In Bullock TH, Heilingenberg W (eds) Electroreception. Wiley, New York, pp 453–464

Brown HR, Andrianov GN, Ilyinsky OB (1974) Magnetic field perception by electroreceptors in Black Sea skates. Nature 249:178–179

Bullock TH (1986) Significance of findings on electroreception for general neurobiology. In Bullock TH, Heilingenberg W (eds) Electroreception. Wiley, New York, pp 651–674

CMACS (2003) A baseline assessment of electromagnetic fields generated by offshore wind farm cables. COWRIE Report1.0 EMF-01-2002 66. 71 pp

CMACS (2011) Gwynt y Môr Offshore Wind Farm. Baseline Fish and EMF Report. J3163v2 prepared by CMACS Ltd for Gwynt y Môr Offshore Wind Farm Ltd. 87 pp

Czech-Damal NU, Liebschner A, Miersch L, Klauer G, Hanke FD, Marshall C, Dehnhardt G et al (2011) Electroreception in the Guiana dolphin (*Sotalia guianensis*). Proc R Soc B 279:663–668

Enger PS, Kirstensen L, Sand O (1976) The perception of weak electric DC currents by the European eel (*Anguilla anguilla*). Comp Biochem Physiol 54A:101–103

Everitt N (2008) Behavioural responses of the shore crab, *Carcinus maenas*, to magnetic fields. MSc thesis, University of Newcastle-upon-Tyne. 94 pp

Freyhof J, Kottelat M (2010) *Anguilla anguilla*. In: IUCN 2012. IUCN red list of threatened species, Version 2012.2. www.iucnredlist.org

Frisk MG, Millar TJ, Dulvy N (2005) Life histories and vulnerability to exploitation of elasmobranchs: inferences from elasticity, perturbation and phylogenetic analyses. J North Atl Fish Organ 35:27–45

Gill AB (2005) Offshore renewable energy: ecological implications of generating electricity in the coastal zone. J Appl Ecol 42:605–615

Gill AB (2011) Removals of the physical resources from the systems: harvesting energy. In Wolanski E, McLusky DS (eds) Treatise on estuarine and coastal science, vol 8. Academic Press, Waltham, pp 217–252

Gill AB, Bartlett M, Thomsen F (2012b) Potential interactions between diadromous fishes of UK conservation importance and the electromagnetic fields and subsea noise from marine renewable energy developments. J Fish Biol 81:664–695

Gill AB, Gloyne-Phillips I, Neal KJ, Kimber JA (2005) The potential effects of electromagnetic fields generated by sub-sea power cables associated with offshore wind farm developments on electrically and magnetically sensitive marine animals—a review. COWRIE Report 1.5 EMF, London. p 90

Gill AB, Huang Y, Gloyne-Philips I, Metcalfe J, Quayle V, Spencer J, Wearmouth V (2009) COWRIE 2.0 Electromagnetic Fields (EMF) Phase 2: EMF-sensitive fish response to EM emissions from subsea electricity cables of the type used by the offshore renewable energy industry. Commissioned by COWRIE Ltd (project reference COWRIE-EMF-1-06)

Gill AB, Huang Y, Spencer J, Gloyne-Philips I (2012a) Electromagnetic fields emitted by high voltage alternating current offshore wind power cables and interactions with marine organisms. Proceedings of the electromagnetics in Current and Emerging Energy and power systems seminar, Institution of Engineering and Technology, London

Gill AB, Kimber JA (2005) The potential for cooperative management of elasmobranchs and offshore renewable energy development in UK waters. J Mar Biol Assoc UK 85:1075–1081

Gill AB, Taylor H (2001) The potential effects of electromagnetic fields generated by cabling between offshore wind turbines upon elasmobranch fishes, Countryside Council for Wales, Contract Science Report 488

Haine OS, Ridd PV, Rowe RJ (2001) Range of electrosensory detection of prey by *Carcharhinus melanopterus* and *Himantura granulata*. Mar Freshwat Res 52:291–296

ICES (2003) Report of the Benthos Ecology Working Group, Fort Pierce, Florida, USA, 28 April–1 May 2003. ICES Document CM 2003/E: 09. 58 pp

Kalmijn AJ (1972) Bioelectric fields in sea water and the function of the ampullae of Lorenzini in elasmobranch fishes. Scripps Institution of Oceanography, La Jolla, CA, Reference Series 72–83. 21 pp

Kalmijn AJ (1982) Electric and magnetic field detection in elasmobranch fishes. Science 218:916–918

Kalmijn AJ (1984) Theory of electromagnetic orientation: a further analysis. In: Bolis L, Keynes RD, Maddrell SHP (eds). Comparative physiology of sensory systems. Crans-sur-Sierre, Switzerland, pp 525–560

Kimber JA (2008) Elasmobranch electroreceptive foraging behaviour: male-female interactions, choice and cognitive ability. PhD thesis, Cranfield University, Bedfordshire, UK. 201 pp

Kimber JA, Sims DW, Bellamy PH, Gill AB (2011) The ability of a benthic elasmobranch to discriminate between biological and artificial electric fields. Mar Biol 158:1–8

Kirschvink JL (1980) South-seeking magnetic bacteria. J Exp Biol 86:345–347

Kirschvink JL (1997) Magnetoreception: homing in on vertebrates. Nature 390:339–340

Kirschvink JL, Dizon AE, Westphal JA (1986) Evidence from strandings for geomagnetic sensitivity in cetaceans. J Exp Biol 120:1–24

Leya T, Rother A, Muller T, Fuhr G, Gropius M, Watermann B (1999) Electro-magnetic antifouling shield (EMAS)—a promising novel antifouling technique for optical systems. Paper presented to the 10th International Congress on Marine Corrosion and Fouling, University of Melbourne, Australia

Lins de Barros HGP, Esquivel DMS, Danon J, de Oliveira LPH (1982) Magneto algae. Academia Brasileira CBPF Notas. FIS 48:104–106

Lohmann KJ, Lohmann CMF (1996) Detection of magnetic field intensity by sea turtles. Nature 380:59–61

Malmivuo J, Plonsey R (1995) Bioelectromagnetism: principles and applications of bioelectric and biomagnetic fields. Oxford University Press, New York

Marine Seen and CMACS (2004) Biology and video surveys of North Hoyle wind turbines, 11–13 August 2004. A report to RWE NPower. 32 pp

Marra LJ (1989) Sharkbite on the SL submarine lightwave cable system: history, causes and resolution. IEEE J Ocean Eng 14:230–237

Metcalfe JD, Holford BH, Arnold GP (1993) Orientation of plaice (*Pleuronectes platessa*) in the open sea — evidence for the use of external directional clues. Mar Biol 117:559–566

Meyer CG, Holland KN, Papastamatiou YP (2005) Sharks can detect changes in the geomagnetic field. J R Soc Interface 2(2):129–130

Myers R, Worm B (2003) Rapid worldwide depletion of predatory fish communities. Nature 423:280–283

New JG, Tricas TC (1998) Electroreceptors and magnetoreceptors: morphology and function. In Sperlakis N (ed) Cell physiology source book, 2nd edn. Academic Press, San Diego, pp 741–758

Nicholls B, Racey PA (2009) The aversive effect of electromagnetic radiation on foraging bats-a possible means of discouraging bats from approaching wind turbines. PLoS ONE 4(7):e6246. doi:10.1371/journal.pone.0006246.

Nielson MR (1986) Test report. Sea electrodes for Konti-Skan 2. ELSAM Report S86/63a. 34 pp

Öhman MC, Sigray P, Westerberg H (2007) Offshore windmills and the effects of electromagnetic fields on fish. Ambio 36:630–633
Pals N, Peters RC, Schoenhage AAC (1982) Local geo-electric fields at the bottom of the sea and their relevance for electrosensitive fish. Neth J Zool 32:479–494
Paris DT, Hurd FK (1969) Basic electromagnetic theory. (McGraw-Hill physical and quantum electronics series). McGraw-Hill, University of Michigan, 591 pp
Patullo BW, Macmillan DL (2007) Crayfish respond to electrical fields. Curr Biol 17:R83–84
Peters RC, Eeuwes LBM, Bretschneider F (2007) On the electrodetection threshold of aquatic vertebrates with ampullary or mucous gland electroreceptor organs. Biol Rev 82:361–373
Poddubny AG (1967) Sonic tags and floats as a means of studying fish response to natural environmental changes to fishing gears. In Conference on Fish Behaviour in Relation to Fishing Techniques and Tactics, Bergen, Norway, pp 793–802. FAO, Rome
Poléo ABS, Harboe M (1996) High voltage direct current (HVDC) sea cables and sea electrodes: effects on marine life. A literature study for the cable projects. Institute of Biology, University of Oslo. 42 pp
Poléo ABS, Johannessen HF, Harboe M (2001) High voltage direct current (HVDC) sea cables and sea electrodes: effects on marine life. Department of Biology, University of Oslo, Norway
Riedman M (1990) The pinnipeds: seals, sea lions and walruses. University of California Press, Berkely, 439 pp
SeaScape (2010) Burbo bank offshore wind farm post-construction 2009 (final) Commercial fish survey. Report prepared by CMACS Ltd
Sheppard JK, Preen AR, Marsh H, Lawler IR, Whiting SD, Jones RE (2006) Movement heterogeneity of dugongs, *Dugong dugon* (Müller), over large spatial scales. J Exp Mar Biol Ecol 334:64–83
Slater M, Jones R, Schultz A (2010) The prediction of electromagnetic fields generated by submarine power cables. Report to Oregon Wave Energy Trust (OWET). 47 pp
Soderberg EF (1969) ELF noise in the sea at depths from 30 to 300 meters. J Geophys Res Space Phys 74:2376–2387
Solov'yov IA, Mouritsen H, Schulten K (2010) Acuity of a cryptochrome and vision based magnetoreception system in birds. Biophys J 99:40–49
Souza JJ, Poluhowich JJ, Guerra RJ (1988) Orientation response of American eels, *Anguilla rostrata*, to varying magnetic fields. Comp Biochem Physiol A 90:57–61
Steullet P, Edwards DH, Derby CD (2007) An electric sense in crayfish? Biol Bull 213:16–20
Stratton JA (2007) Electromagnetic theory. IEEE press series on electromagnetic wave theory, 33. Wiley, New York, 640 pp
Sutherland WJ, Bailey MJ, Bainbridge IP, Brereton T, Dick JTA, Drewitt J, Gilder PM et al (2008) Future novel threats and opportunities facing UK biodiversity identified by horizon scanning. J Appl Ecol 45:821–833
Swedpower (2003) Electrotechnical studies and effects on the marine ecosystem for BritNed Interconnector. Swedpower Ltd, Stockholm
Tricas TC, New JG (1998) Sensitivity and response dynamics of elasmobranch electrosensory primary afferent neurons to near threshold fields. J Comp Physiol A 182:89–101
Tricas TC, Sisneros JA (2004) Ecological functions and adaptations of the elasmobranch electrosense. In von der Emde G, Mogdans J, Kapoor BG (eds) The senses of fishes: adaptations for the reception of natural stimuli. Narosa, New Delhi, pp 308–329
Ugolini A, Pezzani A (1995) Magnetic compass and learning of the y-axis (sea-land) direction in the marine isopod *Idotea baltica* Basteri. Anim Behav 50:295–300
Uhlmann E (1975) Power transmission by direct current. Spinger, New York
Walker TI (2001) Basslink project review of impacts of high voltage direct current sea cables and electrodes on chondrichthyan fauna and other marine life. Report 20 to NSR Environmental Consultants Pty Ltd. 77 pp
Walker MM, Kirschvink JL, Dizon AE, Ahmed G (1992) Evidence that fin whales respond to the geomagnetic field during migration. J Exp Biol 171:67–78
Wardle CS (1993) Fish behaviour and fishing gear. In: Pitcher TJ (ed) Behaviour of teleost fishes. Chapman and Hall, London, pp 609–643
Westerberg H (2000) Effect of HVDC cables on eel orientation. In: Merk T, von Nordheim H (eds) Technische Eingriffe in Marine Lebensräume. Bundesamt für Naturschultz, Lauterbach, Germany, pp 70–76
Willows AOD (1999) Shoreward orientation involving geomagnetic cues in the nudibranch mollusc *Tritonia diomedea*. Mar Freshwat Behav Physiol 32:181–192
Woodruff DL, Ward JA, Schultz IR, Cullinan VI, Marshall KE (2012) Effects of electromagnetic fields on fish and invertebrates. Task 2.1.3: Effects on aquatic organisms. Fiscal Year 2011 Progress Report on the Environmental Effects of Marine and Hydrokinetic Energy. Prepared for the US Department of Energy Contract DE-AC05-76RL01830, Pacific Northwest National Laboratory, Richland, Washington 99352, US
Yano K, Mori H, Minamikawa K, Ueno S, Uchida S, Nagai K, Toda M, Masuda M (2000) Behavioural response of sharks to electric stimulation. Bull Seikai Natl Fish Res Inst 78:13–29
Zimmermann S, Zimmermann AM, Winters WD, Cameron IL (1990) Influence of 60-Hz magnetic fields on sea urchin development. Bioelectromagnetics 11:37–45

Seabirds and Marine Renewables: Are we Asking the Right Questions?

7

Beth E. Scott, Rebecca Langton, Evelyn Philpott and James J. Waggitt

Abstract

The rapid increase in marine renewable energy installations (MREIs) will result in the placing of many novel man-made structures within seabird foraging habitats, and such structures could potentially impact seabird populations directly and indirectly, positively and negatively. However, whether these potential impacts represent real ones, such that they cause detectable trends in population levels, remains unknown. Changes in population dynamics of seabirds are driven primarily by rates of reproduction and adult and juvenile survival, all three of which are impacted by foraging success. Therefore, revealing precisely how MREIs can affect seabird foraging success through changes in foraging behaviour is key to understanding whether large-scale installations could have impacts at a population level. Discussion focuses on how to define foraging habitat and how MREIs might impact those habitats and foraging behaviour indirectly by changes in oceanographic processes and prey characteristics. Foraging behaviours are also likely to be more directly impacted by MREIs, so focus here is also on how changes in foraging behaviour during the more constrained breeding season can influence reproductive output by altering individual energy budgets. A third and more-direct potential impact of MREIs on foraging behaviour is changes in diving behaviour. Throughout, relevant gaps in current knowledge that need to be addressed in order to make robust predictions as to how MREIs might impact seabird populations are highlighted.

Keywords

Diving behaviour · Energetics · Foraging behaviour · Foraging habitat · Marine renewables · Seabirds

Introduction

To predict how the expansion of marine renewable energy installations (MREIs) might impact seabird populations, fundamental understanding of how seabirds use the marine environment to forage needs to be improved. Currently, there has been an expansion of knowledge about the distribution of seabirds at sea with > 30 years of data available from dedicated surveys (Ainley et al. 2012) and more recently, results from the tagging of birds (Montevecchi et al. 2012). There are also long-term data from well-studied colonies that allows understanding of a range of factors that can affect population dynamics (Votier et al. 2009). Most seabirds are long-lived and have a low annual reproductive output, generally producing clutches of between 1 and 3 eggs annually, depending upon species. Population changes will therefore only be significant as a result of variation in adult survival;

B. E. Scott (✉) · R. Langton · E. Philpott · J. J. Waggitt
School of Biological Sciences, Institute of Biological and Environmental Sciences, University of Aberdeen, Tillydrone Avenue, Aberdeen, UK
e-mail b.e.scott@abdn.ac.uk

M. A. Shields, A. I. L. Payne (eds), *Marine Renewable Energy Technology and Environmental Interactions,* Humanity and the Sea,
DOI 10.1007/978-94-017-8002-5_7,

low breeding success and poor juvenile survival will only have a compounding effect if they are repeated regularly (Lack 1968; Ashmole 1971). Foraging behaviour, success or failure, determines how well adult animals maintain their body condition and is the key to adult survival and breeding success. It also dictates whether seabirds will be in direct contact with anthropogenic sources of potential mortality and will influence whether interactions with future MREIs are negative, positive or neutral.

The response of the scientific community to the unknowns of the renewable industry was first to highlight the range of potential effects, both direct and indirect, that wind, tidal and wave developments could have on seabirds (the latest being Grecian et al. 2010; Langton et al. 2011; Witt et al. 2012). Attempts have also been made to quantify and rank the risk from different types of marine renewable devices to various species (Desholm 2009; Garthe and Hüppop 2004; Furness et al. 2012). However, this effort, although focusing attention on what has to be considered, has not increased basic understanding of why seabirds select the areas within which they forage, nor has it increased understanding of the detailed behaviour of seabirds while they are foraging.

This chapter sets out three main areas of scientific study that need to be delivered in order to increase understanding of foraging behaviour to the point where the possible direct and indirect impacts of large-scale marine renewable developments can be quantified with reasonable certainty. The three areas are (i) defining large- and small-scale foraging habitat, (ii) quantifying the energetic constraints of foraging and (iii) understanding diving behaviour.

Defining Foraging Habitat

Large-Scale Habitat

The collection of at-sea seabird data has been increasing over the past 50 years and was mainly achieved via boat surveys (Ainley et al. 2012). In the past decade, however, new techniques such as the use of aircraft and the miniaturization of GPS technology has led to an exponential increase in distributional data (Block et al. 2011). When used in combination (Montevecchi et al. 2012), these data can be used to define the at-sea distributions of seabirds during all seasons. All these techniques have limitations, however. First, many years of repeated sampling by boat surveys are needed before reliable and consistent maps can be created (Maclean et al. 2013). Second, the size and expense of tags has until recently restricted the species and number of individuals that could be tagged. Interestingly, tagging studies indicate that there is great site fidelity for individual birds (Irons 1998; Weimerskirch 2007), but they also show large variation between individuals, implying that a large sample size is needed to define foraging habitat properly (Hamer et al. 2001; Baduini et al. 2006; Kotzerka et al. 2011). Lastly, the use of aircraft, first with observers and more recently using high-definition camera systems, has increased the area covered in short periods of time but comes with the caveat of generally under-representing less visible birds, i.e. smaller species and those diving or foraging close to shore (Thaxter and Burton 2009).

What is encouraging, though, is that the results suggest that species prefer specific locations at a larger scale. Some locations are also sought by multiple species (Kober et al. 2010), and these sites in particular are considered locations where site protection such as a Marine Protected Area (MPA) may be a useful management tool (see the January 2012 Special Issue of *Biological Conservation*). Augmenting knowledge gained from past surveys with known constraints of seabirds' foraging ranges, at least during the breeding season, provides evidence for the most likely foraging areas (Thaxter et al. 2010; Grecian et al. 2012).

There is evidence for interannual differences in large-scale site use, linked to shifts in the boundaries of large-scale surface features of habitats or caused by changes in regional prey abundance and/or distribution (Monaghan et al. 1994; Jahncke et al. 2008; Garthe et al. 2011). What is missing from most of those studies, however, is concurrent data on oceanographic habitat variables and prey characteristics, which might help to determine why seabirds select specific locations. For example, most tagging studies, because of limitations of the loggers, collect few data on the attributes of concurrent habitat in which the seabirds are foraging. Habitat data comparisons, therefore, are usually limited to diving depth and temperature where the birds are foraging (Takahashi et al. 2008; Zavalaga et al. 2010). Generally, the only way to identify habitat preference, and to contrast the areas where seabirds do not forage, is from surface features from satellite information that do not necessarily describe the important subsurface habitat features in which the birds are foraging (Burger 2003; Grémillet et al. 2008; Scott et al. 2010). Not understanding the underlying reasons for seabird use of a foraging location leaves predictions of future use, especially faced with climate change, with great uncertainty.

Fortunately, the number of studies at medium to large spatial scales (10–100 km) that have collected a multiple of environmental variables is increasing (Ladd et al. 2005; Ballance et al. 2006; Scott et al. 2010). Currently, the mechanistic evidence behind the locations of hotspots of marine predator foraging at larger scales points directly to both topographic features such as seamounts and shelf edges (Genin 2004; Yen et al. 2004) and primary productivity, either at surface fronts (Ware and Thomson 2005; Bost et al. 2009) or locations with high subsurface chlorophyll biomass (Scott et al. 2010). Fronts, especially tidal fronts, are spatially and temporally predictable foraging locations for a wide range of seabirds because they tend to support aggregations of prey (Hunt et al. 1999; Weimerskirch 2007; Wakefield et al. 2009). They exist at the location where water depths have increased enough for the frictional effects of tidal mixing to no longer be felt throughout the water column, subjecting the upper section of the water column to warming and thereby stratifying the water column. The mechanisms behind the aggregations of marine animals are driven first by contrasting changes in biweekly mixing (neap to spring tides), which move the location of the frontal regions from close inshore to much farther offshore (neap to spring). This results in increased availability of nutrients from mixed areas of water into the stratified areas and allows high levels of primary production on a predictable biweekly basis. Second, mixing and current characteristics at the front physically support the retention of phyto- and zooplankton, causing predictable aggregations of smaller prey. This level of detail on fronts has been provided to show the level of mechanistic detail needed to allow hypothesis-driven investigations for different types of foraging habitat.

The links between foraging habitat more generally and higher concentrations of chlorophyll and sharp changes in topography point to the level of mixing within the water column being a foraging habitat variable crucial to many seabirds. Levels of water-column mixing will be affected at all spatial scales by many types of marine renewable device: tidal, wave and wind. Therefore, a possible large-scale indirect effect of full-scale developments of marine renewable arrays will be changes in vertical mixing that can influence the behaviour of prey and the entire food chain as levels of mixing ultimately determine the levels, locations and species of primary production. By the very nature of the mixing properties of water columns and fronts, the zones with predictably high and continuous primary production will exist in proximity to tidally energetic areas.

Small-Scale Habitat

The foraging success of seabirds ultimately depends upon the presence of suitable prey, such that the birds will forage predictably in areas where prey is available to them. One needs, therefore, to understand more about the small-scale biophysical attributes of habitats where prey is being captured, so that the reasons for locations (and timing) of foraging sites can be predicted. This knowledge is essential for predicting how foraging success is likely to be influenced by changes to physical conditions attributable to human developments, such as the placement of MREIs.

As with studies at large spatial scales, the studies that have focused on small spatial scales demonstrate that local oceanographic features, in particular the level of vertical mixing, play a major role in determining the small-scale distribution of seabirds because of their effects on prey distribution (Ballance et al. 2006; Embling et al. 2012). What is clear from those studies is that the ability to understand fully the biophysical mechanisms that influence predator–prey overlap requires detailed at-sea surveying and multidisciplinary research with simultaneous collections of seabird distributions, their prey and fine-scale characteristics of the water column. This fact has long been recognized (Haney 1987) and there has been an increase over recent years in the amount of information collected simultaneously and mainly continuously on biophysical variables and other trophic levels (Hunt et al. 1998; Bertrand et al. 2008; Stevick et al. 2008; Montevecchi et al. 2009; Regular et al. 2010; Embling et al. 2012). New technologies such as autonomous gliders (Kahl et al. 2010) and miniature cameras (Takahashi et al. 2004) have also been introduced.

Of multidisciplinary studies focusing on the small scales appropriate for predator–prey encounters, a common finding is again a link to localized, patchy, high biomass of chlorophyll (whether the chlorophyll link is via surface or subsurface abundance) and physical processes such as internal wave activity (Haney 1987; Bertrand et al. 2008; Stevick et al. 2008; Embling et al. 2012). Internal waves can be caused by stratified water flowing over abrupt changes in topography (Moum and Nash 2000), and increases in mixing drives greater fluxes of nutrients into the thermocline subsurface chlorophyll maximum. This has the potential to support higher levels of local primary production. Therefore, it is currently unclear whether the association of predator–prey encounters in areas where internal waves are produced is a causal or correlative link with chlorophyll. The mechanisms linking high levels of predator–prey interactions in these locations therefore can range from complex trophic interactions via bottom–up forcing, with prey available as a result of the greater primary productivity, to the less complex explanation that prey in those locations are easier to catch with internal waves actively aggregating and/or bringing them closer to the surface.

Physical forces driving prey availability provide an obvious mechanism for defining seabird foraging habitat, because increased foraging has been found in a variety of seabird species using fast tidal currents (Schneider et al. 1987; Hunt et al. 1998; Zamon 2003; Peery et al. 2009; Schwemmer et al. 2009). One of the mechanisms suggested is an increase in foraging opportunities caused by zooplankton being accumulated by the effects of tidal currents in an island wake (Alldredge and Hamner 1980). However, there would be great variation between species, with diving species foraging in turbulent, well-mixed water and surface feeders associated with surface convergence associated with tidal features (Ladd et al. 2005; Scott et al. 2010). The use of temperature depth loggers (TDR) on two diving species revealed that common guillemots (*Uria aalge*; midwater divers) foraged in stratified water and that the feeding distribution of

European shags (*Phalacrocorax aristotelis*; benthic feeders) had no association with fronts or the thermocline (Daunt et al. 2003). Intriguingly, interspecies differences in foraging behaviour have also been demonstrated within the types of area targeted for tidal energy installations in regions with strong nearshore tidal currents (Holm and Burger 2002), some species, e.g. ancient murrelets (*Synthliboramphus antiques*) preferentially foraging in fast-flowing water and diving ducks foraging in slack water. There is even evidence of differences within the same species where different foraging behaviours were used in stratified vs. mixed waters, Takahashi et al. (2008) finding that Brünnich's guillemots (*Uria lomvia*) dive deeper in mixed water.

Foraging Habitat Discussion

It is clear from the research findings presented above that to understand and define the conditions, timing and locations of seabird foraging habitat will require a fully multidisciplinary approach to be instigated. If the effects of multiple large-scale marine renewable developments are to be predicted with certainty, not only must the biophysical cues and mechanisms seabirds use to forage be understood, but one also needs to know how prey species are influenced by the fine-scale physical processes. What is also clear is that understanding the differences between seabird species is crucial. The effect of MREIs on seabird foraging will depend on species-specific aspects such as the energetic cost of foraging and the type of foraging behaviour in which they engage, particularly for seabirds that forage by diving. These will be taken up respectively in the two sections below.

Quantify the Energetic Constraints of Foraging Seabirds

During their breeding season, most species of seabird are constrained by having to commute between offshore areas used for resting and feeding, and the colonies where they care for their offspring. Adults have to manage their time and energy budgets in such a way that they can cope with the cost of regular flights and return to the colony frequently enough, with food, to rear a chick successfully while minimizing the amount of time the chick spends alone. There is now evidence that at least some species of seabird exhibit avoidance behaviour around offshore windfarms, whereas others may be attracted to such developments (Larsen and Guillemette 2007; Masden et al. 2009; Lindeboom et al. 2011), and similar effects may be observed around wave and tidal energy sites. The abundances of prey species are also modified by the developments (Wilhelmsson et al. 2006; Perrow et al. 2011; Wilhelmsson and Langhamer 2014, Chap. 5). Modelling studies have shown that these effects could have implications for the survival of individual seabirds as a consequence of changes in their rates of energy expenditure and intake (Kaiser et al. 2005). Below, we provide a description of the type of information that needs to be considered when thinking about the possible implications of non-lethal effects of renewable energy developments (modified after Langton et al. 2011).

The Energy Expenditure of Foraging Seabirds

A seabird's rate of metabolism at a particular time will depend on the activities in which it is engaged. As a result, the total daily energy expenditure (DEE) of an individual seabird will be determined by the energy costs of its activities and the amount of time spent performing each activity. Changes in behaviour and habitat caused by marine renewable developments may alter the amount of time birds need to spend engaged in different activities and reduce the time available for other activities, so data on the energetic costs of these different behaviours are needed to be able to calculate total energy expenditures. The mass-independent metabolic rates vary between activities and species. The major characteristics that differentiate the DEE between species are the mode of foraging and flight.

Laboratory studies have shown that, on average, the cost of diving for a common guillemot is 13.05 W kg^{-1} (Croll and McLaren 1993) and for a European shag, 22.66 W kg^{-1} (Enstipp et al. 2005). These measurements, however, were obtained from captive birds diving in a tank 8 m deep for the guillemot and 1 m deep for the shag, so do not provide reliable estimates of the dive costs of free-ranging birds because of the uncharacteristically shallow nature of the dives. Foraging common guillemots regularly dive deeper than 40 m (Hedd et al. 2009; Thaxter et al. 2009). In addition, swimming in water with fast currents, similar to the waters experienced by birds foraging in tidally active areas, increases energy costs (Heath and Gilchrist 2010); such conditions are not experienced by seabirds diving in tanks.

In contrast to guillemots and shags, black-legged kittiwakes (*Rissa tridactyla*) are plunge-divers, diving into the water from height but only penetrating the topmost layer. The energy expenditure recorded for plunge-diving by free-ranging black-legged kittiwakes is 431.24 W kg^{-1}, an extremely high value compared with pursuit-diving, although plunge-dives are much shorter in duration. This could be related to the cost of having to take off vertically from the water surface with wet plumage after a dive (Jodice et al. 2003). The reduced energy efficiency associated with this foraging method may suggest that it is only viable when feeding on high-density prey patches with a good probability of a successful dive. Compared with plunge-diving, surface-feeding by kittiwakes is not very energy expensive, just 17.56 W kg^{-1} (Jodice et al. 2003).

Birds also differ in their methods of flight. Some flap their wings continually whereas others intersperse flapping with gliding; energy costs will be related to wing-beat frequency. There is a trade-off between the characteristics of wings for efficient flying and wing-propelled diving. Wing-propelled divers such as auks typically have reduced wingspan and area compared with other birds of the same mass. Consequently, in order to remain airborne, auks need to flap their wings faster (Pennycuick 1987), meaning that their flight mode is more energy-intensive than species such as kittiwakes.

The implications of these differences in flight mode and foraging is that any changes in behaviour caused by the deployment of MREIs will impact the total DEE of bird species differently, depending on species and the activity affected. For example, because of their different energy costs of diving, the impact on a common guillemot of changing the proportion of time it spends diving could be less than for a European shag. Similarly, if birds have to commute farther to a feeding patch, by avoiding devices, for example, the proportional change in total energy expenditure of a black-legged kittiwake will be less than for more-inefficient fliers such as common guillemots.

Foraging Efficiency

The rate of energy gain of a seabird will be a function of the calorific content of the prey and the rate of prey capture. Understanding how feeding rates relate to prey density or availability is necessary to predicting how changes in prey populations or distributions caused by marine renewable developments will alter a seabird's energy intake. There are now results published from studies that have quantified successfully the relationship between broad-scale prey abundance and population level variables (e.g. chick diet, nest attendance and breeding success) over an entire season (Piatt et al. 2007; Buren et al. 2012). However, because of the difficulty in observing free-ranging foraging birds and prey abundances at sea simultaneously, there is still little information on the nature of the functional relationship between prey density and instantaneous intake by seabirds at shorter time-scales of individual foraging bouts (Grémillet et al. 2004).

One laboratory study on double-crested cormorants (*Phalacrocorax auritus*) revealed that the relationship between feeding rate (g min^{-1}) and fish density (g m^{-3}) was best described by a type III functional response (S-shaped) curve (Enstipp et al. 2007). This means that increases in prey density would only enhance a bird's energy intake up to a maximum, after which further rises would have no impact. Moreover, any reduction in prey density below a certain threshold would lead to a lower feeding rate, meaning either a decrease in the energy consumption of the adult and/or the chick, or the adult having to expend more time and energy obtaining the same quantity of food. For the double-crested cormorant, this threshold was ~4 g m^{-3} (Enstipp et al. 2007), but that value is likely to be species-specific. Any impacts of renewable devices on the energy balance of a species arising from changes in the density of available prey will depend on the positions of the new and old prey densities relative to the threshold value.

At a smaller scale, changes in the density of prey within a patch, e.g. in the aggregation of prey at subsurface structures, could alter the instantaneous foraging efficiency, the quantity of food consumed per unit of time spent diving (Chimienti 2012). Changes at a larger scale, such as distribution of prey patches in space, could change the quantity of food consumed over a whole day, because the time taken to commute to a prey patch may increase or decrease. Therefore, the number of foraging trips an adult bird has time to perform in a day may be modified.

Seabird Time Budgets

The rates of energy expenditure and assimilation during different behaviours are not the only factors that contribute to a seabird's energy budget; time spent on those activities is important too. The variation in metabolic rates and foraging efficiencies described above are set by a bird's physiology and morphology, and they will not be altered by marine renewable developments, but the latter may alter the time seabirds spend performing different behaviours.

Observed activity budgets vary between species, possibly as a result of differences in body mass and the relative costs of each activity. For example, northern gannets (*Morus bassanus*) and common guillemots, of which the latter is the less efficient flier, spend 50 % and 3 %, respectively, of their foraging trip in flight (Monaghan et al. 1994; Hamer et al. 2007). One aspect of a time budget that many species do seem to have in common is to be at the nest for at least half the day (Monaghan et al. 1994; Jodice et al. 2003), attributable to obligate biparental care patterns. Despite this, though, individual time budgets can vary greatly to allow birds to cope with changes in the environment and reproductive demands.

Breeding adult birds forage more frequently at locations nearer their colony and use feeding patches that are closer together than those used by non-breeders (Peery et al. 2009). This would reduce the time spent commuting and allow more time to be spent catching prey. Similarly, using closer foraging sites can increase the feeding rate to the chick. Black-legged kittiwakes halve their trip duration between incubation periods and post-hatching

(Hamer et al. 1993), probably because the adult is constrained by having to transport food to the nest at regular intervals when it is rearing a chick. The extra demands of a chick and the increase in total time at sea (Cairns et al. 1987) suggest that any impacts of renewable devices will be greater during chick-rearing than during incubation. The time adults spend foraging in years with reduced prey availability will increase (Hamer et al. 1993; Monaghan et al. 1994; Uttley et al. 1994), and despite the added demands of rearing chicks, their foraging range has to expand (Hamer et al. 1993; Monaghan et al. 1994; Uttley et al. 1994; Burke and Montevecchi 2009). This results in a decrease in chick-feeding rates when food is scarce (Uttley et al. 1994). Shifts in the distribution of available prey caused by renewable energy developments are therefore likely to have similar impacts on foraging time and chick provisioning rates as do natural fluctuations. Along with breeding status and environmental conditions, the number of trips a seabird undertakes in a day also depends on species; a northern gannet, for example, may perform an average of just one trip a day (Hamer et al. 2001) whereas a common tern (*Sterna hirundo*) may perform as many as 12 (Pearson 1968). This variation in number of trips individual seabirds perform will alter the frequency with which they encounter MREIs, so will influence the magnitude of any impacts (Masden et al. 2010).

Altering the time spent engaged in foraging is a long proposed and supported theory relating to how breeding seabirds cope with environmental change without jeopardizing reproductive success (Cairns et al. 1987; Hamer et al. 1993; Monaghan et al. 1994; Uttley et al. 1994). Time spent at the colony is usually the part sacrificed if parents need to increase foraging time (Hamer et al. 1993; Monaghan et al. 1994). The extent of this buffering capacity does vary between species, however. Common guillemots usually have sufficient spare time in their budget to decrease the proportion of the day at the nest without the chicks being left unattended for too long (Monaghan et al. 1994). Kittiwakes, on the other hand, often have to leave chicks alone in times of food shortage (Hamer et al. 1993; Kitaysky et al. 2000), elevating the risk of chick mortality. For example, black-legged kittiwakes at Shetland experienced complete breeding failure when, because of a shortage of sandeels, they left chicks unattended for 17 % of the time (Hamer et al. 1993). This suggests that kittiwakes are already struggling to meet the demands of chick rearing in a good year and cannot cope with any additional pressure, a hypothesis supported by studies that have shown that the total DEE of black-legged kittiwakes does not vary regardless of fluctuations in food resources (Welcker et al. 2010). Therefore, marine renewable developments may have greater population impacts for some seabird species than for others.

The Energetic Constraint of Foraging

Any changes to time-activity budgets of seabirds are limited by the energy individuals have available to use. There may be a point at which the energy gained from extra foraging is less than the additional cost, so in that situation the birds will eventually be unable to support their offspring and themselves. Under such circumstances, the individual seabird would have to make a decision whether to increase its effort to attempt to successfully fledge existing offspring, potentially incurring a fitness cost, or risk sacrificing current reproduction in favour of adult survival and future reproductive prospects. Being long-lived with low clutch sizes, seabirds likely take up the latter option. This could have population level impacts if the choice to abandon reproductive effort is made repeatedly. Even if the energy required to boost foraging effort is always less than the energy gained, increasing the time engaged in foraging cannot continue *ad infinitum* and a foraging seabird will eventually be constrained by the time available and possibly also a physiological upper limit of energy expenditure (Weiner 1992).

Understanding Seabird Diving Behaviour

Marine renewable energy installations exploiting wave or wind resources, as well as some tidal energy sites, will have large and conspicuous structures both above and below the sea surface. Most research focusing upon interactions between seabirds and MREIs has concentrated on devices which are visible above the water. For example the consequences of seabirds avoiding or colliding with windfarms during flight have been subject to increased study (Desholm and Kahlert 2005; Drewitt and Langston 2006). However, most MREIs will have underwater structures such as moorings and foundations, and some tidal and wave-energy devices will also have underwater structures that move, e.g. turbine blades and undulating hinges. Therefore, species that forage throughout the water column, including auks, northern gannets and cormorants *Phalacrocorax* spp., might interact with devices both above and below the sea surface. For those species, diving behaviour represents an important link between an individual deciding where to forage and successfully capturing prey. Therefore, because efficient foraging is essential for both breeding success and adult survival, the impacts of the underwater components of MREIs on seabird diving behaviours requires careful consideration.

Because of the obvious technical challenges, subsurface interactions between MREIs and seabirds have yet to receive much research attention. Reviews have focused on the maximum diving depths that appear relevant when discussing tidal energy devices that have underwater components near the seabed (see Fig. 7.1 in Langton et al. 2011; Furness et al. 2012). However, wave- and wind-energy devices often have moorings or foundations stretching between the seabed and the sea surface, so individual seabirds would encounter underwater structures regardless of their maximum diving depth. Therefore, when evaluating the impacts of underwater components of MREIs in a broader sense, it is important to focus on how those underwater components might influence seabird diving behaviour through changes to their foraging habitat, before determining whether the changes could have negative or positive consequences on foraging efficiency. An approach to this issue might be: (i) outline seabird diving methods and how they differ among species; (ii) describe how underwater components may alter adjacent foraging habitat; (iii) speculate how these changes might impact seabird diving behaviour and foraging efficiency.

Seabird Diving

Among diving species of seabird in UK waters, there are three main diving methods, plunge-diving, pursuit-diving with wings and pursuit-diving with feet. Plunge-diving species include gannets, although they penetrate farther into the water column than black-legged kittiwakes, which are also plunge-divers. Individual gannets detect prey from the air and then make short and rapid dives to ambush it within the upper water column; on entering the water, gannets generally travel at ~6 m s^{-1}, although this decreases to ~1 m s^{-1} at maximum depth (Ropert-Coudert et al. 2009). Typical plunge-dives of gannets rarely last more than 10 s or exceed 10 m of depth. Sometimes, however, individual gannets use underwater wingbeats to move through the water column immediately following their initial plunge dive. These most likely reflect them pursuing prey initially located within upper surface layers rather than them searching for prey underwater (Ropert-Coudert et al. 2009); such extended plunge-dives may last up to 40 s and reach depths >10 m. This mixed diving strategy generally produces bimodal dive depths and durations among gannets, suggesting that shallow and deep dives are discrete forms of behaviour most likely representing different approaches to varying situations (Hamer et al. 2009).

In contrast, specialist pursuit-divers include auks and cormorants, which undertake dives of much longer duration that start from a static position on the water surface. However, although often classified as having the same diving technique, auks and cormorants have different diving behaviours associated with their contrasting methods of propulsion through the water column, perhaps reflecting trade-offs between speed and manoeuvrability. Auks are pursuit-divers that beat their wings to move through the water column. In most cases, they descend from the water surface at ~1.5 m s^{-1} (Watanuki and Sato 2008), then increase speed at maximum depth to >2 m s^{-1} (Swennen and Duiven 1991). At that depth, auks may adopt a horizontal position similar to penguins, to maintain buoyancy (Kato et al. 2006). Wing propulsion might allow auks to swim faster during dives, but at the expense of manoeuvrability, so favouring prey pursuit in open water (Lovvorn and Liggins 2002). In contrast, cormorants are pursuit-divers that use footbeats to move through the water column, descending from the water surface generally at ~1.5 m s^{-1} but often slowing down at maximum depth to <1 m s^{-1} (Ropert-Coudert et al. 2006). At depth, cormorants tend to adopt a vertical position to maintain buoyancy, often facing directly downwards (Kato et al. 2006). Foot-propulsion probably provides cormorants with greater manoeuvrability during diving at the expense of speed, favouring their foraging in structurally complex habitats (Lovvorn and Liggins 2002). Indeed, cormorants appear to detect and ambush prey primarily at close distances, using tactile cues rather than pursuing prey in open water (Martin et al. 2008).

Despite there being three distinctly defined diving methods, species described as having the same diving method often show subtle differences in diving behaviour. Of the auks found in the UK, razorbills (*Alca torda*) tend to make shallower dives than closely related common guillemots, and may also descend from the water surface at more oblique angles (Watanuki et al. 2006; Thaxter et al. 2010), whereas another alcid, the Atlantic puffin (*Fratercula arctica*), may prefer to dive even shallower and for shorter periods than both these species (Wanless et al. 1988). Similar differences are also found between two species of cormorant found in the UK. Grémillet et al. (1999) noted that great cormorants (*Phalacrocorax carbo*) undertook shallower and shorter dives than European shags at similar locations. They also descended at a more oblique angle than European shags, which dive almost vertically from the water surface (Watanuki et al. 2005).

How Might Underwater Components Impact Foraging Habitats and Efficiencies?

Underwater components of MREIs have the potential to impact foraging habitats through two mechanisms; prey characteristics and environmental conditions. First, underwater components could act as fish aggregation devices (FADs) by providing shelter from currents/predators or new foraging opportunities. They could also function as artificial reefs and attract new species into an area. As a result, fish species and abundance could change following the construction of MREIs (Inger et al. 2009). This subject is discussed further in Wilhemsson and Langhamer (2014, Chap. 5). Moreover, MREI design could well influence exactly how fish communities change around the devices. For example, MREIs with substantial components near the seabed could cause aggregations of fish at deeper depths than those that have a large part of their subsurface construction near the sea surface, altering the vertical distribution of forage fish. Further, underwater components could change local hydrodynamics including current speeds, direction and turbulence characteristics (Shields et al. 2011). Finally, the removal of wave and tidal energy from the water column could change sedimentation processes including deposition, transport and removal, with consequences for water turbidity (Shields et al. 2011).

Changes in foraging habitats following the installation of MREIs could impact seabird diving behaviour and foraging efficiency through several mechanisms, namely changes in prey characteristics, prey response to underwater structures, changes in hydrodynamic conditions and changes in water turbidity.

Seabird dive duration and depth depend primarily upon prey characteristics. Seabirds taking pelagic prey from the water column could perform relatively short dives, perhaps detecting prey items from the water surface before commencing their dive. In contrast, those taking demersal prey would likely dive for much longer in locating and pursuing prey near or on the seabed (Elliott et al. 2008). However, the vertical distribution of pelagic prey and their escape response within the water column is important. For example, when pelagic prey is deep in the water column, seabirds may approach from above using shallow dives to herd the shoals of fish towards the seabed. In contrast, when the pelagic prey is shallower, seabirds may approach the targets from below using deep dives to force shoals of fish towards the surface (Benoit-Bird et al. 2011). The size and behaviour of prey items may also be important. For example, seabirds could perform long and extended dives in gathering small, slow-moving, shoaling species (Hedd et al. 2009), but short and rapid dives to ambush large, fast-moving, solitary species (Garthe et al. 2000). As MREIs may impact both the species composition of prey communities and their vertical distribution, they have the potential to impact seabird diving depth and duration and hence the energy costs of the dives (Chimienti 2012).

The ability of seabirds to capture prey might depend too upon the manner in which fish behave around underwater structures. For example, should the fish seek refuge within complex structures, then capture rates might decrease; this situation applies particularly to auks and gannets that seem unsuited for foraging in such habitats. Cormorants may be less susceptible to losing prey among such refuges. In contrast, if fish merely aggregate around structures, capture rates could increase because they would become easier to catch in dense shoals (see above; Enstipp et al. 2007). Therefore, the design of underwater components (complex vs. simple), in conjunction with the natural tendencies of the fish species present either to remain in open water or to seek refuge, could determine whether and which seabirds may experience reduced or increased capture rates of prey.

It also seems a reasonable assumption that the energy costs of dives would depend on current speed and direction and turbulence. For example, the hydrodynamics could influence buoyancy or drag, with consequences on the effort required for a seabird to move through the water. Therefore it is possible that the removal or dispersal of energy in the water column around underwater components could reduce the energy costs of dives around them (Heath and Gilchrist 2010). The counter-argument would then be that fast currents or turbulence could restrict prey movement and its ability to evade capture. As a result, the energy costs of seabird diving in such conditions may be outweighed by increased rates of capture (Hunt et al. 1999). However, precisely how the hydrodynamics impact the energy costs of dives or the rates of prey capture remains largely unknown (Heath and Gilchrist 2010), but despite the uncertainty, it seems reasonable to assume that the overall effects from changes in the hydrodynamics could vary among species given their different methods of diving and prey capture.

Finally, the ability of seabirds to detect and capture prey could depend upon the level of turbidity. In this respect, some species seem more sensitive to changes than others, and differences are most likely associated with the different dive methods. For example, auks that tend to detect and pursue prey by sight through the water may be less successful in foraging when turbidity is high (Regular et al. 2011). By contrast, cormorants that most likely detect and ambush prey at close distances may not suffer decreased rates of capture, unless the levels of turbidity become very high (Enstipp et al. 2007). In contrast to auks and cormorants, though, gannets depend solely on detecting prey from the air (Hamer et al. 2009), so reduced visibility might negatively impact their foraging success simply by discouraging them from diving rather than directly changing their actual diving behaviour.

Diving Behaviour

The fundamental differences in seabird diving behaviour make it likely that underwater components of MREIs will influence seabird species in different ways. Moreover, the wide variety of MREI designs and locations mean that the effects on species could differ between installations, so the influence of underwater components on seabird diving behaviour needs to be predicted for different designs. Predicting how underwater components might affect seabird dives requires a good understanding of their diving methods, the way changes in prey characteristics and behaviour, hydrodynamics and turbidity varies in the presence of MREIs, and finally how these changes could negatively or positively influence foraging efficiency through either changing the energy costs of the dives or the rates of prey capture.

Comprehensive understanding of the final two elements above is still a long way off, because thus far, detailed knowledge of seabird diving behaviour has been gathered primarily with biologging devices. These are attached directly onto seabirds and can record information including dive depth, time and the horizontal and vertical orientation of individual birds during their dives (Ropert-Coudert and Wilson 2005). Currently, however, time–depth recorders (TDRs) cannot record hydrodynamic conditions, turbidity or prey behaviour during diving events, so our knowledge of how the hydrodynamics influence seabird dives remains limited (Heath and Gilchrist 2010). The increasing miniaturization of biologging devices means that TDRs could be deployed with cameras that may be able to establish prey characteristics and water conditions during seabird dives (see Watanuki et al. 2008). However, such new devices will initially be restricted to larger species (gannets and cormorants). Another potential solution lies in the use of sonar equipment, which could allow seabird dive data to be collected *in situ* from research vessels or from moored devices (Brierley and Fernandes 2001; Benoit-Bird et al. 2011). By recording dives *in situ*, foraging behaviour and prey characteristics can be qualified and quantified more accurately than at present. The provision of simultaneously collected oceanographic data would allow quantification of physical conditions and therefore better define the characteristics of foraging habitats. Such an increase in understanding, coupled with mathematical modelling that uses observational data on the rates of prey capture, dive characteristics and water column (foraging) habitat to test for changes in foraging efficiencies through changes in energy costs of dives, is what is needed to address the question of whether and how MREIs may influence seabird populations through changes in their diving behaviour.

Conclusions

The title of this chapter suggests that current research studies may not be focusing on the type of mechanistic understanding of seabird foraging needed to be able to understand and predict if MREIs will influence seabird populations significantly. It is hoped that the evidence-based argument presented here demonstrates that to understand with any degree of certainty or predictability the effect of MREIs on seabird populations, the focus needs to be on evaluating why seabirds choose to forage where and when they do. Therefore, future research focus relating to MREIs surely needs to be on the collection of information that, first, better defines the physical characteristics of the foraging habitats of seabird species and second, generates better knowledge of species-specific energy and time constraints of foraging ranges. This second focus also needs to determine how different seabird species interact with potential changes in prey availability and concentrations in the vicinity of the MREIs. The third focus needs to be improvements in knowledge of the energetic and efficiency of species-specific diving behaviour and the relationship between rates of prey capture and the physical oceanography, comparing the results with information from areas with and without MREIs. Exploring this level of multidisciplinary and biophysical detail is needed to be able to understand and predict the impacts of large-scale marine renewable developments on seabird populations.

References

Ainley DG, Ribic CA, Woehler EJ (2012) Adding the ocean to the study of seabirds: a brief history of at-sea seabird research. Mar Ecol Prog Ser 451:231–243

Alldredge AL, Hamner WM (1980). Recurring aggregation of zooplankton by a tidal current. Estuar Coast Mar Sci 10:31–37

Ashmole NP (1971) Sea bird ecology and the marine environment. In: Farner DS, King JR, Parkes K (eds) Avian biology, 1. Academic Press, New York, pp 223–286

Baduini CL, Hunt GL, Pinchuk AI, Coyle KO (2006) Patterns in diet reveal foraging site fidelity of short-tailed shearwaters in the southeastern Bering Sea. Mar Ecol Prog Ser 320:279–292

Ballance LT, Pitman RL, Fiedler PC (2006) Oceanographic influences on seabirds and cetaceans of the eastern tropical Pacific: a review. Prog Oceanog 69:360–390

Benoit-Bird KJ, Kuletz K, Heppell S, Jones N, Hoover B (2011) Active acoustic examination of the diving behavior of murres foraging on patchy prey. Mar Ecol Prog Ser 443:217–235

Bertrand A, Gerlotto F, Bertrand S, Gutiérrez M, Alza L, Chipollini A, Diaz E et al (2008) Schooling behaviour and environmental forcing in relation to anchoveta distribution: an analysis across multiple spatial scales. Prog Oceanog 79:264–277

Block BA, Jonsen ID, Jorgensen SJ, Winship AJ, Shaffer SA, Bograd SJ, Hazen EL et al (2011) Tracking apex marine predator movements in a dynamic ocean. Nature 475:86–90

Bost CA, Cotté C, Bailleul F, Cherel Y, Charrassin JB, Guinet C, Ainley DG et al (2009) The importance of oceanographic fronts to marine birds and mammals of the southern oceans. J Mar Syst 78:363–376

Brierley AS, Fernandes PG (2001) Diving depths of northern gannets: acoustic observations of *Sula bassana* from an autonomous underwater vehicle. Auk 118:529–534

Buren AD, Koen-Alonso M, Montevecchi WA (2012) Linking predator diet and prey availability: common murres and capelin in the Northwest Atlantic. Mar Ecol Prog Ser 445:25–35

Burger AE (2003) Effects of the Juan de Fuca Eddy and upwelling on densities and distributions of seabirds off southwest Vancouver Island, British Columbia. Mar Ornithol 31:113–122

Burke CM, Montevecchi WA (2009) The foraging decisions of a central place foraging seabird in response to fluctuations in local prey conditions. J Zool 278:354–361

Cairns DK, Bredin KA, Montevecchi WA (1987) Activity budgets and foraging ranges of breeding common murres. Auk 104:218–224

Chimienti M (2012) Modelling the foraging behaviour of diving predators in complex and heterogeneous landscapes: the impact of renewable tidal devices on diving seabirds. MRes thesis, University of Aberdeen

Croll DA, McLaren E (1993) Diving metabolism and thermoregulation in common and thick-billed murres. J Comp Physiol B 163:160–166

Daunt F, Peters G, Scott BE, Grémillet D, Wanless S (2003) Rapid-response recorders reveal interplay between marine physics and seabird behaviour. Mar Ecol Prog Ser 255:283–288

Desholm M (2009) Avian sensitivity to mortality: prioritizing migratory bird species for assessment at proposed wind farms. J Environ Manag 90:2672–2679

Desholm M, Kahlert J (2005) Avian collision risk at an offshore wind farm. Biol Lett 1:296–298

Drewitt AL, Langston RHW (2006) Assessing the impacts of wind farms on birds. Ibis 148:29–42

Elliott KH, Woo K, Gaston AJ, Benvenuti S, Dall'Antonia L, Davoren GK (2008) Seabird foraging behaviour indicates prey type. Mar Ecol Prog Ser 354:289–303

Embling CB, Illian J, Armstrong E, van der Kooij J, Sharples J, Camphuysen KCJ, Scott BE (2012) Investigating fine-scale spatio-temporal predator–prey patterns in dynamic marine ecosystems: a functional data analysis approach. J App Ecol 49:481–492

Enstipp MR, Grémillet D, Jones DR (2007) Investigating the functional link between prey abundance and seabird predatory performance. Mar Ecol Prog Ser 331:267–279

Enstipp MR, Grelmillet D, Lorentsen S (2005) Energetic costs of diving and thermal status in European shags (*Phalacrocorax aristotelis*). J Exp Biol 208:3451–3461

Furness RW, Wade HM, Robbins AMC, Masden EA (2012) Assessing the sensitivity of seabird populations to adverse effects from tidal stream turbines and wave energy devices. ICES J Mar Sci 69:1466–1479

Garthe S, Benvenuti S, Montevecchi WA (2000) Pursuit plunging by northern gannets (*Sula bassana*) feeding on capelin (*Mallotus villosus*). Proc Roy Soc London. B. Biol Sci 267:1717–1722

Garthe S, Hüppop O (2004) Scaling possible adverse effects of marine wind farms on seabirds: developing and applying a vulnerability index. J App Ecol 41:724–734

Garthe S, Montevecchi WA, Davoren GK (2011) Inter-annual changes in prey fields trigger different foraging tactics in a large marine predator. Limnol Oceanog 56:802–812

Genin A (2004) Bio-physical coupling in the formation of zooplankton and fish aggregations over abrupt topographies. J Mar Syst 50:3–20

Grecian WJ, Inger R, Attrill MJ, Bearhop S, Godley BJ, Witt MJ, Votier SC (2010) Potential impacts of wave-powered marine renewable energy installations on marine birds. Ibis 152:683–697

Grecian WJ, Witt MJ, Attrill MJ, Bearhop S, Godley BJ, Grémillet D, Hamer KC et al (2012) A novel projection technique to identify important at-sea areas for seabird conservation: an example using northern gannets breeding in the north-east Atlantic. Biol Conserv 156:43–52

Grémillet D, Kuntz G, Delbart F, Mellet M, Kato A, Robin J, Chaillon P et al (2004) Linking the foraging performance of a marine predator to local prey abundance. Funct Ecol 18:793–801

Grémillet D, Lewis S, Drapeau L, van der Lingen CD, Huggett JA, Coetzee JC, Verheye HM et al (2008) Spatial match–mismatch in the Benguela upwelling zone: should we expect chlorophyll and sea-surface temperature to predict marine predator distributions? J App Ecol 45:610–621

Grémillet D, Wilson RP, Storch S, Yann G (1999) Three-dimensional space utilization by a marine predator. Mar Ecol Prog Ser 183:263–273

Hamer KC, Humphreys EM, Garthe S, Hennicke J, Peters G, Grémillet D, Phillips RA et al (2007) Annual variation in diets, feeding locations and foraging behaviour of gannets in the North Sea: flexibility, consistency and constraint. Mar Ecol Prog Ser 338:295–305

Hamer KC, Humphreys EM, Magalhães MC, Garthe S, Hennicke J, Peters G, Grémillet D et al (2009) Fine-scale foraging behaviour of a medium-ranging marine predator. J Anim Ecol 78:880–889

Hamer KC, Monaghan P, Uttley JD, Walton P, Burns MD (1993) The influence of food supply on the breeding ecology of kittiwakes *Rissa tridactyla* in Shetland. Ibis 135:255–263

Hamer KC, Phillips RA, Hill JK, Wanless S, Wood AG (2001) Contrasting foraging strategies of gannets *Morus bassanus* at two North Atlantic colonies: foraging trip duration and foraging area fidelity. Mar Ecol Prog Ser 224:283–290

Haney JC (1987) Ocean internal waves as sources of small-scale patchiness in seabird distribution on the Blake Plateau. Auk 104:129–133

Heath JP, Gilchrist HG (2010) When foraging becomes unprofitable: energetics of diving in tidal currents by common eiders wintering in the Arctic. Mar Ecol Prog Ser 403:279–290

Hedd A, Regular PM, Montevecchi WA, Buren AD, Burke CM, Fifield DA (2009) Going deep: common murres dive into frigid water for aggregated, persistent and slow-moving capelin. Mar Biol 156:741–751

Holm KJ, Burger AE (2002) Foraging behavior and resource partitioning by diving birds during winter in areas of strong tidal currents. Waterbirds 25:312–325

Hunt GL, Mehlum F, Russell RW, Irons D, Decker MB, Becker PH (1999) Physical processes, prey abundance, and the foraging ecology of seabirds. Proc Int Ornithol Congr 22:2040–2056

Hunt GL, Russell RW, Coyle KO, Weingartner T (1998) Comparative foraging ecology of planktivorous auklets in relation to ocean physics and prey availability. Mar Ecol Prog Ser 167:241–259

Inger R, Attrill MJ, Bearhop S, Broderick AC, Grecian WJ, Hodgson DJ, Mills C et al (2009) Marine renewable energy: potential benefits to biodiversity? An urgent call for research. J App Ecol 46:1145–1153

Irons DB (1998) Foraging area fidelity of individual seabirds in relation to tidal cycles and flock feeding. Ecology 79:647–655

Jahncke J, Vlietstra LS, Decker MB, Hunt GL (2008) Marine bird abundance around the Pribilof Islands: a multi-year comparison. Deep-Sea Res II: Top Stud Oceanog 55:1809–1826

Jodice PGR, Roby DD, Suryan RM, Irons DB, Kaufman AM, Turco KR, Visser GH (2003) Variation in energy expenditure among black-legged kittiwakes: effects of activity-specific metabolic rates and activity budgets. Physiol Biochem Zool 76:375–388

Kahl LA, Schofield O, Fraser WR (2010) Autonomous gliders reveal features of the water column associated with foraging by Adélie penguins. Integ Comp Biol 50:1041–1050

Kaiser MJ, Elliott AJ, Galanidi M, Rees EIS, Caldow RWG, Stillman RA, Sutherland WJ et al 2005. Predicting displacement of common scoter *Melanitta nigra* from benthic feeding areas due to offshore windfarms. University of Bangor Report COWRIE-BEN-03-2002. 266 pp

Kato A, Ropert-Coudert Y, Grémillet D, Belinda C (2006) Locomotion and foraging strategy in foot-propelled and wing-propelled shallow-diving seabirds. Mar Ecol Prog Ser 308:293–301

Kitaysky AS, Hunt GL, Flint EN, Rubega MA, Decker MB (2000) Resource allocation in breeding seabirds: responses to fluctuations in their food supply. Mar Ecol Prog Ser 206:283–296

Kober K, Webb A, Win I, Lewis M, O'Brien S, Wilson LJ, Reid JB (2010) An analysis of the numbers and distribution of seabirds within the British Fishery Limit aimed at identifying areas that qualify as possible marine SPAs. JNCC Report, 431. ISSN 0963–8091

Kotzerka J, Hatch SA, Garthe S (2011) Evidence for foraging-site fidelity and individual foraging behavior of pelagic cormorants rearing chicks in the Gulf of Alaska. Condor 113:80–88

Lack D (1968) Ecological adaptations for breeding in birds. Methuen, London

Ladd C, Jahncke J, Hunt GL, Coyle KO, Stabeno PJ (2005) Hydrographic features and seabird foraging in Aleutian Passes. Fish Oceanog 14:178–195

Langton R, Davies IM, Scott BE (2011) Seabird conservation and tidal stream and wave power generation: information needs for predicting and managing potential impacts. Mar Pol 35:623–630

Larsen JK, Guillemette M (2007) Effects of wind turbines on flight behaviour of wintering common eiders: implications for habitat use and collision risk. J App Ecol 44:516–522

Lindeboom HJ, Kouwenhoven HJ, Bergman MJN, Bouma S, Brasseur S, Daan R, Fijn RC et al (2011) Short-term ecological effects of an

offshore wind farm in the Dutch coastal zone; a compilation. Environ Res Lett 6, 035101. doi:10.1088/1748-9326/6/3/035101
Lovvorn JR, Liggins GA (2002) Interactions of body shape, body size and stroke-acceleration patterns in costs of underwater swimming by birds. Funct Ecol 16:106–112
Maclean IMD, Rehfisch MM, Skov H, Thaxter CB (2013) Evaluating the statistical power of detecting changes in the abundance of seabirds at sea. Ibis 155:113–126
Martin GR, White CR, Butler PJ (2008) Vision and the foraging technique of great cormorants *Phalacrocorax carbo*: pursuit or close-quarter foraging? Ibis 150:485–494
Masden EA, Haydon DT, Fox AD, Furness RW (2010) Barriers to movement: modelling energetic costs of avoiding marine wind farms amongst breeding seabirds. Mar Poll Bull 60:1085–1091
Masden EA, Haydon DT, Fox AD, Furness RW, Bullman R, Desholm M (2009) Barriers to movement: impacts of wind farms on migrating birds. ICES J Mar Sci 66:746–753
Monaghan P, Walton P, Wanless S, Uttley JD, Burns MD (1994) Effects of prey abundance on the foraging behaviour, diving efficiency and time allocation of breeding guillemots *Uria aalge*. Ibis 136:214–222
Montevecchi WA, Benvenuti S, Garthe S, Davoren GK, Fifield D (2009) Flexible foraging tactics by a large opportunistic seabird preying on forage- and large pelagic fishes. Mar Ecol Prog Ser 385:295–306
Montevecchi WA, Hedd A, McFarlane Tranquilla L, Fifield DA, Burke CM, Regula, PM, Davoren, GK et al (2012) Tracking seabirds to identify ecologically important and high risk marine areas in the western North Atlantic. Biol Conserv 156:62–71
Moum JN, Nash JD (2000) Topographically induced drag and mixing at a small bank on the continental shelf. J Phys Oceanog 30:2049–2054
Pearson TH (1968) Feeding biology of sea-bird species breeding on Farne Islands, Northumberland. J Anim Ecol 37:521–552
Peery MZ, Newman SH, Storlazzi CD, Beissinger SR (2009) Meeting reproductive demands in a dynamic upwelling system: foraging strategies of a pursuit-diving seabird, the marbled murrelet. Condor 111:120–134
Pennycuick CJ (1987) Flight of auks (Alcidae) and other northern seabirds compared with southern Procellariiformes: Ornithodolite observations. J Exp Biol 128:335–347
Perrow MR, Gilroy JJ, Skeate ER, Tomlinson ML (2011) Effects of the construction of Scroby Sands offshore wind farm on the prey base of little tern *Sternula albifrons* at its most important UK colony. Mar Poll Bull 62:1661–1670
Piatt JF, Harding AMA, Shultz M, Speckman SG, van Pelt TI, Drew GS, Kettle AB (2007) Seabirds as indicators of marine food supplies: Cairns revisited. Mar Ecol Prog Ser 352:221–234
Regular PM, Davoren GK, Hedd A, Montevecchi WA (2010) Crepescular foraging by a pursuit-diving seabird: tactics of common murres in response to the diel vertical migration of capelin. Mar Ecol Prog Ser 415:295–304
Regular PM, Hedd A, Montevecchi WA (2011) Fishing in the dark: a pursuit-diving seabird modifies foraging behaviour in response to nocturnal light levels. PLoS ONE 6:e26763. doi:10.1371/journal.pone.0026763
Ropert-Coudert Y, Daunt F, Kato A, Ryan PG, Lewis S, Kobayashi K, Mori Y et al (2009) Underwater wingbeats extend depth and duration of plunge dives in northern gannets *Morus bassanus*. J Avian Biol 40:380–387
Ropert-Coudert Y, Grémillet D, Kato A (2006) Swim speeds of free-ranging great cormorants. Mar Biol 149:415–422
Ropert-Coudert Y, Wilson RP (2005) Trends and perspectives in animal-attached remote sensing. Frontiers Ecol Environ 3:437–444
Schneider D, Harrison NM, Hunt GL (1987) Variation in the occurrence of marine birds at fronts in the Bering Sea. Estuar Coast Shelf Sci 25:135–141
Schwemmer P, Adler S, Guse N, Markones N, Garthe S (2009) Influence of water flow velocity, water depth and colony distance on distribution and foraging patterns of terns in the Wadden Sea. Fish Oceanog 18:161–172
Scott BE, Sharples J, Ross ON, Wang J, Pierce GJ, Camphuysen CJ (2010) Sub-surface hotspots in shallow seas: fine-scale limited locations of top predator foraging habitat indicated by tidal mixing and sub-surface chlorophyll. Mar Ecol Prog Ser 408:207–226
Shields MA, Woolf DK, Grist EPM, Kerr SA, Jackson AC, Harris RE, Bell MC et al (2011) Marine renewable energy: the ecological implications of altering the hydrodynamics of the marine environment. Ocean Coast Manag 54:2–9
Stevick PT, Incze LS, Kraus SD, Rosen S, Wolff N, Baukus A (2008) Trophic relationships and oceanography on and around a small offshore bank. Mar Ecol Prog Ser 363:15–28
Swennen C, Duiven P (1991) Diving speed and food-size selection in common guillemots, *Uria aalgae* (sic). Neth J Sea Res 27:191–196
Takahashi A, Matsumoto K, Hunt GL, Shultz MT, Kitaysky AS, Sato K, Iida K et al (2008) Thick-billed murres use different diving behaviors in mixed and stratified waters. Deep-Sea Res II: Top Stud Oceanog 55:1837–1845
Takahashi A, Sato K, Naito Y, Dunn MJ, Trathan PN, Croxall JP (2004) Penguin-mounted cameras glimpse underwater group behaviour. Proc Roy Soc B: Biol Sci 271(Suppl. 5):S281–S282
Thaxter CB, Burton NHK (2009) High definition imagery for surveying seabirds and marine mammals: a review of recent trials and development of protocols. British Trust for Ornithology Report Commissioned by Cowrie Ltd
Thaxter CB, Daunt F, Hamer KC, Watanuki Y, Harris MP, Grémillet D, Peters G et al (2009) Sex-specific food provisioning in a monomorphic seabird, the common guillemot *Uria aalge*: nest defence, foraging efficiency or parental effort? J Avian Biol 40:75–84
Thaxter CB, Wanless S, Daunt F, Harris MP, Benvenuti S, Watanuki Y, Grémillet D et al (2010) Influence of wing loading on the trade-off between pursuit-diving and flight in common guillemots and razorbills. J Exp Biol 213:1018–1025
Uttley JD, Walton P, Monaghan P, Austin G (1994) The effects of food abundance on breeding performance and adult time budgets of guillemots *Uria aalge*. Ibis 136:205–213
Votier SC, Hatchwell BJ, Mears M, Birkhead TR (2009) Changes in the timing of egg-laying of a colonial seabird in relation to population size and environmental conditions. Mar Ecol Prog Ser 393:225–233
Wakefield ED, Phillips RA, Matthiopoulos J (2009) Quantifying habitat use and preferences of pelagic seabirds using individual movement data: a review. Mar Ecol Prog Ser 391:165–182
Wanless S, Morris JA, Harris MP (1988) Diving behaviour of guillemot *Uria aalge*, puffin *Fratercula arctica* and razorbill *Alca torda* as shown by radio-telemetry. J Zool 216:73–81
Ware DM, Thomson RE (2005) Ecology: bottom–up ecosystem trophic dynamics determine fish production in the Northeast Pacific. Science 308:1280–1284
Watanuki Y, Daunt F, Takahashi A, Newell M, Wanless S, Sato K, Miyazaki N (2008) Microhabitat use and prey capture of a bottom-feeding top predator, the European shag, shown by camera loggers. Mar Ecol Prog Ser 356:283–293
Watanuki Y, Sato K (2008) Dive angle, swim speed and wing stroke during shallow and deep dives in common murres and rhinoceros auklets. Ornithol Sci 7:15–28
Watanuki Y, Takahashi A, Daunt F, Wanless S, Harris M, Sato K, Naito Y (2005) Regulation of stroke and glide in a foot-propelled avian diver. J Exp Biol 208:2207–2216
Watanuki Y, Wanless S, Harris M, Lovvorn JR, Miyazaki M, Tanaka H, Sato K (2006) Swim speeds and stroke patterns in wing-propelled divers: a comparison among alcids and a penguin. J Exp Biol 207:1217–1230

Weimerskirch H (2007) Are seabirds foraging for unpredictable resources? Deep-Sea Res II: Top Stud Oceanog 54:211–223
Weiner J (1992) Physiological limits to sustainable energy budgets in birds and mammals: ecological implications. Trends Ecol Evol 7:384–388
Welcker J, Moe B, Bech C, Fyhn M, Schultner J, Speakman JR, Gabrielsen GW (2010) Evidence for an intrinsic energetic ceiling in free-ranging kittiwakes *Rissa tridactyla*. J Anim Ecol 79:205–213
Wilhelmsson D, Langhammer O (2014) The influence of fisheries exclusion and addition of hard substrata on fish and crustaceans (Chap. 5 of this volume)
Wilhelmsson D, Malm T, Öhman MC (2006) The influence of offshore windpower on demersal fish. ICES J Mar Sci 63:775–784
Witt MJ, Sheehan EV, Bearhop S, Broderick AC, Conley DC, Cotterell SP, Crow E, et al (2012) Assessing wave energy effects on biodiversity: the Wave Hub experience 2012. Philos Trans Roy Soc (A) Math Phys Eng Sci 370:502–529
Yen PPW, Sydeman WJ, Hyrenbach KD (2004) Marine bird and cetacean associations with bathymetric habitats and shallow-water topographies: implications for trophic transfer and conservation. J Mar Syst 50:79–99
Zamon JE (2003) Mixed species aggregations feeding upon herring and sandlance schools in a nearshore archipelago depend on flooding tidal currents. Mar Ecol Prog Ser 261:243–255
Zavalaga CB, Halls JN, Mori GP, Taylor SA, Dell'omo G (2010) At-sea movement patterns and diving behavior of Peruvian boobies *Sula variegata* in northern Peru. Mar Ecol Prog Ser 404:259–274

Marine Renewable Energy and Environmental Interactions: Baseline Assessments of Seabirds, Marine Mammals, Sea Turtles and Benthic Communities on the Oregon Shelf

8

Sarah K. Henkel, Robert M. Suryan and Barbara A. Lagerquist

Abstract

The wave climate along the west coast of North America presents great opportunities for the development of offshore renewable energy, yet initial assessments of the potential ecological effects of wave energy development have only just started. An enhanced regional understanding of the biological resources in the area is needed, and a key information gap is the distribution of both physical substrata and important biological communities. An initial renewable energy project targeted for Oregon is a mobile Ocean Test Facility developed by the Northwest National Marine Renewable Energy Center (NNMREC), led by Oregon State University (OSU), for testing wave energy converters. In addition, a number of wave and wind energy projects have been proposed for the Pacific Northwest of the US. In this chapter, an overview of the oceanographic characteristics of the region is presented, summarizing some of the interactions of concern, and highlighting baseline research projects focused on seabirds, marine mammals and benthic ecology in preparation for siting and deploying the NNMREC Ocean Test Facility and offshore renewable structures generally in the region.

Keywords

Benthos · Cetaceans · Flatfish · Habitats · Invertebrates · ROV · Seabirds · Wave · Wind

Introduction

The wave climate along the west coast of North America provides a great opportunity for developing offshore renewable energy. The Electric Power Research Institute estimates the amount of wave energy potential along the US West Coast to be 440 TW-h year^{-1} (Bedard et al. 2005), or some 10% of US energy demands for the year 2010. In addition to the relatively consistent and predictable wave energy produced across the long fetch in the North Pacific, the region possesses the coastal infrastructure and demand for electrical power generation (Bedard et al. 2005). Wind resources in the United States are also available offshore in the Pacific Northwest, with significant potential in 60–900 m of water depth (Thresher and Musial 2010). Interest in the development of renewable energy projects on the outer continental shelf (OCS) of the Pacific Northwest continues to increase as technologies develop and states increase renewable energy portfolios.

The long fetch of the Pacific Ocean and the prevailing westerly winds generally drive the high wave energy flux on the Oregon coast (Boehlert et al. 2008). However, seasonally changing patterns of wind stress result in variable ocean surface water circulation on Oregon's continental shelf. During summer, offshore high pressure systems and associated northerly or northwesterly winds drive the upwelling of deep, dense, cold water towards the ocean surface; at that time, circulation of surface waters on the continent shelf is dominated by the south-flowing California Current. In contrast, low offshore pressure systems during winter drive southwesterly storm winds that result in surface circulation

S. K. Henkel (✉) · R. M. Suryan · B. A. Lagerquist
Hatfield Marine Science Center, Oregon State University,
2030 SE Marine Science Drive, Newport, OR 97365, USA
e-mail: sarah.henkel@oregonstate.edu

M. A. Shields, A. I. L. Payne (eds), *Marine Renewable Energy Technology and Environmental Interactions,* Humanity and the Sea,
DOI 10.1007/978-94-017-8002-5_8,

being dominated by the north-flowing Davidson Current. Decade-scale shifts in the California Current can affect biological communities in the ecosystem, ranging from benthic infauna (Oliver et al. 2008) to demersal fish (Robinette et al. 2007) and indirectly affecting predators such as seabirds and whales (Thompson et al. 2012) when warm regimes and associated declines in plankton production cause stress or degradation of the assemblages. On shorter time-scales, *El Niño* events, which increase wave activity and storms (leading to sedimentation), can cause major, but short-term, disturbances to the communities. Hence, evaluation of potential effects of marine renewable energy on the ecosystem have to be made in the context of seasonal and climate trends.

On the inner continental shelf (depths <40 m), the bottom sediments are transported by a combination of wind-driven currents, wind waves, and tidal and estuarine-induced currents. Typically, waves influence bottom currents at depths of up to 50 m (Largier et al. 2008), so the reduction of wave energy in that zone could influence bottom currents, which on the inner continental shelf can transport sand-sized sediment. Therefore, changes to benthic communities and the potential for scaling-up effect on higher trophic levels mostly arise within that zone. On the mid-continental shelf (depths 40–100 m), water circulation is mainly influenced by wind-driven currents, whereas on the outer continental shelf (100–200 m), shoaling waves and regional currents control water circulation seasonally.

The marine life of coastal Oregon is dominated by cold-temperate species accustomed to broad seasonal changes in the energetics and water properties of the system. Although specific impacts from offshore wind or wave facilities need to be determined by pre- and post-installation surveys, it is clear that any renewable ocean energy project will have direct interactions with the seafloor and associated benthic communities as well as potentially serving as aggregators or collision risks to foraging and migrating species. Here, research conducted to date on the groups of organisms that will likely have the greatest potential for interactions with devices in the region are described, i.e. seabirds, marine mammals and benthic communities.

Current and Proposed Projects in Oregon

The Northwest National Marine Renewable Energy Center (NNMREC), led by Oregon State University (OSU) in partnership with the University of Washington, was established through the US Department of Energy Water Power Program with state and local funding to support wave and tidal energy development for the United States. The OSU-NNMREC Ocean Test Facility (OTF) is a pioneering effort to deliver a mobile capability for testing wave energy conversion (WEC) devices. The project is the first application of its kind in the world and highlights OSU's role in leading development of marine renewable technology and accelerating its commercialization in a manner compatible with ocean and coastal environments and coastal users. The mobile, floating capability to test wave energy technologies without a connection to an electrical grid allows for data to be collected under different wave conditions, at various depths depending on device requirements, and, in future applications, at a number of sites. The OTF's lack of a grid connection precludes environmental impacts associated with a cable to shore and allows for relatively expeditious removal of project components from a site.

The OTF was initially located some 2 miles off the coast of central Oregon, USA, near the city of Newport. The study area for the project measures 3 miles from north to south and 2 miles from east to west. The OTF itself is limited to a <1 mile^2 site (the project site) located within the 6 mile^2 project area. OSU-NNMREC conducted a number of environmental studies for baseline characterization of the OTF site and is currently monitoring a variety of factors at the site, many described in this chapter, and will continue to monitor during and after the testing of different devices in future. The studies are designed to increase knowledge of the potential effects the project, and wave energy projects in general, may have on specific ecological components.

In addition to the OSU-NNMREC OTF operating in summer 2012, another project in Oregon has obtained Federal Regulatory Commission licensing to date. Ocean Power Technologies was issued an original license in August 2012 for a wave energy project with an installed capacity of 1.5 MW to be located in Oregon territorial waters, about 2.5 nautical miles off the coast of Reedsport, in Douglas County. Wave-generating buoys will be installed there, and some baseline environmental surveys have been conducted. Various other wind and wave projects have been discussed for Oregon but are still in the early stages.

Seabirds

Potential Interactions of Concern

All US seabirds are protected under the US Migratory Bird Treaty Act, making it unlawful to pursue, hunt, capture, kill, take (disturb) or sell them. Several species regularly found along the Oregon coast are given additional protection through the US Endangered Species Act, including marbled murrelet (*Brachyramphus marmoratus*), short-tailed albatross (*Phoebastria albatrus*) and snowy plover (*Charadrius alexandrinus nivosus*). Other species are internationally listed on the IUCN Red List as vulnerable or endangered, including black-footed albatross (*P. nigripes*) and pink-footed shearwater (*Puffinus creatopus*). Other flying animals such as hoary bat (*Lasiurus cinereus*) and silver haired bat (*Lasionycteris noctivagans*) are also sometimes encountered offshore.

For seabirds, there is a potential collision risk with any structure above the water's surface (Boehlert et al. 2008), with the greatest concern for wind turbines associated with their moving blades (Grecian et al. 2010). The collision concern is greatest for dense arrays and less so for a single or few devices. Diving species dominate the marine avifauna community numerically year-round off the coast of Oregon, and the potential collision risk for seabirds with subsurface structures is considered to be low given their agility and visual acuity underwater. One abundant species of seabird in Oregon's waters, the common murre (*Uria aalge*), can dive as deep as 150 m (Hedd et al. 2009), although most dive to within 50 m of the surface. Therefore, the upper water column portion of submerged structures poses the greatest potential for interaction. An equal concern, however, is if subsurface structures function as fish-aggregating devices and also accumulate marine debris, attracting birds for foraging that can then become entangled. The positive effect of foraging opportunities could, however, offset potential negative subsurface interactions (Boehlert et al. 2008; Inger et al. 2009). Unless discouraged from landing, some birds (especially gulls and cormorants) will use above-water structures for roosting while resting between foraging bouts and also occasionally for nest building during the breeding season.

Wind turbines pose a potentially great risk for seabirds, regardless of whether they are monopile structures fixed to the seafloor or floating platforms. Although the rotating blades are of greatest concern, stationary structures below and above the water's surface pose the same potential risks as many WECs. There are four broad categories of WEC; attenuators, point absorbers, overtopping devices and terminators. Attenuators are typically multi-segmented, articulating surface structures that generally have a relatively low profile above and below the surface so pose the least collision risk. Point absorbers can have large structures >10 m above the water's surface and sometimes extend >30 m below. With associated mooring lines (or bridles with multiple anchors per device), such devices have a large potential surface area for interaction, especially if aggregated in a dense array of tens to hundreds of devices. Overtopping devices and terminators capture water movement through wave action to power a turbine, and the potential for them to impact seabirds negatively tends to relate to entrapment of the birds in the reservoir, either unintentionally or intentionally entering to pursue prey within it.

All seabirds have the potential to interact with structures at the water's surface. Whether a species could potentially encounter the upper structures of a tall wind turbine (60–100 m above the ocean for some) or the 30 m subsurface structure of a point absorber depends on the flight capabilities and foraging modes of the species in question. Wind turbine blades pose the greatest potential threat to seabirds and, although they rotate at relatively slow speeds (e.g. <22 rpm), the tip of a single blade >30 m long can exceed 250 km h^{-1}, so is not easily avoided. Depending on the size of the turbine, the blades may not extend closer than 30 m to the water, but species that tend to fly higher than that or that use dynamic soaring have the potential to collide with all portions of the turbine, include the blades at their highest point. Heavier-bodied diving species with their high wing loading (high body weight, low wing surface area), however, most often fly at altitudes of <30 m (Day et al. 2004), likely below the potential blade impact area, so in that case, the turbine presents a similar structural threat to that of a WEC. Structures above the water's surface raise a risk of collision during poor visibility periods such as at night or in fog, particularly for species that fly in flocks and fast and that remain 10–30 m from the water's surface. In addition to concerns about impact during poor visibility, there is concern too for light attraction of some phototactic species. Whereas navigation lighting can be minimized within the limits of safety standards and preferably include non-continuous lighting (Gehring et al. 2009), bright, continuous white lighting on vessels and structures during device installation and maintenance is a key concern.

Increasingly, diverse interests in commercial and recreational use of marine resources are creating new challenges for coastal ocean management. One concern of increased offshore use and development off the coast of Oregon is the potential impact on marine bird populations. Recently, Suryan et al. (2012) summarized the primary surveys of seabird breeding colonies and at-sea distribution along the Oregon coast to describe spatial patterns in species distribution and to identify gaps where additional data were needed.

Baseline Seabird Surveys

NNMREC has supported the compilation of information on seabird colony and at-sea distribution along the Oregon coast. The abundance of breeding birds during summer (>1 million in total, primarily common murre and Leach's storm-petrel, *Oceanodroma leucorhoa*) is greatest in northern and southern Oregon because of availability there of suitable breeding habitat on large offshore rocks and islands. There are fewer breeding colonies along sandy shores, but adjacent coastal waters are still frequented by breeding birds and nonbreeding migrants, generally in lesser densities during summer. Seabird density, and likely potential interaction with offshore structures, is greatest near the shore and steadily declines to low levels beyond the outer continental shelf (Fig. 8.1). Dynamic soaring species, however, which have a greater potential to interact with taller structures such as wind turbines, tend to be more common on the mid- to outer shelf. Species composition also changes dramatically among seasons (Fig. 8.2). Low-flying (<30 m above the sea) diving species dominate in most seasons off Oregon, however, which has potential conservation implications for interactions with structures above and below the water's surface. Given the abundance of storm-petrels, increased light

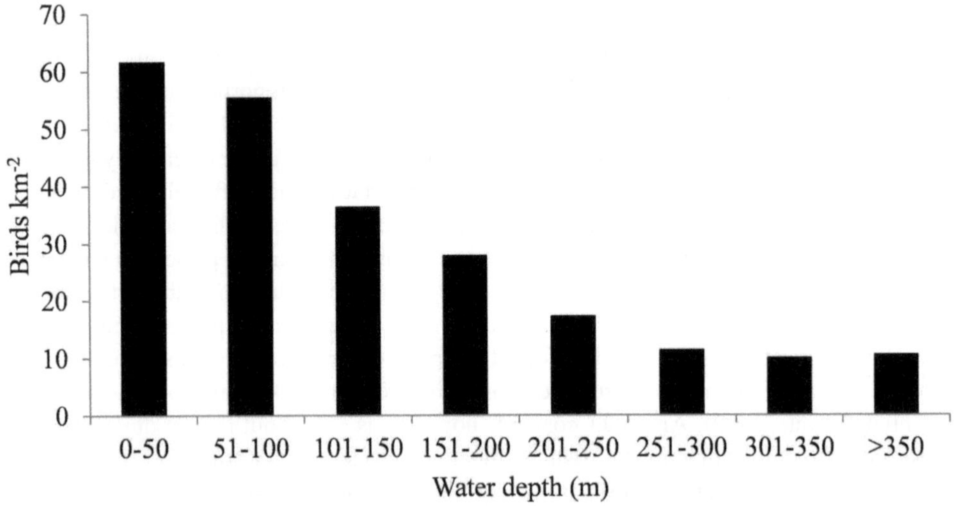

Fig. 8.1 Variation in seabird densities by depth across the continental shelf from the shore to beyond the shelf break (200 m). The overall density of birds decreases, but species composition changes, with more dynamic soaring species, such as the Procellariiformes (albatrosses, petrels, and relatives), being found farther offshore. Dynamic soaring species have greater potential to interact with tall structures such as wind turbines (from Suryan et al. 2012)

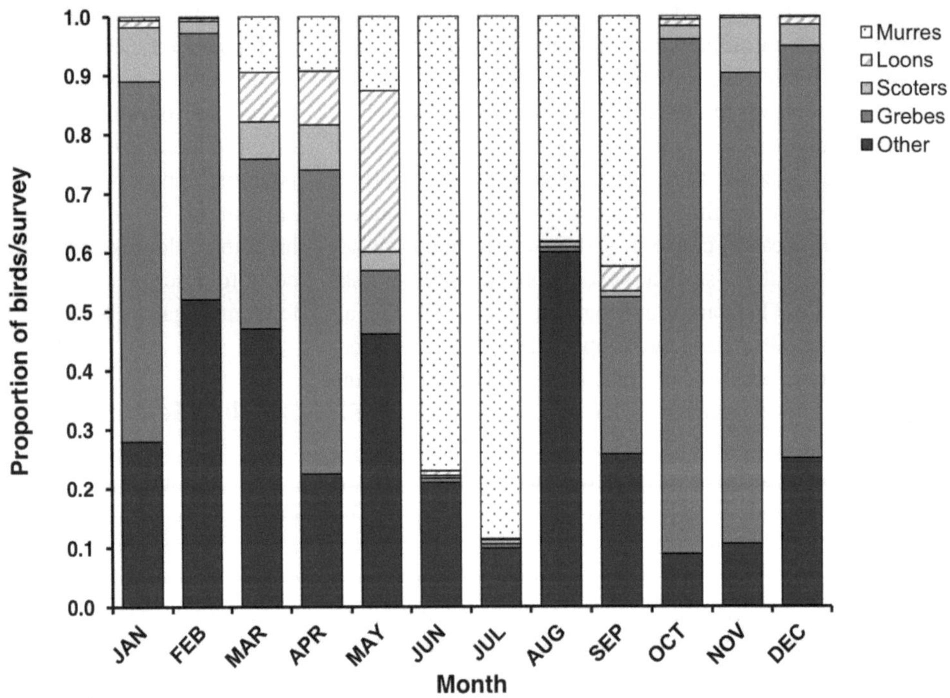

Fig. 8.2 Seasonal variation in nearshore seabird species composition and abundance observed during the years 2004–2009 near the Oregon–Washington border. In nearly all months, diving species that typically fly near the surface of the water were the most abundant (from Phillips et al. 2011)

pollution, especially during construction and maintenance, is also a concern for these and other nocturnal, phototactic species. There have been dramatic declines or redistributions at some breeding colonies, suggesting that long-term planning needs to consider changing habitat requirements on land and at sea.

It is important to keep in mind the location on the shelf and the importance of pre-installation surveys, to avoid migration corridors and otherwise high-use areas. For instance, 50 m depth is considered optimal location for wave energy devices (Boehlert et al. 2008), but it also corresponds with the highest densities of seabirds (Suryan et al. 2012). Efforts have generally involved the compilation of existing sources of seabird colony and at-sea data (Suryan et al. 2012), and the use of at-sea data to model seabird distribution and to identify high-use areas (Nur et al. 2011). The State of Oregon has made publically available in summarized form many of the layers of data used in their offshore marine spatial planning efforts (Oregon.marinemap.org). Moreover, US Geological Survey and US Bureau of Ocean Energy Management are conducting year-round aerial surveys throughout much of the northern California Current, Washington to northern California. Dedicated survey effort could supplement the rather spatially and temporally coarse aerial surveys, such as bird observations during oceanographic sampling cruises. Much can be gained from studies tracking individual birds, and data are available for some larger seabird species, such as albatrosses (*Phoebastria* spp.; Fernández et al. 2001) and shearwaters (*Puffinus* spp.; Adams et al. 2012). Fortunately, micro-electronic technology can now track small-bodied seabird species off Oregon and provide information crucial for identifying migration corridors, residence time (transitory vs. resident) and continuously/repeatedly used high-use areas (hotspots). Additional information needed includes flight altitude, which can be obtained using individual tracking devices, and marine radar studies.

Development of New Tools to Help Survey Organisms and Their Interactions with Devices

Investigators at NNMREC are developing a synchronized sensor array for remotely monitoring avian and bat interactions with energy devices. The integrated array is being designed to continuously monitor the interactions (including impacts) of birds and bats on the blades and structures of wind turbines. In contrast to land-based wind facilities, animal casualties in offshore energy installations can only be assessed efficiently over the long term using an on-board detection system with data transmitted remotely to shore-based data-processing centres. The synchronized array of sensors will initially include accelerometers, contact microphones, visual and infrared spectrum cameras and bioacoustic recorders. On-board, custom-designed data post-processing and statistical-based software will detect impacts from accelerometers and contact microphones or targets within the field of view of cameras and trigger an event record using sensor-specific detection algorithms. The monitoring system is being designed to run continuously so that information stored in a ring buffer from all sensors arising before and after the event will be transmitted to investigators for analyses. Remote access to the recorded images and sensor data will make it possible to quantify interactions, including collisions, and to identify organisms involved to the lowest taxon possible. The sensor array can be installed to monitor a single device during testing or multiple turbines in parallel. It is anticipated that the integrated observing array can be applied in various configurations to a variety of marine WEC devices, in addition to wind turbines.

Marine Mammals and Sea Turtles

Potential Interactions of Concern

The waters off the coast of Oregon hold at least 33 species of marine mammal (Table 8.1), including seven mysticetes (baleen whales), 21 odontocetes (toothed whales), three otariids (eared seals) and two phocids (true seals). Of these, seven are listed as endangered under the Endangered Species Act (blue, fin, sei, humpback, North Pacific right, sperm, and southern resident killer whale) and one as threatened (Steller sea lion). The most common marine mammal in nearshore Oregon Territorial Sea waters (within 3 nautical miles, or 5.6 km, of shore) are grey whales, harbour porpoises, harbour seals, northern elephant seals, California sea lions, and Steller sea lions. Minke, humpback, blue, fin and killer whales and Dall's porpoise are also found, but at lesser frequency than the more common species. The other marine mammal species listed are found over the outer continental shelf or even farther from shore.

Marine mammal presence in Oregon waters may be year-round (e.g. harbour seals) or seasonal (e.g. killer whales and most baleen whales) and, within a species, may also vary in terms of distance from shore. It has been suggested that harbour porpoises occupy deeper offshore water in late winter (Dohl et al. 1983). During summer and autumn, grey whales feed off Oregon in shallow water often < 1 km from shore, whereas during winter and spring, migrating grey whales are farther offshore (average > 5 km; Ortega-Ortiz and Mate 2008). Presence may also vary by sex and reproductive status. For example, it is typically just male California sea lions that move north to Oregon in autumn and winter from breeding areas in California. Adult male Steller sea lions migrate north to Washington, British Columbia and Alaska during late summer and autumn to feed, returning to Oregon to breed in spring and summer. Many adult female Stellers, with and without pups, and juveniles also disperse north to wintering areas outside Oregon in late summer and autumn although some remain at haul-outs in central and southern Oregon during winter (Scordino 2006).

Four species of sea turtle have been documented through sightings or strandings off the Oregon coast: leatherback, loggerhead, olive Ridley and green; all four are listed as endangered under the Endangered Species Act. The presence of olive Ridley, loggerhead and green turtles is irregular and generally linked to unusual oceanic conditions. In the eastern North Pacific, loggerheads and green sea turtles have been sighted as far north as Alaska, but are most common off California and farther south (NOAA 2012), with few sightings off Oregon. Leatherbacks, by contrast, regularly feed off Oregon in summer and autumn when sea surface temperatures are highest (Green et al. 1992; Bowlby et al. 1994; Benson et al. 2011) and densities of gelatinous zooplankton greatest (Suchman and Brodeur 2005). Waters off Oregon and Washington are one of the main eastern Pacific feeding destinations for leatherbacks nesting in the western Pacific (Benson et al. 2011), and critical habitat for leatherbacks has been designated from Cape Flattery, WA, to Cape Blanco, OR, east of the 2,000 m depth contour (Federal Register, 77 FR 4170, 26 January 2012).

Table 8.1 Species of marine mammal and sea turtle found off Oregon

Common name	Scientific name	Endangered species act status
	Mysticetes	
Blue whale	*Balaenoptera musculus*	Endangered
Fin whale	*Balaenoptera physalus*	Endangered
Grey whale	*Eschrichtius robustus*	
Eastern North Pacific		Not listed
Western North Pacific		Endangered
Humpback whale	*Megaptera novaeangliae*	Endangered
Minke whale	*Balaenoptera acutorostrata*	Not listed
North Pacific right whale[a]	*Eubalaena japonica*	Endangered
Sei whale	*Balaenoptera borealis*	Endangered
	Odontocetes	
Baird's beaked whale	*Berardius bairdii*	Not listed
Blainville's beaked whale	*Mesoplodon densirostris*	Not listed
Common bottlenose dolphin[b]	*Tursiops truncatus truncatus*	Not listed
Cuvier's beaked whale	*Ziphius cavirostris*	Not listed
Dall's porpoise	*Phocoenoides dalli*	Not listed
Dwarf sperm whale	*Kogia sima*	Not listed
Gingko-toothed beaked whale	*Mesoplodon gingkodens*	Not listed
Harbour porpoise	*Phocoena phocoena*	Not listed
Hubb's beaked whale	*Mesoplodon carlhubbsi*	Not listed
Killer whale	*Orcinus orca*	
Southern resident		Endangered
Transient		Not listed
Offshore		Not listed
Lesser beaked whale	*Mesoplodon peruvianus*	Not listed
Northern right whale dolphin	*Lissodelphis borealis*	Not listed
Pacific white-sided dolphin	*Lagenorhynchus obliquidens*	Not listed
Perrin's beaked whale	*Mesoplodon perrini*	Not listed
Pygmy sperm whale	*Kogia breviceps*	Not listed
Risso's dolphin	*Grampus griseus*	Not listed
Short-beaked common dolphin	*Delphinus delphis delphis*	Not listed
Short-finned pilot whale	*Globicephala macrorhynchus*	Not listed
Sperm whale	*Physeter macrocphalus*	Endangered
Stejneger's beaked whale	*Mesoplodon stejnegeri*	Not listed
Striped dolphin[c]	*Stenella coeruleoalba*	Not listed
	Pinnipeds	
Steller sea lion	*Eumatopias jubatus*	Threatened
California sea lion	*Zalophus californianus*	Not listed
Northern fur seal	*Callorhinus ursinus*	Not listed
Harbour seal	*Phoca vitulina*	Not listed
Northern elephant seal	*Mirounga angustirostris*	Not listed
	Sea turtles	
Leatherback turtle	*Dermochelys coriacea*	Endangered
Green turtle	*Chelonia mydas*	Endangered
Loggerhead turtle	*Caretta caretta*	Endangered
Olive Ridley turtle	*Lepidochelys olivacea*	Endangered

[a] No North Pacific right whales have been seen off Oregon, but there have been sightings in California and Washington, so they may occur off Oregon

[b] Common bottlenose dolphins may range into Oregon and Washington waters during warm-water periods (Carretta et al. 2011)

[c] No sightings of striped dolphins have been reported for Oregon and Washington, but some have stranded in both states (Carretta et al. 2011)

The potential impacts of wave and offshore wind energy development on marine mammals and sea turtles in Oregon shelf waters may depend on time of year because of the seasonal distribution of many species. Impacts may be direct or indirect (affecting marine mammal and sea turtle prey), negative (collision/entanglement) or positive (more haul-out sites for pinnipeds, increased feeding opportunities as a result of fish-aggregating properties). They will also depend on the type of technology, location (nearshore vs. offshore, latitude), and development phase (installation, operation or decommission) along with the behavioural state of the animals. Numerous workshops and publications have addressed the range of potential impacts to marine mammals, identifying information gaps, and providing recommendations for evaluating and mitigating impacts (Gill 2005; Madsen et al. 2006; Thomsen et al. 2006; Wilson et al. 2007; Boehlert et al. 2008; Ortega-Ortiz and Lagerquist 2008; Dolman and Simmonds 2010).

Direct Effects

Collision

For both marine mammals and sea turtles, there is a potential collision risk with various components of wave and offshore wind energy structures. Animals may collide with submerged structures, surface structures and/or the mooring lines holding structures in place. Animals may collide with objects when they are swimming through the water, when they come to the surface to breathe or as a result of structures being pushed down on them in times of heavy ocean swell (Wilson et al. 2007). During installation, maintenance and decommissioning there is risk of ship strikes on animals or their collision with vessels, but these risks are likely small because of the animals' ability to detect engine noise and to avoid the vessels.

Collision risk will vary depending on the type of technology and the behavioural state or age of the animal. An animal in active pursuit of fast-swimming prey or avoiding a predator may be at more risk than another engaged in other activities, and a newborn or young animal with little experience navigating through the water may be at greater risk of collision than juveniles or adults. Collisions may result in minor or major injuries, or even mortality. Risks are believed to be greater for baleen whales than for highly manoeuvrable pinnipeds, echolocating odontocetes or sea turtles, but they increase for all in times of reduced visibility, such as during storms when turbidity increases. Also, when surface waves are sufficient to entrain air bubbles into the water column, such as during times of high sea states or in surf zones, the echolocation abilities of odontocetes may be reduced, increasing their collision risk (Wilson et al. 2007).

Mooring lines may pose a particular risk for baleen whales, because they will likely be even harder to detect than larger structures. Collision with mooring lines could result in lacerations as well as blunt trauma (Boehlert et al. 2008), depending on line diameter and tension.

Entanglement

Mooring lines or cables associated with marine renewable energy technology can pose an entanglement risk for marine mammals and sea turtles if the lines/cables are slack or capable of forming loops; thin lines tend to pose more of a risk than thick ones. Moreover, derelict fishing line or nets may become entangled on mooring lines/cables and may themselves add to the entanglement risk to marine mammals and sea turtles.

Noise Effects

Sound in the ocean may affect marine mammals in a variety of ways, ranging from no effect to acute lethal effects, depending on the characteristics of the sound and the sensitivity of an animal's hearing (Richardson et al. 1995; Southall et al. 2007). If marine mammals cannot detect sound above ambient noise, however, there will likely be no effect, and sounds may be detected and still fail to elicit a response, behavioural or physiological. Whether or not an animal responds behaviourally to a sound in its environment and the type of that response depends on species, context, the properties of the sound and its novelty or whether the animal has prior experience of the sound (Southall et al. 2007). Species such as harbour porpoises and beaked whales are fairly sensitive to a range of human sounds at very low levels of exposure (Southall et al. 2007; Tyack et al. 2011), and they may exhibit a behavioural response before other species do. Behavioural responses to sound can range from the interruption of normal activities (such as resting, feeding or social interactions) to short- or long-term displacement from an area (Richardson et al. 1995). If sounds have overlapping frequencies with marine mammal vocalizations or that of their predators/prey, vocalizations may be masked, hence affecting a marine mammal's ability to communicate with others, navigate, feed or avoid predation.

Sound exposure may also induce physiological effects on marine mammals, including low to severe levels of stress responses (changes in heart rate and/or respiration, release of stress hormones), which in turn may affect metabolism, immune response and reproductive success (Southall et al. 2007). Sound exposure may also result in temporary or permanent reduction in hearing sensitivity or tissue damage, all of which have potential ultimately to affect an animal's survival.

Sound generated from wave and offshore wind energy development may be produced by the energy conversion devices themselves, wave action against structures, vibration from mooring lines/cables or shipping and construction activities during installation, maintenance or decommissioning. With the exception of pile-driving, which may or may not be required for anchoring floating offshore wind platforms, or explosions potentially associated with decommissioning, most of the sounds produced will likely be of insufficient intensity to induce physical injury to or permanent hearing loss of marine mammals. There is concern, however, that sounds from such developments may disturb and displace marine mammals from critical habitat, including feeding, breeding, resting and transiting areas or migration corridors, the extent of which will depend on the size and the location of the facilities, and the nature of the sounds they generate. Such displacement may be temporary, perhaps only during construction or decommissioning, or long term, e.g. throughout the operational life of a facility. The population impact of noise disturbance is largely unknown (Gilles et al. 2009), but would likely depend on the scale of the disturbance and the species. Genetically distinct populations of harbour porpoise along the Oregon coast, for instance, may suffer reduced foraging success, which could lead to adverse health and reproductive effects if the animals are displaced from crucial foraging habitat.

Recordings of pile-driving noise have been made at offshore windfarm construction sites in Denmark, Sweden and the UK and have yielded calculated source levels of 226–262 dB re 1 μPa at 1 m (Nedwell et al. 2003; Thomsen et al. 2006; Bailey et al. 2010). Such source levels can cause physical damage or hearing impairment in marine mammals at close range (<100 m; Nedwell et al. 2003; Thomsen et al. 2006; Bailey et al. 2010) and may disturb/displace marine mammals if farther away. Recorded reductions in echolocation clicks of harbour porpoises during construction of offshore windfarms in northern Europe (including pile-driving) indicate a decrease in density of harbour porpoises out to distances >15 km from the construction site (Tougaard et al. 2005; Carstensen et al. 2006; Brandt et al. 2011). As well as decreased acoustic activity during pile-driving, Tougaard et al. (2003) documented changes in surface behaviour, with more non-directional swimming (presumably associated with feeding) during days where there was no pile-driving than during days where there was. Both these effects were significant at ranges of up to 15 km from the construction site.

Tougaard et al. (2009) measured the underwater noise of wind turbines in Denmark and Sweden under normal operation and discovered that the noise was only measurable above ambient noise at frequencies <500 Hz with total sound pressure in the range 109–127 dB re 1 μPa. They concluded that the noise would be unlikely to cause injury, hearing loss or sound masking for seals or porpoises and that behavioural disturbance was unlikely for harbour porpoises unless they were close to the turbine foundations (within 70 m), whereas harbour seals might react at up to a few hundred metres away. Koschinski et al. (2003) recorded distinct reactions at close range to simulated wind turbine noise by both harbour porpoises and harbour seals during a playback experiment in Canada. However, the results may have been confounded by the introduction of high-frequency artefacts into the signal to which the porpoises and seals may have been responding (Madsen et al. 2006). Lucke et al. (2007) showed that, at short range in the open sea, the operational sound from wind turbines could have a masking effect on harbour porpoises.

No studies to date have documented the effects on baleen whales of noise from wave or offshore wind energy development. Noises associated with petroleum industry activities have been shown, however, to cause avoidance reactions in migrating grey whales, with 50 % of the animals studied avoiding exposure to continuous sounds (engine or drilling noise) at received levels of 120 dB re 1 μPa (Malme et al. 1983) and impulses from airguns at received levels of 170 dB re 1 μPa (Malme et al. 1984). Given these results and the fact that the hearing sensitivity of baleen whales extends to low frequency sounds (Southall et al. 2007), where much of the energy from pile-driving and wind turbine noise is centred, it is likely that baleen whales would be disturbed/displaced by such development. Based on their recordings of pile-driving for two deep-water turbines in Scotland, Bailey et al. (2010) suggested that minke whales and other cetaceans with hearing focused on mid- and low frequencies may exhibit behavioural disturbance to pile-driving up to 50 km from the source. The distance at which baleen whales can hear an operating wind turbine is likely significantly <50 km, and behavioural effects are likely to be minor and at ranges of up to a few kilometres only (Madsen et al. 2006). Turbines may also have a masking effect on baleen whale communication, but likely only over distances of kilometres in areas with no other significant sound sources, such as motorized shipping (Madsen et al. 2006).

Little is known about the hearing ability of sea turtles or their response to or use of sound in their environment (Bartol and Ketten 2006). Their hearing sensitivity appears, though, to be limited to low frequencies, with optimum sensitivity at 100–1,000 Hz and an upper limit of 2,000 Hz (Lenhardt 1994; Bartol et al. 1999; Ketten and Bartol 2005). There are, however, differences in hearing ranges between different sizes and age classes of loggerhead and green sea turtles, with smaller, younger animals having a greater range of hearing than older ones (Ketten and Bartol 2005). In-water hearing thresholds of green sea turtles range from 160 to 200 dB re 1 μPa for frequencies of 100–1,000 Hz (Lenhardt 1994). No hearing data are available for leatherback sea turtles, but based on the hearing capabilities of other sea turtles (NTRC 2008), they likely also hear low frequency sounds.

As with cetaceans that also hear low frequencies, sea turtles may be disturbed/displaced by the low-frequency operational noise of wind turbines, but likely only when they are very close to the devices, given the animals' generally high hearing thresholds. Shipping noise frequencies associated with deployment, maintenance or decommissioning of wave or wind energy facilities may overlap with those of sea turtle hearing and also lead to disturbance or displacement.

Pile-driving or underwater detonations associated with decommissioning could disturb or displace sea turtles, but they may also cause physical damage or hearing impairment at close range. Nothing has been published on the effects of pile-driving on sea turtles, but studies examining the effects of airgun exposure demonstrate that turtles do avoid such pulsed sounds. In trials with green and loggerhead sea turtles, McCauley et al. (2000) showed that turtle swimming speed increased noticeably at airgun sound levels of 166 dB re 1 μPa rms, and that the animals behaved erratically (possibly indicating agitation) at levels >175 dB re 1 μPa rms. O'Hara (1990) and Moein et al. (1994) also demonstrated avoidance by loggerhead sea turtles exposed to airgun signals with estimated received levels of ≥175 dB. In the study by Moein et al. (1994), turtles exposed to airgun sound several days after first being exposed to them failed to show statistically significant avoidance, and the authors attributed this either to habituation or a temporary shift in hearing caused by the first exposure.

Viada et al. (2008) provide a review of studies describing the effects of underwater explosions on sea turtles. The types of explosion trialled were those used in the removal of oil platforms and in US Navy exercises and ordinance detonations, and the effects ranged from no apparent effect to behavioural disturbance, mild physical discomfort, injury to the auditory system and other internal organs, and death. Underwater explosions can cause death through extensive lung haemorrhaging, gastrointestinal injury, concussive brain trauma, fractures to the cranium, skeleton and shell, and massive inner ear trauma (Ketten 1995; Viada et al. 2008). The effects of underwater explosions depend upon several factors, including the size and the depth of the animal in the water, overall water column depth, size, type and depth of the explosive charge, and the distance between the animal and the explosion (Viada et al. 2008).

Electromagnetic Effects

The magnetic component of the electromagnetic field from energy-generating devices and associated transmission cables or undersea pods may affect cetaceans, but the extent of the effect is not known. Some cetaceans and sea turtles are thought to detect and perhaps orientate themselves in relation to the Earth's magnetic field (Putman et al. 2011), so the fields from renewable energy developments may have a temporary effect on those animals as they come within the vicinity of a development (Gill 2005).

Chemical Pollution

Marine animals are exposed to not just one, but a whole cocktail of chemical contaminants in the ocean (Montie et al. 2009), and many such contaminants can accumulate in the body tissues of marine vertebrates, owing to their environmental persistence, hydrophobicity and resistance to metabolism (Montie et al. 2010). Developing marine mammal offspring may be at particular risk because of the transfer of accumulated toxins from the mother during gestation and lactation (Krahn et al. 2010). Chemical contaminants, specifically organochlorines, have been associated with reproductive problems and effects on endocrine and immune function in marine mammals (O'Shea et al. 1999), and they may have the potential to affect neurodevelopment (Montie et al. 2009). Pollutants have also been associated with the disease fibropapillomatosis, disruption of endocrine function and immune system suppression in sea turtles (Hamann et al. 2010).

Marine renewable energy development can add to the suite of contaminants already found in the ocean, through several pathways. Increased vessel traffic during installation, maintenance and decommissioning may introduce chemical pollution into the marine environment in the form of fuel or oil leakages/spills. Renewable energy devices themselves may also contribute to chemical pollution in the event of leaking hydraulic fluid or other unexpected damage, or the release of biofouling agents during maintenance or decommissioning (Dolman and Simmonds 2010). Chemical spills may cause temporary displacement from surrounding areas and also indirectly affect marine mammals and sea turtles via their influence on prey or habitat.

Attraction

Depending on the shape and size of surface structures, wave and offshore wind energy technology may attract pinnipeds by providing haul-out sites. This may of course offer positive benefit, in terms of resting, thermoregulation and predator avoidance, but it may place the animals at enhanced risk to illegal shooting (Boehlert et al. 2008). Pinnipeds may also be at risk of injury from contact with exposed moving or articulated parts while attempting to haul-out or leave the structures (Wilson et al. 2007).

Indirect Effects

Increased or Decreased Foraging Opportunities

Wave and offshore wind energy developments may indirectly influence marine mammals and sea turtles through effects on their prey. Surface and/or subsurface structures have the potential to attract fish, jellyfish and other aquatic organisms by acting as fish-aggregating devices (FADs) or artificial reefs, in turn attracting marine mammals and sea turtles by offering increased foraging opportunity. Scheidat et al. (2011) reported increased numbers of harbour porpoises in a windfarm area off the Dutch coast between baseline periods and operational periods, suggesting that this may be due to increased food availability (a reef effect) or decreased vessel traffic (a sheltering effect). Conversely, changes to the benthic environment may affect species assemblages and result in decreased prey for some marine mammals or sea turtles. For example, if benthic amphipods or epibenthic mysids are displaced or otherwise negatively affected by the placement of hard structures or transmission cables on the seafloor, there may be temporary or localized reductions in foraging opportunities for grey whales.

Increased Predation

If sharks are attracted to marine renewable energy development through either their aggregating properties or electromagnetic field effects, predation pressure on marine mammals and sea turtles in the vicinity may increase. Predation pressure by killer whales may increase too if such predators associate marine renewable energy facilities with enhanced presence of marine mammals. Depending on the size and extent of renewable energy facilities, their potential to exclude marine mammals from particular areas (as described above) may place animals at increased risk of predation. For instance, if grey whale cow/calf pairs are prevented from travelling close to shore by facilities extending along long stretches of the coast, they may be more vulnerable to predation by killer whales. Similarly, if a stretch of installations offshore cause a sufficient barrier effect, then whales might be trapped effectively in a corridor.

Assessment of Effects

In order to assess the effects of marine renewable energy development on marine mammals off Oregon, information gaps such as baseline distribution, abundance, seasonality, migration routes, habitat preference, behaviour and diet of many species need to be addressed (Boehlert et al. 2008). For example, densities have been modelled for 12 species of cetacean along the US west coast based on line-transect surveys conducted from 1986 to 2006 (Barlow et al. 2009), but the surveys were not conducted during winter or spring and did not cover the nearshore area well. Nearshore seabird surveys from 1997 to 2007 along the Oregon coast included sightings of grey whales and harbour porpoises, but were seasonal, only between May and August each year (ODFW 2011). Further visual observations (from shore, vessel or aircraft), telemetry studies and passive acoustic monitoring might all help fill some of these information gaps, and they need to be conducted prior to development and during construction, operation and decommissioning in order to assess better the effects of a development on marine mammals.

As part of the effects assessment, the acoustic output from wave and offshore wind energy facilities need to be characterized to help predict the effect of sound on marine mammals. Underwater recordings can be collected for different phases of development (installation, operation and maintenance, and decommissioning) and for different types of energy-generation technology. Recordings ought also be made year-round to account for seasonal variation in sound-speed properties, differences in weather conditions and signal-to-noise ratio issues that might affect detection and response characteristics.

Baseline Marine Mammal Studies

Three studies have been initiated to date by Oregon State University's Marine Mammal Institute that help address some of the baseline information gaps for grey whales, a shore-based study of grey whale migration, satellite tracking of "resident" grey whales, and a preliminary test of the effectiveness of an acoustic deterrent as a mitigation tool for collision or entanglement risk.

Shore-Based Migration Study

The shore-based observational study was conducted during 2007/2008 to evaluate the potential exposure of grey whales to wave energy development in the Oregon Territorial Sea (Ortega-Ortiz and Mate 2008). Its objective was to generate accurate, up-to-date baseline information on grey whale behaviour and distribution relative to shore in an area where installation of wave-driven electricity generators had been proposed. Three observers surveyed grey whales from an observation station at Yaquina Head on the central Oregon coast (44.67675°N 124.07956°W; 25.39 m above mean sea level) using 7 × 50 binoculars and a theodolite (2 s resolution, 30 × scope). Observations were conducted from the start of January to the end of May during daylight whenever environmental conditions permitted. Locations were recorded of all whales seen during scan surveys of the 200° field of view of the ocean, and of individual groups tracked during focal follows. The average distance from shore, the median depth of locations and the average speed were all significantly different between southbound and two northbound phases of migration, with southbound animals farthest from shore, in deeper water, travelling fastest. The cow/calf migration pattern was opposite, slower than the other phases of migration and closer to shore. Overall, 61 % of all locations and 78 % of locations during the cow/calf migration were within the Oregon Territorial Sea (OTS), where wave energy development has initially been proposed.

Satellite Tagging of Resident Grey Whales

Satellite tags were applied to 18 eastern grey whales (EGW) off the coasts of Oregon and northern California from September to December 2009. These whales are members of the Pacific Coast Feeding Group (PCFG), a subset of EGW that feeds from northern California to southeastern Alaska during summer/autumn. Tracking periods for these animals ranged from 3.3 to 382.9 days and revealed movements to Baja Mexico and as far north as Icy Bay, Alaska. The vast majority of the locations of the tagged whales were inside the OTS, and funding is currently being sought to analyse how the movements of these whales relate to proposed offshore energy development.

Testing the Effectiveness of an Acoustic Deterrent for Grey Whales

During winter and spring 2012, a preliminary study tested the effectiveness of a low-powered acoustic deterrent for grey whales. The objective of the study was to determine whether the sound source would deflect migratory grey whale movements by 500 m. If so, a deterrent could be used to keep grey whales, and potentially other baleen whales, away from marine renewable energy facilities, if such facilities proved risky in terms of collision or entanglement. In this study, an acoustic device was moored in 50 m of water ~4.6 km west of Yaquina Head in the migration path of grey whales. The device emitted a 1-s sound every 20 s during a predetermined experimental period each day. Observers conducted concurrent observations from shore using a theodolite to track whale locations. The study continued during the 2012/2013 grey whale migration to assure a sufficient sample size, after which whale behaviour and distribution will be compared between experimental and control (no sound) periods to determine whether whales respond to the sound.

Benthos

Potential Interactions of Concern

Rocky reefs in the north Pacific Northwest are diverse ecosystems, hosting a variety of fish and invertebrate species. In shallower waters, large, canopy-forming kelps colonize rocks in certain areas, and others are covered with smaller algal species and invertebrates. The structure and function of hard-bottom ecosystems farther offshore and below the photic zone depends on the presence of habitat-forming sessile invertebrates such as sea anemones (e.g. *Metridium*), barrel, vase and shelf sponges, corals (e.g. *Lophelia*), crinoids (*Florometra serratissima*), basket stars (*Gorgonocephalus eucnemis*) and bryozoans. As all these species are slow-growing and fragile, activities that contact the seabed or impact the environment that shapes the communities can harm species which may take a long time to recover. The placement of anchors or burying cable through the habitats can be particularly destructive, so it is imperative that the locations of the habitats be identified in order for the effects of device installation and, where relevant, cable-laying be minimized. Over the long term, these habitats are most likely to be indirectly affected by changes to sedimentation patterns resulting from the installation or operation of nearby wind or wave arrays. For instance, increased siltation during anchor deployment and cable-laying may reduce light availability for seaweeds. Moreover, disturbance during installation and changes to sediment processes attributable to wave-energy capture or scour around static device components can cause smothering or burying of sessile invertebrates and the displacement of mobile invertebrates.

Sedimentary (soft bottom) habitat is the dominant habitat on the continental shelf and slope throughout the Pacific Northwest. Although sandy or muddy habitats are sometimes thought to be barren, they are in fact highly dynamic and full of life. Coastal sedimentary habitats experience high levels of nutrient cycling and are a nursery ground for many ecologically and economically important species, such as sole and crabs. Coastal sedimentary habitats also serve as feeding grounds because they are home to an abundance of invertebrate epibenthic organisms, an important food source for marine animals such as flatfish (Hogue and Carey 1982) and grey whales (Newell 2009). Infaunal invertebrates modify the sediment and structure the habitat, making them key species despite their individual small size. As grain size often determines which animals can live in the sediment, changes to sediment movement as a consequence of ocean energy extraction or alterations of flow around large devices and arrays may affect the distribution of soft-bottom organisms. Finally, scyphozoan jellies, which are important prey for leatherback sea turtles (Graham 2009), also live in these habitats and have benthic life history stages that may be impacted by installation of marine renewable energy facilities. Further, the free-swimming stages of these and other gelatinous zooplankton may be entrained or damaged by moving, subsurface components of renewable energy devices.

During construction and decommissioning, sedimentary habitats may be directly disturbed by the movement of anchors, substations and underwater power cables. It is hypothesized that these effects will be similar to those associated with dredging-related disturbance, so a local loss of sedentary infauna and reef-builders can be expected, and mobile benthic organisms would be displaced temporarily (Gill 2005).

Changes in local hydraulics (current patterns and water mixing) may result from energy capture by large arrays of wave-energy devices (Cada et al. 2007) or by the installation of large, hard structures in previously "featureless" habitats. Such changes can affect organisms by altering the patterns or rates of food delivery, the mixing of eggs and sperm, the dispersal of spores and/or larvae and how temperature varies throughout the water column, all of which may impact benthic species distributions and/or abundances. Changes in water movement resulting from wave-energy capture or the presence of devices inducing scour also may affect how sand is moved within and among coastal areas, impacting benthic species.

Wave-energy capture and offshore wind devices, their anchors and mooring lines will introduce hard material that may attract fish (Nelson 2003), be colonized by a variety of invertebrates, including jellyfish polyps (Holst and Jarms 2007) and/or become settlement sites for non-native species. The fish, invertebrates and seaweeds that colonize hard structures will be different from those typically found in sandy habitats, so a new biological community will arise, which may result in novel food or novel predators for the resident organisms (Gill 2005). Additionally one may anticipate the formation of shell mounds if the devices are deployed for the long term. Shell mounds are a feature of the seabed around offshore oil platforms in California (Page et al. 1999) and around wind turbines in Europe (Hiscock et al. 2002) where structures have been colonized by fouling organisms, which then fall or are scraped off the devices as part of regular maintenance. It is anticipated that similar mounds will develop beneath wave-energy devices. In California, sea stars and rock crabs dominate the megafauna on shell mounds associated with oil platforms, with rock crab abundance up to 24 times greater than Dungeness crab (Goddard and Love 2010).

Initially, most wave-energy devices being considered in Oregon were buoy-type devices targeted for installation in 40–70 m of water, primarily on areas of sediment, so that concrete blocks could be used as anchors. In that case, the actual area of seafloor expected to be converted from soft bottom to hard substratum by the deployment of concrete anchors is <5 % of the overall buoy-array footprint. More recently, devices that operate on the seabed have been proposed, such that the amount of benthic habitat directly exposed to the device is greater. However, in many cases the overall project footprint of installations using the devices is smaller, potentially resulting in a smaller but more intense impact area. Bottom-mounted devices tend to dominate nearer the shore in shallower water, but in Oregon too there has been increased interest in offshore, floating wind devices. These devices would use mooring lines and anchors similar to those for floating wave devices but be located farther offshore in deeper water and typically with a broad footprint of widely separate devices. Therefore, it is necessary for scientists and engineers to consider the dynamics of benthic populations and the potential impact area of renewable energy installations on the inner, middle and outer continental shelf, at a variety of spatial scales.

In order to conduct analyses of the effects of marine renewable energy installations, it is necessary to determine the spatial and temporal extent of the interaction between the devices and benthic invertebrates. As the devices are placed in sedimentary habitat, there will be 100 % overlap between the area of device deployment and the potential area of effects to sediment-associated benthic invertebrates. The effects of seabed installations and/or energy removal may go beyond the spatial extent of the installation, so the extent of potential effects on the sediment and associated invertebrates and fish likely will be greater than the project footprint. Although benthic changes will be observed mainly near installations in shallower water with effects likely to be highly localized, sand adjacent to an artificial reef installed off La Jolla, CA, in 13 m of water was scoured as far as 15 m from the reef (Davis et al. 1982), and grain-size analysis of sediment collected along a transect from Oil Platform "Eva" off Huntington Beach, CA, in 18 m of water indicated coarse sand out to 20 m from the platform and very fine sand beyond that (Wolfson et al. 1979). Hence, sedimentary changes may be observed up to 20 m away from renewable energy device installations. Such differences in grain size are important to benthic biological communities in these zones because depth and median grain size are the major drivers of species distributions; Wolfson et al. (1979) noted changes in epifaunal and infaunal invertebrates with distance from the platform. Studies of offshore platforms in the Mediterranean similarly observed that benthic infaunal assemblages varied with distance from a platform and that the spatial extents of the differences varied with depth of the platform (30 vs. 90 m water depth; Terlizzi et al. 2008) and over time (Manoukian 2010). Energy removal itself will be another consideration in predicting the effects on the benthic infaunal communities, but it is not yet known what the spatial extent of energy removal will be, i.e. how far the energy removal shadow might persist.

The temporal scale of the stressor will likely be temporary and short term for the potential effects of cable-laying and other installation activities. The temporal scale of the stressor of the presence of device components on the seafloor [cables (if exposed over hard bottom substratum), anchors and some styles of device] and of the indirect effects of devices and cables in the water column (by attracting new species and shell-mound development) on benthic habitats and organisms is for the duration of the project. For wave-energy devices, the temporal scale of stress attributable to energy removal is for the duration of the project, but the amount of energy removal and the scale of potential effects may vary based on the operation of the wave devices and sea state. Alternatively, for wind-energy devices, stress to benthic organisms associated with energy removal is not anticipated.

Baseline Benthic Surveys

To evaluate the effect of devices and/or energy removal on benthic invertebrates and fish in a project area, information about their distribution, habitat association, behaviour and food habit needs to be collected. The extent of temporal and spatial variability in species or assemblages of interest as well as habitat needs to be characterized before project-related changes can be evaluated, so baseline data have to be obtained across seasons, depths and latitude. Reference sites need to be established to evaluate temporal changes at locations reasonably distant from a project site. There are a variety of different techniques that can be employed to survey physical properties of the water or substrata, whole assemblages of organisms and/or to target specific species of interest.

In 1988, the US Minerals Management Service (MMS) conducted benthic reconnaissance surveys for invertebrates on soft and hard bottom substrata of central and northern California's outer continental shelf areas in four geological basins, in preparation for oil and gas exploration in those regions. On hard-bottom substrata generally, observed invertebrate communities changed with depth and substratum parameters, and basin differences were secondary factors, a result consistent with previous MMS- and industry-sponsored studies (Lissner 1989). Similarly, with the soft-bottom invertebrate assemblages, the main patterns were related to depth; sediment size characteristics had a secondary influence and other inter-basin differences appeared to have only a minor influence (Lissner 1989). These results from California suggest that there are few regional differences in benthic invertebrate assemblages and that communities could potentially be predicted if depth and sediment type were well characterized at a site.

In September of 1988, 1989 and 1990, the MMS sponsored surveys of fish assemblages of three rocky banks in the Pacific Northwest (Washington and Oregon) using the manned submersible "Delta". Differences in fish assemblages were observed across the three sampling years and even greater differences among the three sites (Pearcy et al. 1989; Hixon et al. 1991; Stein et al. 1992). These results underscore the need for site-specific baseline and effects surveys and the need to develop an understanding of temporal trends before investigating the potential effects of project installations.

The US Environmental Protection Agency (EPA) initiated an Environmental Monitoring and Assessment Program (EMAP) in 1990 to develop, test and validate environmental monitoring methods for sampling benthic macrofaunal invertebrates. Originally the EMAP protocol required 3–5 replicate samples per station, but studies have shown since that a single sample per station is sufficient (Summers et al. 1992; Macauley et al. 1993). The number of sampled stations per site varies based on the degree of expected heterogeneity of the site. One station per 2 km^2 is often used. To maximize cost efficiency and minimize small-scale endpoint variability in future comparative studies, the EPA recommends taking one 0.1 m^2 benthic macrofaunal sample and each station and sieving through 1 mm mesh (Ferraro et al. 2006), with stations initially sampled seasonally to assess baseline temporal variability. To assess spatial variability, a grid or a random distribution of sampling stations (e.g. determined using a randomized, tessellated, stratified sampling design used by EMAP; Stevens and Olsen 2004) needs to be established such that several stations of varying distance from the proposed installation are sampled.

To collect epifaunal invertebrates, including gelatinous zooplankton, by day (at night, many migrate towards the surface) and fish in sedimentary habitat, a bottom trawl tends to be used, and a 2-m beam trawl is usually employed to collect epifaunal samples from various sediment types by sampling at and just above the seabed. It performs reliably on soft and coarse sediment, its small size makes it easy to deploy, and it usually delivers a sample of manageable size (Ware and Kenny 2011), although multiple tows may be required to achieve adequate statistical power. For each tow, an average towing speed of 1.5 knots needs to be maintained for 5–10 min, depending on the density of organisms. The sample also needs to be sufficiently large to characterize the resident epifaunal assemblage adequately.

Numbers and/or biomass of epibenthic fish and invertebrates, including scyphozoan jellies that are important prey for leatherback turtles, presence, density, size and temporal distribution all can be ascertained using visual and/or hydroacoustic survey methods (Georgakarakos and Kitsiou 2008; Trenkel et al. 2008). Specific visual methods used in practice include SCUBA or diver-operated video transects (Martin and Lowe 2010), towed video transects using sled-mounted cameras (Sheehan et al. 2010), manned submersibles (Yoklavich and O'Connell 2008) and remotely operated vehicles (ROVs; Pacunski et al. 2008).

In Oregon, two major baseline survey projects have been undertaken to characterize benthic habitats and species that could be affected by future energy-generating installations. The first project has been repeat sampling of the site of the Wave Energy OTF operated by OSU-NNMREC. The benthic surveys have included sampling for both invertebrates and fish in the sedimentary habitat found at the test site. The second baseline project is a broader, alongshelf survey of benthic invertebrates only in both sedimentary and rocky habitats at selected sites on the outer continental shelf. Survey areas for both projects initially were mapped using multibeam sonar and acoustic backscatter to classify the habitat prior to biological sampling. Both are discussed below.

Baseline Surveys of the NNMREC OTF Site

Pre-installation baseline sampling of benthic habitats and species was conducted at and around the NNMREC OTF location from May 2010 to December 2011. After exploratory video sled surveys in May 2010, 12 sample stations were established on a regular grid of four transects with stations at approximately 30, 40 and 50 m on each. Properties of the full water column were sampled on each visit with a SeaBird CTD profiler with additional sensors at every station. Stations were sampled bimonthly, weather permitting, in order to determine the temporal intensity of sampling required to quantify seasonal variability in habitat and organisms at the location.

All 12 stations were sampled for sediment and infaunal organisms using a 0.1 m^2 modified Gray–O'Hare box corer, and a subsample of sediment from the undisturbed surface collected and preserved for grain size and total organic carbon analysis. The sample was then sieved on board through a 1 mm mesh screen; samples were stained and preserved for identification to the lowest possible taxon (species in most cases) and enumeration.

For beam trawl surveys, nine stations (three transects) were sampled on each visit. The central transect aligns with the OTF site with reference transects located approximately two minutes of latitude to the north and south. The beam trawl is 2 m wide and 0.5 m high, with wall netting of 20 mm mesh and a liner of 3 mm. Tow duration was 10 min from contact with the seabed to retrieval. Fish and most epifaunal invertebrates were collected and preserved. Dungeness crab (*Metacarcinus magister*), seastars (*Pisaster brevispinus*, *Luidia foliolata*) and scyphomedusae (*Chrysaora fuscescens*) were enumerated, sexed (crabs only), measured (crabs and seastars), then released. In the laboratory, all fish were identified and analysed morphometrically for body condition, and the gut contents of selected flatfish species identified. Invertebrates, mostly mysids and *Crangon* shrimp, were sorted to species and counted.

Two distinct sediment types were found in the OTF area near Newport, OR: silty sand at ~30 m and potentially shallower and nearly pure sand at 40 m and deeper. A single uniform infaunal invertebrate assemblage was found across the silty sand stations (all 30 m stations) that differed from the assemblage found at deeper, sand stations. At the deeper (50 m) stations, infaunal invertebrate assemblages differed between the two northern and the two southern stations, suggesting that infaunal invertebrate assemblages vary significantly over relatively small spatial scales. Hence, spatial sampling at greater intensity is required to characterize a site and to investigate potential near- and far-field effects of renewable energy installations.

There was no seasonal variability in either grain size or infaunal invertebrate assemblage at the site in the 2 years of bimonthly sampling. However, comparisons with similar studies conducted by the US Army Corps of Engineers indicate that there have been changes in relative abundances of species over longer time-scales, suggesting that baseline or effects monitoring may be needed on an annual basis for a sandy site such as that off Newport, OR. Contrastingly, fish species presence in the area varied with season over the 18-month sampling period, so baseline or effects monitoring of fish assemblages would need to be conducted across seasons to encompass the full variability present at a site. Scyphomedusae were collected only in summer and early autumn (June, August and October) with abundances in 2010 twice what they were in 2011. In 2011, ctenophores (*Pleurobranchia bachei*) were more abundant, concurrent with the decrease in scyphomedusae. These fish and gelatinous zooplankton data, indicating seasonal variability, suggest that effects monitoring must, at a minimum, ensure that samples are taken in the same season across years so that seasonal changes that arise naturally are not ascribed to project effects. Further, multiple years of baseline data collection are needed to characterize interannual variability. Results of the benthic baseline surveys conducted for the NNMREC OTF are fully reported in Henkel (2011).

Outer Continental Shelf Invertebrate Survey

This was conducted on both sedimentary and rocky reef habitats. Sedimentary habitats were investigated for infaunal invertebrates and rocky habitats surveyed for epifaunal invertebrates with a remotely operated vehicle (ROV). Identification and enumeration of fish from the ROV is at the time of writing under way.

For infaunal invertebrate collections, the study used a stratified random sampling design, known as a Generalized Random Tessellation Stratified (GRTS) survey design that distributed 118 sampling stations across six sites from northern California to Washington State. At each station, benthic grab samples were obtained using a modified Gray–O'Hare 0.1 m^2 box corer. After collection of a sediment subsample, samples were sieved on board through a 1.0 mm screen and preserved. At each station, vertical water-column profiles of conductivity, temperature, dissolved oxygen and depth were obtained with a Sea-Bird Electronics CTD unit with additional sensors. Benthic infauna were sorted and identified as described above and median grain size, percentage silt/clay and percentage total organic carbon determined.

Unique infaunal invertebrate assemblages were identified in sedimentary habitats at each of the six Pacific Northwest shelf sites. Hence, for the siting of renewable energy devices, it does not appear that baseline surveys conducted at one site can necessarily serve as a proxy for other sites. Shallower sites had greater spatial heterogeneity in infaunal invertebrate assemblages, so a recommendation for monitoring would be that shallower sites require more sampling in order to characterize them adequately.

A modified deep-ocean *Phantom* remotely operated vehicle (ROV), was used to survey substratum type and macroinvertebrates in hard-bottom habitats. The ROV is equipped with two cameras, one facing downwards (used for accurate species density counts) and perpendicular to the seafloor, the other outwards (used mostly for species ID) and angled roughly 30° from the dorsal surface of the ROV. The ROV was equipped with a CTD that measured depth (m), temperature (°C) and salinity (psu) and with a navigation instrument that measured latitude and longitude, all every second. Each survey station consisted of three parallel transects, each ~250 m long and 250 m apart. Two outer continental shelf sites were surveyed in this manner in 2011 and another in 2012.

In rocky habitats, two major groups of substrata were observed to host different macroinvertebrate communities: high-relief substrata (flat and ridge rock) and finer-sediment substrata (various combinations of mud, gravel, pebble, cobble and boulders). High-relief substrata were associated with a diverse array of sessile taxa, including crinoids, sponges and gorgonians, and communities of these sessile invertebrates sometimes differed between ridge- and flat-dominated rocky, elevated substrata. Low-relief finer sediment habitat was generally associated with motile invertebrates. Fine-sediment substrata composed mainly of mud, mud plus boulders and mud plus gravel each yielded unique macroinvertebrate assemblages.

Development of New Tools to Help Survey Organisms and their Interactions with Devices

A video lander (also referred to as a drop-video camera) is a sampling tool designed to survey the colonization of organisms on devices and to assess potential fish-aggregating effects of anchors and other benthic device components. The video lander consists of an aluminium frame with two sets of video cameras with lights mounted on the frame. A similar device with a single camera is described in Hannah and Blume (2012). The two cameras are orientated 180° from each other so that they face opposite directions. The lander is deployed at the 40 and 50 m stations on two of the established transect lines as reference locations as well as being dropped near each anchor of the Ocean Sentinel (n=3; ~45 m depth) and at each anchor of the WEC under test (as appropriate for each device type), for a total of six reference and up to six anchor drops per visit. It is left on the seabed for a total of 15 min at each drop station. The number of each species or taxa of fish observed over time by each camera is counted and the primary (mostly sand) and secondary (potentially anchor) substratum observed is recorded. Counts are compared to determine whether there are more fish at anchor locations than reference locations and whether the camera facing the anchor records more fish than the camera facing away from the anchor. This sampling method will also provide for observation of derelict gear that may become tangled on the anchors and result in subsequent animal entanglement.

Plans for Post-Installation Studies

Plans for post-installation benthic surveys at the NNMREC OTF are as described in the baseline monitoring section above, with box cores taken at each of the 12 established stations and the beam trawl used at each of the nine stations established. Each year of testing, samples are scheduled to be taken in spring, summer and autumn, allowing analysis of the spatial and temporal distributions of organisms, the condition of fish, and whether their feeding habits have changed from the situation pre-installation. With a video lander also available, whereas it was not available for baseline monitoring, it will be possible too to assess the potential fish-aggregating effects of anchors.

Conclusions

In order to evaluate the potential impacts of renewable energy installations on the environment and associated organisms, the habitat, the resident organisms and their dynamics must first be characterized. Strong seasonal, interannual and decade-scale patterns in ocean conditions are known for the Pacific Northwest, so an understanding needs to be developed of how organisms respond to this changing environment before specific project effects can be determined. Researchers at OSU and their collaborators are engaged in myriad studies to understand organism distributions in areas targeted for marine renewable energy in the Pacific Northwest and their potential for interaction with renewable energy installations. The OSU-NNMREC OTF is an innovative effort to deliver a mobile capability for testing the output of WEC devices and represents the first installation of an offshore renewable energy device in Oregon, providing an opportunity to observe expected environmental interactions. In addition to modelling seabird distributions, surveying grey whale migration and sampling benthic habitats and organisms as described in this chapter, NNMREC researchers have investigated other interactions of potential concern. This includes characterizing the ambient noise field, modelling sound propagation in the area and recording acoustic outputs from the test platform and a wave-energy device under test.

An objective of all these baseline efforts is to reduce the uncertainty associated with the variety of potential effects of marine renewable energy installations, with the goal of identifying priority interactions that need to be monitored post-installation. These studies represent a pioneering effort to support development of marine renewable technologies in a manner compatible with ocean and coastal ecosystems. The research aims to contribute to monitoring standards for marine renewable energy installations such that potential environmental impacts can be assessed across device types and different regions of the world.

References

Adams J, MacLeod C, Suryan RM, Hyrenbach KD, Harvey JT (2012) Summer-time use of west coast US National Marine Sanctuaries by migrating sooty shearwaters (*Puffinus griseus*). Biol Conserv 156:105–116

Bailey H, Senior B, Simmons D, Rusin J, Picken G, Thompson PM (2010) Assessing underwater noise levels during pile-driving at an offshore windfarm and its potential effects on marine mammals. Mar Poll Bull 60:888–897

Barlow J, Ferguson MC, Becker EA, Redfern JV, Forney KA, Vilchis IL, Fiedler PC et al (2009) Predictive modeling of cetacean densities in the eastern Pacific Ocean. NOAA Technical Memorandum NOAA-TM-NMFS-SWFSC-444. http://swfsc.noaa.gov/submenu.aspx?ParentMenuId=32

Bartol SM, Ketten DR. (2006) Turtle and tuna hearing. In: Swimmer Y, Brill R (ed) Sea turtle and pelagic fish sensory biology: developing techniques to reduce sea turtle bycatch in longline fisheries. NOAA Technical Memorandum, NMFS-PIFSC-7, pp 98–105

Bartol SM, Musick JA, Lenhardt ML. (1999) Auditory evoked potentials of the loggerhead sea turtle (*Caretta caretta*). Copeia 1999:836–840

Bedard R, Hagerman G, Previsic M, Siddiqui O, Thresher R, Ram B (2005) Offshore wave power feasibility demonstration project: final summary report, project definition study, EPRI Global WP 009–US Rev 1. Palo Alto, CA

Benson SR, Eguchi T, Foley DG, Forney KA, Bailey H, Hitipeuw C, Samber BP et al (2011) Large-scale movements and high-use areas of western pacific leatherback turtles, *Dermochelys coriacea*. Ecosphere 2(7):27 article 84

Boehlert GW, McMurray GR, Tortorici CE (eds) (2008) Ecological effects of wave energy in the Pacific Northwest. NOAA Technical Memorandum, NMFS-F/SPO-92, p 174

Bowlby CE, Green GA, Bonnel ML (1994) Observations of leatherback turtles offshore of Washington and Oregon. Northwest Natur 75:33–35

Brandt MJ, Diederichs A, Betke K, Nehls G (2011) Responses of harbour porpoises to pile driving at the Horns Rev II offshore wind farm in the Danish North Sea. Mar Ecol Prog Ser 421:205–216

Cada G, Ahlgrimm J, Bahleda M, Bigford T, Stavrakas SD, Hall D, Moursund R et al (2007) Potential impacts of hydrokinetic and wave energy conversion technologies on aquatic environments. Fisheries 32:174–181

Carretta JV, Forney KA, Oleson E, Martien K, Muto MM, Lowry MS, Barlow J et al (2011) US Pacific marine mammal stock assessments: 2011. NOAA Technical Memorandum, NMFS-SWFSC-488

Carstensen J, Henriksen OD, Teilmann J (2006) Impacts of offshore wind farm construction on harbour porpoises: acoustic monitoring of echolocation activity using porpoise detectors (T-PODS). Mar Ecol Prog Ser 321:295–308

Davis N, Vanblaricom GR, Dayton PK (1982) Man-made structures on marine-sediments—effects on adjacent benthic communities. Mar Biol 70:295–303

Day RH, Rose JR, Prichard AK, Blaha RJ, Cooper BA (2004) Environmental effects on the fall migration of eiders at Barrow, Alaska. Mar Ornithol 32:13–24

Dohl TP, Guess RC, Duman ML, Helm RC (1983) Cetaceans of central and northern California, 1980–1983: status, abundance, and distribution. Final report to the Minerals Management Service, Contract 14-12-0001-29090, 284 pp

Dolman S, Simmonds M (2010) Towards best environmental practice for cetacean conservation in developing Scotland's marine renewable energy. Mar Policy 34:1021–1027

Fernández P, Anderson DJ, Sievert PR, Huyvaert KP (2001) Foraging destinations of three low-latitude albatross (*Phoebastria*) species. J Zool 254:391–404

Ferraro SP, Cole FA, Olsen AR (2006) A more cost-effective EMAP benthic macrofaunal sampling protocol. Environ Monit Assess 116:275–290

Gehring J, Kerlinger P, Manville AM (2009) Communication towers, lights, and birds: successful methods of reducing the frequency of avian collisions. Ecol Appl 19:505–514

Georgakarakos S, Kitsiou D (2008) Mapping abundance distribution of small pelagic species applying hydroacoustics and co-kriging techniques. Hydrobiologia 612:155–169

Gill AB (2005) Offshore renewable energy: ecological implications of generating electricity in the coastal zone. J Appl Ecol 42:605–615

Gilles A, Scheidat M, Siebert U (2009) Seasonal distribution of harbour porpoises and possible interference of offshore wind farms in the German North Sea. Mar Ecol Prog Ser 383:295–307

Goddard JHR, Love MS (2010) Megabenthic invertebrates on shell mounds associated with oil and gas platforms off California. Bull Mar Sci 86:533–554

Graham TR (2009) Scyphozoan jellies as prey for leatherback sea turtles off central California. Master's thesis. San José State University, San José, CA. Paper 3692. http://scholarworks.sjsu.edu/etd_theses/3692

Grecian WJ, Inger R, Attrill MJ, Bearhop S, Godley BJ, Witt MJ, Votier SC (2010) Potential impacts of wave-powered marine renewable energy installations on marine birds. Ibis 152:683–697

Green GA, Brueggeman JJ, Grotefendt RA, Bowlby CE, Bonnell ML, Balcomb KC (1992) Cetacean distribution and abundance off Oregon and Washington, 1989–1990. Chapter 1. In: Brueggeman JJ (ed) Oregon and Washington marine mammal and seabird surveys. Minerals Management Service Contract Report 14-12-0001-30426

Hamann M, Godfrey MH, Seminoff JA, Arthur K, Barata PCR, Bjorndal KA, Bolten AB et al (2010) Global research priorities for sea turtles: informing management and conservation in the 21st century. Endanger Species Res 11:245–269

Hannah RW, Blume MTO (2012) Tests of an experimental unbaited video lander as a marine fish survey tool for high-relief deepwater rocky reefs. J Exp Mar Biol Ecol 430/431:1–9

Hedd A, Regular PM, Montevecchi WA, Buren AD, Burke CM, Fifield DA (2009) Going deep: common murres dive into frigid water for aggregated, persistent and slow-moving capelin. Mar Biol 156:741–751

Henkel SK (2011) Baseline characterization and monitoring of the OSU mobile ocean test berth site: benthic habitat characteristics and organisms on the Central Oregon Coast. A report prepared for the Oregon Wave Energy Trust. 31 pp. http://www.oregonwave.org

Hiscock K, Tyler-Walters H, Jones H (2002) High level environmental screening study for offshore wind farm developments—marine habitats and species project. Report from the Marine Biological Association to the Department of Trade and Industry New and Renewable Energy Programme (AEA Technology, Environment Contract W/35/00632/00/00)

Hixon MA, Tissot BN, Pearcy WG (1991) Fish assemblages of rocky banks of the Pacific Northwest. A final report by the Department of Zoology and College of Oceanography of Oregon State University for the US Department of the Interior, Minerals Management Service Pacific OCS Office, Camarllo, CA. Contract 14-12-0001-30445

Hogue EW, Carey AG (1982) Feeding ecology of 0-age flatfishes at a nursery ground on the Oregon coast. Fish Bull US 80:555–565

Holst S, Jarms G (2007) Substrate choice and settlement preferences of planula larvae of five Scyphozoa (Cnidaria) from German Bight, North Sea. Mar Biol 151:863–871

Inger R, Attrill MJ, Bearhop S, Broderick AC, Grecian WJ, Hodgson DJ, Mills C et al (2009) Marine renewable energy: potential benefits to biodiversity? An urgent call for research. J Appl Ecol 46:1145–1153

Ketten DR (1995) Estimates of blast injury and acoustic trauma zones for marine mammals from underwater explosions. In: Kastelein RA,

Thomas JA, Nachtigall PE (eds) Sensory systems of marine mammals. De Spil Publishers, Woerden, pp 391–407

Ketten DR, Bartol SM (2005) Functional measures of sea turtle hearing. Final report to the Office of Naval Research. ONR Award N00014-02-1-0510; Organization Award 13051000

Koschinski S, Culik BM, Henriksen OD, Tregenza N, Ellis G, Jansen C, Kathe G (2003) Behavioural reactions of free-ranging porpoises and seals to the noise of a simulated 2 MW windpower generator. Mar Ecol Prog Ser 265:263–273

Krahn MM, Hanson B, Schorr GS, Emmons CK, Burrows DG, Bolton JL, Baird RW et al (2009) Effects of age, sex and reproductive status on persistent organic pollutant concentrations in "Southern Resident" killer whales. Mar Poll Bull 58:1522–1529

Largier J, Behrens D, Robart M (2008) The potential impact of WEC development on nearshore and shoreline environments through a reduction in nearshore wave energy. California Energy Commission, PIER Energy-Related Environmental Research Program and California Ocean Protection Council

Lenhardt M (1994) Seismic and very low frequency sound induced behaviors in captive loggerhead marine turtles (*Caretta caretta*). In: Bjorndal KA, Bolten AB, Johnson DA, Eliazar PJ (eds) Proceedings of the 14th annual symposium on sea turtle biology and conservation. NOAA Technical Memorandum, NMFS-SEFSC-351. pp 238–248

Lissner A (1989) Benthic reconnaissance of central and northern California OCS areas. Final report volume 1: technical report. Submitted by SAIC and MEC Analytical Systems to US Department of the Interior, Minerals Management Service Pacific OCS Office, Los Angeles, CA. Contract 14-12-0001-30388

Lucke K, Lepper PA, Hoeve B, Everaarts E, van Elk N, Siebert U (2007) Perception of low-frequency acoustic signals by a harbor porpoise (*Phocoena phocoena*) in the presence of simulated offshore wind turbine noise. Aquat Mamm 33:55–68

Macauley JM, Summers JK, Engle VD, Heitmullert PT, Adams AM (1993) Annual statistical summary: EMAP Estuaries Louisianian Province—1993. EPA/620/R-96/003, US Environmental Research Laboratory, Gulf Breeze, FL

Madsen PT, Wahlberg M, Tougaard J, Lucke K, Tyack P (2006) Wind turbine underwater noise and marine mammals: implications of current knowledge and data needs. Mar Ecol Prog Ser 309:279–295

Malme CI, Miles PR, Clark CW, Tyack P, Bird JE (1983) Investigations of the potential effects of underwater noise from petroleum industry activities on migrating gray whale behavior. Bolt Beranek and Newman Inc. Report 5366 for US Minerals Management Service, Anchorage, AK. NTIS PB86-174174

Malme CI, Miles PR, Clark CW, Tyack P, Bird JE (1984) Investigations of the potential effects of underwater noise from petroleum industry activities on migrating gray whale behavior/Phase II: January 1984 migration. Bolt Beranek and Newman Inc. Report 5586 for US Minerals Management Service, Anchorage, AK. NTIS PB86-218377

Manoukian S, Spagnolo A, Scarcella G, Punzo E, Angelini R, Fabi G (2010) Effects of two offshore gas platforms on soft-bottom benthic communities (northwestern Adriatic Sea, Italy). Mar Environ Res 70:402–410

Martin CJB, Lowe CG (2010) Assemblage structure of fish at offshore petroleum platforms on the San Pedro Shelf of southern California. Mar Coast Fish Dynam Manag Ecosyst Sci 2:180–194

McCauley RD, Fewtrell J, Duncan AJ, Jenner C, Jenner M-N, Penrose JD, Prince RIT et al (2000) Marine seismic surveys—a study of environmental implications. Aust Pet Prod Explor Assoc 40:692–708

Moein SE, Musick JA, Keinath JA, Barnard DE, Lenhardt M, George R (1994) Evaluation of seismic sources for repelling sea turtles from hopper dredges. Report for US Army Corps of Engineers, from Virginia Institute of Marine Science, VA, USA

Montie EW, Letcher RJ, Reddy CM, Moore MJ, Rubinstein B, Hahn ME (2010) Brominated flame retardants and organochlorine contaminants in winter flounder, harp and hooded seals, and North Atlantic right whales from the Northwest Atlantic Ocean. Mar Poll Bull 60:1160–1169

Montie EW, Reddy CM, Gebbink WA, Touhey KE, Hahn ME, Letcher RJ (2009) Organohalogen contaminants and metabolites in cerebrospinal fluid and cerebellum gray matter in short-beaked common dolphins and Atlantic white-sided dolphins from the western North Atlantic. Environ Poll 157:2345–2358

Nedwell J, Langworthy J, Howell D (2003) Assessment of sub-sea acoustic noise and vibration from offshore wind turbines and its impact on marine wildlife; initial measurements of underwater noise during construction of offshore windfarms, and comparison with background noise. Subacoustech Report 544R0424 to COWRIE, The Crown Estate, 16 Carlton House Terrace, London, SW1Y 5AH

Nelson PA (2003) Marine fish assemblages associated with fish aggregating devices (FADs): effects of fish removal, FAD size, fouling communities, and prior recruits. Fish Bull US 101:835–850

Newell CL (2009) Ecological interrelationships between summer resident gray whales (*Eschrichtius robustus*) and their prey, mysid shrimp (*Holmesimysis sculpta* and *Neomysis rayi*) along the Central Oregon Coast. Oregon State University, Corvallis

NOAA (National Oceanic and Atmospheric Administration) (2012) Office of protected resources. Species. Sea turtles. http://www.nmfs.noaa.gov/pr/species/turtles/

NTRC (Northwest Training Range Complex) (2008) Draft environmental impact statement/overseas environmental impact statement. http://nwtrangecomplexeis.com/Documents.aspx

Nur N, Jahncke J, Herzog MP, Howar J, Hyrenbach KD, Zamon JE, Ainley DG et al (2011) Where the wild things are: predicting hotspots of seabird aggregations in the California Current System. Ecol Appl 21:2241–2257

ODFW (Oregon Department of Fish and Wildlife) (2011) ODFW's Ecological Atlas—Science Workshop, 20–21 September 2011. http://www.oregonocean.info/index.php?option=com_content & task=view & id=381&Itemid=12

O'Hara J (1990) Avoidance responses of loggerhead turtles (*Caretta caretta)* to low frequency sound. Copeia 1990:564–567

Oliver JS, Kim SL, Slattery PN, Oakden JA, Hammerstrom KK, Barnes EM (2008) Sandy bottom communities at the end of a cold (1971–1975) and warm (1997–1998) regime in the California Current: impacts of high and low plankton production. Nature Precedings, http://dx.doi.org/10.1038/npre.2008.2103.1

Ortega-Ortiz JG, Lagerquist BA (2008) Report of the workshop on potential effects of wave energy buoys on marine mammals of the Oregon coast. Report submitted to the Oregon Wave Energy Trust, Portland, OR

Ortega-Ortiz JG, Mate BR (2008) Distribution and movement patterns of gray whales off central Oregon: shore-based observations from Yaquina Head during the 2007/2008 migration. Report submitted to the Oregon Wave Energy Trust, Portland, OR

O'Shea TJ, Reeves RR, Long AK (1999) Marine mammals and persistent ocean contaminants: Proceedings of the Marine Mammal Commission Workshop, Keystone, CO, 12–15 October 1998

Pacunski RE, Palsson WA, Greene HG, Gunderson D (2008) Conducting visual surveys with a small ROV in shallow water. In: Reynolds JR, Greene HG (eds) Marine habitat mapping technology for Alaska. Alaska Sea Grant College Program, University of Alaska, Fairbanks, pp 109–128

Page HM, Dugan JE, Dugan DS, Richards JB, Hubbard DM (1999) Effects of an offshore oil platform on the distribution and abundance of commercially important crab species. Mar Ecol Prog Ser 185:47–57

Pearcy WG, Stein DL, Hixon MA, Pikitich EK, Barss WH, Starr RM (1989) Submersible observations of deep-reef fishes of Heceta Bank, Oregon. Fish Bull US 87:955–965

Phillips EM, Zamon JE, Nevins HM, Gibble CM, Duerr RS, Kerr LH (2011) Summary of birds killed by a harmful algal bloom along the south Washington and north Oregon coasts during October 2009. Northwest Natur 92:120–126

Putman NF, Endres CS, Lohmann CMF, Lohmann KJ (2011) Longitude perception and bicoordinate magnetic maps in sea turtles. Curr Biol 21:463–466

Richardson WJ, Greene CR, Malme CI, Thompson DH (1995) Marine mammals and noise. Academic Press, New York

Robinette DP, Howar J, Sydeman WJ, Nur N (2007) Spatial patterns of recruitment in a demersal fish as revealed by seabird diet. Mar Ecol Prog Ser 352:259–268

Scheidat M, Tougaard J, Brasseur S, Carstensen J, van Polanen Petel T, Teilmann J, Reijnders P (2011) Harbour porpoises (*Phocoena phocoena*) and wind farms: a case study in the Dutch North Sea. Environ Res Lett 6:1–10

Scordino J (2006) Steller sea lions (*Eumetopias jubatus*) of Oregon and northern California: seasonal haulout abundance patterns, movements of marked juveniles, and effects of hot-iron branding on apparent survival of pups at Rogue Reef. Masters thesis, Oregon State University, 113 pp

Sheehan EV, Stevens TF, Attrill MJ (2010) A quantitative, non-destructive methodology for habitat characterization and benthic monitoring at offshore renewable energy developments. PLoS ONE 5(12):e14461

Southall BL, Bowles AE, Ellison WT, Finneran JJ, Gentry RL, Greene CR, Kastak D et al (2007) Marine mammal noise exposure criteria: initial scientific recommendations. Aquat Mamm 33:411–521

Stein DL, Tissot BN, Hixon MA, Barss WH (1992) Fish-habitat associations on a deep reef at the edge of the Oregon continental shelf. Fish Bull US 90:540–551

Stevens DL, Olsen AR (2004) Spatially balanced sampling of natural resources. J Am Statist Assoc 99:262–278

Suchman CL, Brodeur RD (2005) Abundance and distribution of large medusae in surface waters of the Northern California current. Deep-Sea Res II: Top Stud Oceanog 52:51–72

Summers JK, Macauley JM, Heitmuller PT, Engle VD, Adams AM, Brooks GT (1992) Annual Statistical Summary: EMAP–Estuaries Louisianian Province—1991, EPA/600/R-93/001, US Environmental Protection Agency, Office of Research and Development, Environmental Research Laboratory, Gulf Breeze, FL

Suryan RM, So KJ, Phillips EM, Zamon JE, Lowe RW, Stephensen SW (2012) Seabird colony and at-sea distribution along the Oregon coast: implications for offshore energy facility placement and information gap analysis. Northwest National Marine Renewable Energy Center Report 2, Oregon State University, Corvallis, OR, NNMREC, p 26. http://hdl.handle.net/1957/30569

Terlizzi A, Bevilacqua S, Scuderi D, Fiorentino D, Guarnieri G, Giangrande A, Licciano M et al (2008) Effects of offshore platforms on soft-bottom macro-benthic assemblages: a case study in a Mediterranean gas field. Mar Poll Bull 56:1303–1309

Thompson SA, Sydeman WJ, Santora JA, Black BA, Suryan RM, Calambokidis J, Peterson WT et al (2012) Linking predators to seasonality of upwelling: using food web indicators and path analysis to infer trophic connections. Prog Oceanog 101:106–120

Thomsen F, Lüdemann K, Kafemann R, Piper W (2006) Effects of offshore wind farm noise on marine mammals and fish. Biola, Hamburg, Germany, on behalf of COWRIE Ltd

Thresher R, Musial W (2010) Ocean renewable energy's potential role in supplying future electrical energy needs. Oceanography 23:16–21

Tougaard J, Carstensen J, Henrikksen OD, Skov H, Teilmann J (2003) Short-term effects of the construction of wind turbines on harbor porpoises at Horns Reef. Technical report to Techwise A/S, HME/362-02662. Hedeselskabet, Roskilde. http://www.hornsrev.dk

Tougaard J, Carstensen J, Teilmann J, Bech NI (2005) Effects of the Nysted offshore wind farm on harbor porpoises. Technical Report to Energi E2 A/S. NER, Roskilde. http://uk.nystedhavmoellepark.dk

Tougaard J, Henriksen OD, Miller LA (2009) Underwater noise from three types of offshore wind turbines: estimation of impact zones for harbor porpoises and harbor seals. J Acoust Soc Am 125:3766–3773

Trenkel VM, Mazauric V, Berger L (2008) The new fisheries multibeam echosounder ME70: description and expected contribution to fisheries research. ICES J Mar Sci 65:645–655

Tyack PL, Zimmer WMX, Moretti D, Southall BL, Claridge DE, Durban JW, Clark CW et al (2011) Beaked whales respond to simulated and actual navy sonar. PLoS ONE 6(3):e17009. doi:10.1371/journal.pone.0017009

Viada ST, Hammer RM, Racca R, Hannay D, Thompson MJ, Balcom BJ, Phillips NW (2008) Review of potential impacts to sea turtles from underwater explosive removal of offshore structures. Environ Impact Assess Rev 28:267–285

Ware SJ, Kenny AJ (2011) Guidelines for the conduct of benthic studies at marine aggregate extraction sites, 2nd edn. Marine Aggregate Levy Sustainability Fund, London, 80 pp

Wilson BR, Batty S, Daunt F, Carter C (2007) Collision risks between marine renewable energy devices and mammals, fish and diving birds. Report to the Scottish Executive. Scottish Association for Marine Science, Oban, Scotland, PA37 1QA

Wolfson A, Vanblaricom GR, Davis D, Lewbel GS (1979) The marine life of an offshore oil platform. Mar Ecol Prog Ser 1:81–89

Yoklavich MM, O'Connell V (2008) Twenty years of research on demersal communities using the Delta submersible in the Northeast Pacific. In: Reynolds JR, Greene HG (eds) Marine habitat mapping technology for Alaska. Alaska Sea Grant College Program, University of Alaska, Fairbanks, pp 143–155

Rethinking Underwater Sound-Recording Methods to Work at Tidal-Stream and Wave-Energy Sites

9

Ben Wilson, Paul A. Lepper, Caroline Carter and Stephen P. Robinson

Abstract

Commercial-scale devices to extract energy from tidal streams and waves may be new, but an associated industry is developing fast. In most countries, device introduction will require investigation and some level of proof that they do not unduly harm local wildlife. Of the impacts that they might have, the emission of acoustic energy (noise) into the marine environment is important. In operation, it is possible, though unlikely, that they will emit sufficient noise to cause auditory damage to sensitive species, but some level of area avoidance/attraction and masking is likely. Nevertheless, all such devices will require perceivable acoustic signatures for animals to detect and avoid colliding with them. To understand these issues, information on operational device acoustic characteristics is required along with information on existing background noise levels at sites suitable for extraction of marine energy. However, the energetic features of these locations with intense lateral, vertical or oscillatory motion mean that conventional methods of underwater sound recording are unsuitable. Here new methods for sound measurement specifically tailored to tidal-stream and wave-energy sites are introduced. The methods are illustrated following performance tests and real measurements at the European Marine Energy Centre tidal test site in Orkney, UK.

Keywords

Acoustics · Hydrokinetic energy · Hydrophone · Marine renewable energy · Measurement · Noise

With the rapid development of offshore renewable energy, there is a parallel need to quantify their interactions with the marine environment and to seek ways to mitigate any negative impacts (Boehlert and Gill 2010). An area of concern that is frequently raised is the potential acoustic impacts of such devices on wildlife (Inger et al. 2009). Having direct equivalents to onshore structures, the acoustic properties of typical offshore wind turbines in operation are relatively well understood and because of ongoing development, have been relatively well quantified (Madsen et al. 2006; Tougaard et al. 2008). As the wind-harvesting rotors, gearing and generator machinery are suspended above the sea, the bulk of the operational sound introduced into the marine environment is derived from noise conducted down and out of the tower walls (Betke et al. 2004). Where measured, this is generally at lower frequencies (pure tones below 500 Hz—1 kHz) and intensities [<145 dB re 1 μPa at 1 m (rms)] and is therefore not considered a significant acoustic threat to species such as coastal marine mammals (Madsen et al. 2006; Lucke et al. 2007; Tougaard et al. 2009). For fish, the sound is also more likely to be restricted to masking communication and orientation signals rather than hearing damage or area exclusion

B. Wilson (✉) · C. Carter
Scottish Association for Marine Science (SAMS), Oban, Argyll PA37 1QA, UK
e-mail: ben.wilson@sams.ac.uk

P. A. Lepper
School of Electronic, Electrical and Systems Engineering, Loughborough University, Loughborough, Leicestershire LE11 3TU, UK

S. P. Robinson
National Physical Laboratory (NPL), Teddington, Middlesex TW11 0LW, UK

M. A. Shields, A. I. L. Payne (eds), *Marine Renewable Energy Technology and Environmental Interactions,* Humanity and the Sea,
DOI 10.1007/978-94-017-8002-5_9,

(Wahlberg and Westerberg 2005). In contrast, the construction phase, particularly piling of the turbine tower foundation into the seabed, introduces relatively short-term but high-intensity impulsive noise (Bailey et al. 2010; Casper et al. 2012). Such activity is at sound levels where harm to sensitive marine organisms is possible (Madsen et al. 2006), so there has been and continues to be significant effort to understand and mitigate construction-related noise. With wind turbines being deployed in increasingly deeper water, it is likely that pile-driving will give way to other potentially less noisy fixing methods (pin-piled jackets, then anchored floating structures, etc), but whether the construction of these is quieter remains to be seen (Norro et al. 2013).

The acoustic properties of marine renewables (Tidal-Stream Generators, TSGs, and Wave Energy Converters, WECs) are less well understood, and arguably because they gather kinetic energy from the seawater itself during their operational phase, they have a greater potential to present underwater acoustic issues than operational offshore wind turbines. This is because, in most cases, the energy harvesting and conversion machinery floats on or is entirely submerged within the sea. The devices therefore interact with the motion of the water and, being in direct contact with the water, transmit sound better into the surrounding medium. To complicate matters, without the long period of onshore evolution experienced by the wind turbine industry, a far greater diversity of device concepts is being progressed simultaneously by the tidal stream and wave sectors (>170, according to the Scottish Renewables Forum held at Inverness in 2011). For TSGs, current concepts range through horizontal axis turbines, vertical axis turbines, reciprocating hydrofoils, venturi-effect devices, tidal kites and Archimedes screws (www.aquaret.com). WECs similarly range through attenuators, point absorbers, oscillating wave surge converters, oscillating water column, overtopping, submerged pressure differential, bulge wave and rotating mass devices (www.aquaret.com). For each of these types too, there are several variants and a variety of novel devices that do not fit into any of the currently known energy extraction categories.

Despite the large number of device concepts being progressed simultaneously, most are currently in the conceptual or experimental phase and do not yet exist at full scale or in final operational configuration. In parallel to the diversity of devices and sites being targeted for tidal-stream and wave-energy extraction, there is also an assortment of fixing concepts being developed and tested (Huang and Aggidis 2008). Currently, pin-piled substructures, cable/chained anchor moorings, monopiles and gravity bases dominate the options (www.aquaret.com). Unlike wind turbines, the windows of time available for construction and fixing of tidal- and wave-energy devices tend to be limited, so elements of construction noise associated with device placement are likely to be both more diverse and more restricted to brief weather opportunities and periods of low tidal flow. For obvious reasons therefore, there is focus on rapid-drop deployment methods with potentially less acoustic impacts than pile-driving.

Why the Interest in Noise from Marine Renewables?

As with other marine anthropogenic activities, there is growing interest in the sound produced by wave- and tidal-stream energy devices (Inger et al. 2009; IWC 2013). Primary attention in other forms of marine industry noise has focused on its potential to harm marine fauna, particularly marine mammals and fish and to a lesser extent birds and turtles (Yelverton and Richmond 1981; Popper 2003; Nowacek et al. 2007; Piniak et al. 2012). Intense noise, such as underwater explosions, seismic pulses, sonar or pile-driving, can cause physical trauma to tissues surrounding internal airspaces (especially lungs) in animals in the immediate environment (Richardson et al. 1998). For offshore renewables, however, more attention has been paid to auditory damage associated with Permanent Threshold Shift (PTS) and Temporary Threshold Shift (TTS) in hearing sensitivity (Madsen et al. 2006). These arise when acoustic energy damages the hair cells used in auditory reception (Ketten 2012). Although there is uncertainty about precisely what levels of sound are required to cause auditory damage in marine species, it is clear that intensity (and therefore proximity) coupled with duration or repetition are key (Southall et al. 2007). PTS and TTS are key physiological issues in terms of degraded sound perception, but additional effects are also likely at lower intensities of sound. At levels from perception upwards and frequencies where animals have sensitivity, anthropogenic sounds may cause a broad range of behavioural responses, including increased vigilance, avoidance, attraction, startle, activity switching and communication masking (Richardson et al. 1998; Wahlberg and Westerberg 2005).

Precisely what responses animals show are likely to depend on intrinsic factors such as the species and the mode of sound perception (pressure or particle-motion detection). More subtle factors such as age, gender, current activity, experience, prior exposure, motivation and precise features of the sound stimulus may also play a role (NRC 2005; Götz and Janik 2010). In addition, extrinsic features such as geographic location and water depth may play a part. Responses may, of course, be complex (Miksis-Olds et al. 2007) and subject to behavioural compensation (New et al. 2013), so to begin to understand them it is essential to gain a good grasp of the acoustic emissions of the sound sources of concern, in this instance marine renewable energy devices. Moreover, to understand the more subtle issues of disturbance and masking, information is required on local levels of background noise, so that potential zones of perception can be integrated. This is particularly pertinent because manoeuvring animals need some warning if they are to avoid colliding with devices in the limited visibility of temperate coastal waters. It is the acoustic cues that devices produce that animals (particularly large marine vertebrates) likely need for safe passage. The precise extent of response will determine whether animals use the information to avoid collision or are excluded from the whole area (and its associated habitat).

Coupled with a scientific need to understand the impacts of industrial sound on biota, there are drivers from the regulatory side. In Europe, current guidelines/regulations typically require developers to produce an environmental impact assessment (EIA) to determine potential effects on key marine species (Harvey and Clarke 2012). Within the UK, the Joint Nature Conservation Committee (JNCC) currently recommends the use of impact criteria such as those proposed by Southall et al. (2007) for marine mammals. This practice is becoming widely established, e.g. in association with the latest round of offshore windfarm developments in the UK. Various additional criteria exist in other European countries such as Germany and the Netherlands. In addition, indicators of good environmental status (GES) are being developed under EU regulations.

All of these criteria have common features. Typically, the process will involve initial assessment of the soundfield in the vicinity of a noise-generating system using appropriate metrics, particularly the intensities at frequencies of relevance to the species of concern. Propagation of these levels is then modelled to determine their spatial extent and then compared with pre-defined impact thresholds. Mitigation actions are then set out if required or possible.

The EIA process often requires prediction of likely soundfields in advance of system construction. Actual sound levels at a fixed distance from a source are likely to depend upon the source characteristics and the local acoustic environment, which itself depends on bottom topography, seabed type, water depth, water properties and sea state. Sound propagation can be complex in shallow waters because of interactions with the sea surface and seabed and within the water column. To predict the soundfield one needs to understand the characteristics of the sound source and then apply propagation models.

To determine the characteristics of the sound source (i.e. the source level), an assessment of the propagation loss between the source and the measurement location (the received level) is needed, because sound has to be measured at some distance from the device of interest. First, being a large structure, sound emanates from many places simultaneously, so measuring too close to a device will over-represent those parts nearest the recorder. Hence, only at a distance can the device as a whole be summarized into a single signature. For convenience, this is usually expressed as if the entire device was a single infinitely small point in space (termed a monopole). Second, soundfields near the source are typically complex and recorded levels can vary greatly, with only small changes in range. Recording in the farfield (at a distance) therefore produces more robust and repeatable measures. The characteristics of the sound source, once encapsulated into a broadband or frequency-dependent source level, need to be described as independently of the environment as possible.

How is Underwater Sound from Renewables Generally Measured?

Underwater sound is measured in a variety of ways, depending on the frequency requirements relative to receptive species, the nature of the sound source and the platforms available. Most of the measurements undertaken are made with equipment that detects the pressure-change component of the acoustic field and typically use piezoelectric transducer hydrophones (underwater microphones). Such receivers are generally positioned at a known distance from the source, sufficiently far away to be clear of the acoustic nearfield elements of the source signal (Wahlberg and Westerberg 2005). Recordings of farfield pressure changes are then made with, and preferably without, the object of target interest being present or active. There is also increasing interest in the particle motion associated with marine renewables, because this is the medium of perception for most fish and invertebrates. However, the equipment needed to measure this is more complicated and the receptive species have traditionally been of less conservation concern. This, however, is a situation that is likely to change in future as suitable measurement equipment becomes available.

Historically and of relevance here, there have been many forms of acoustic assessment associated with offshore wind developments, to document pile-driving during construction and the noise of turbine operation. Examples include those described by Nedwell et al. (2003), De Jong and Ainslie (2008), Lepper and Robinson (2008), Tougaard et al. (2009) and Bailey et al. (2010). Although the details of these at-sea assessments vary, their generalities are similar and are described below.

Studies typically used wide-bandwidth, high-sensitivity hydrophones with low-noise preamplifiers and recorders sampling at up to 500 kHz, thus allowing sound frequencies of up to at least 200 kHz to be documented. Hydrophones are either mounted from seabed-moored buoys or hung over the side of a boat. For boat deployments, one or several hydrophones are tethered using an anti-heave buoy that is allowed to drift several metres away from the vessel to reduce recording-boat noise. The hydrophone elements are suspended or weighted at a range of depths from 5 to 15 m below the surface or, for shallow sites, simply hung in midwater.

Moorings are generally positioned between 1 and 22 km from the source, although one study recorded 40 m or less from the foundations. Boats tend to be stationary or engaged in sprint-stop-measure manoeuvres along transects usually arranged as spokes running out perpendicular to the source. For comparison, the transect spokes pass close to the static buoys and run from a starting distance to the source from some 100 m out to 8–20 km or even as far away as 60 km. During the 30–120 s recording periods from a vessel, its engine(s), echosounder and generators are extinguished and the underwater sound signal monitored using headphones for electrical interference and excessive wave-slap on the boat's hull. In addition to the recorders intended for sound-level estimation, an additional "comparative" hydrophone buoy set-up has been moored within 2 km of the sound source to provide a general recording of temporal variations in sound levels.

For all these studies, good information on the range to the source has been crucial for back-calculating source levels, with GPS universally used and notes being taken of vessel drift during measurements. Recordings tend also to be time-stamped so that different measurement platforms running concurrently can be correlated. Information on other relevant variables has also been collected, including wind speed, weather, other manoeuvring vessels and the site's sound-velocity profiles logged using conductivity, temperature and salinity (CTD) probes. With the rapid development of electronics, the sound cards and recording systems used have varied depending on the study. For widely applicable results, all data acquisition electronics have to be calibrated, usually before and after the trials, so that measured voltages can be converted to absolute sound pressure levels.

The use of hydrophones to measure underwater sound must not be confused with passive acoustic monitoring (PAM) techniques, which are a suite of tools developed to detect and log the sounds produced by vocalizing animals (Van Parijs et al. 2009) rather than sound energy within the water *per se*.

The Challenge of Wet Renewable Energy Habitats

Wave and tidal-stream devices are designed to extract renewable kinetic energy from the environment, as do offshore wind turbines, but by the very nature of the energy they extract they are placed in very different marine environments from offshore wind devices. Tidal Stream Generators (TSGs) are placed in areas of strong, full-water column lateral flow that will periodically reach speeds of 3–4 m s^{-1}, i.e. at sites typically in channels between land masses or around notable headlands. The level of water flow usually means that bottom sediments are moved or removed, leaving rock, boulders and pockets of coarse gravel. Wave-energy sites are most often off exposed coastlines that are subject to wind and swell coming mainly from offshore. Items on the surface are subject to strong vertical motion, whereas objects in the water column or on the seabed experience oscillatory motion or surge. Sediments can vary depending on depth, but sediment removal leaving scoured rock is common.

Although tidal-stream and wave-energy sites do experience periods of low kinetic energy (i.e. slack tides or settled weather) which may facilitate making recordings of ambient sound or device noise, such periods are unfortunately not when the relevant underwater sound will be generated. For tidal streams, slack water would halt sediment relocation or in-water turbulence and TSGs would be stationary. Likewise, in settled weather, wave sites would not have breaking waves, shoreline surf or surge, sediment will not be moved and WECs will not operate.

With the addition of significant horizontal, vertical or oscillatory water motion, the methods typically used to document the acoustic outputs from offshore wind turbines (described above) are not applicable. As a result, new methods are needed, or existing ones will have to be adapted, so that both ambient noise and TSG/WEC acoustic outputs can be investigated. Below, we outline two new methods specifically developed for the purpose, both trialled at the European Marine Energy Centre (EMEC) in Orkney, Scotland, a facility established in 2003 to test tidal-stream and wave-energy devices and currently the world's leading site for testing full-scale, grid-connected machines in seawater.

Measurement of Underwater Sound in Tidal Streams and from TSGs

As outlined above, water moving laterally presents notable hurdles in terms of working at tidal-stream sites. Simply fixing bottom-mounted recorders in these energetic areas is logistically complex and requires sufficient weight for heavy mooring equipment deployed using large boats to keep equipment stationary and free from vibration on the bottom. Likewise, anchoring boats as recording platforms is problematic and potentially dangerous. More significantly though, recording from a fixed platform in moving water will expose a hydrophone element to substantial water shear over its surface, resulting in spurious (often termed "parasitic") flow noise not actually present in the ambient environment. In most studies of marine acoustics, this problem is negligible because current speeds are low, but in tidal streams, this is clearly not the case. Worse, the effect will scale with the flow such that an artefact will be correlated with the parameters being measured.

Traditional solutions to the problem include (i) attempting to shield the hydrophone, (ii) calculating, then subtracting,

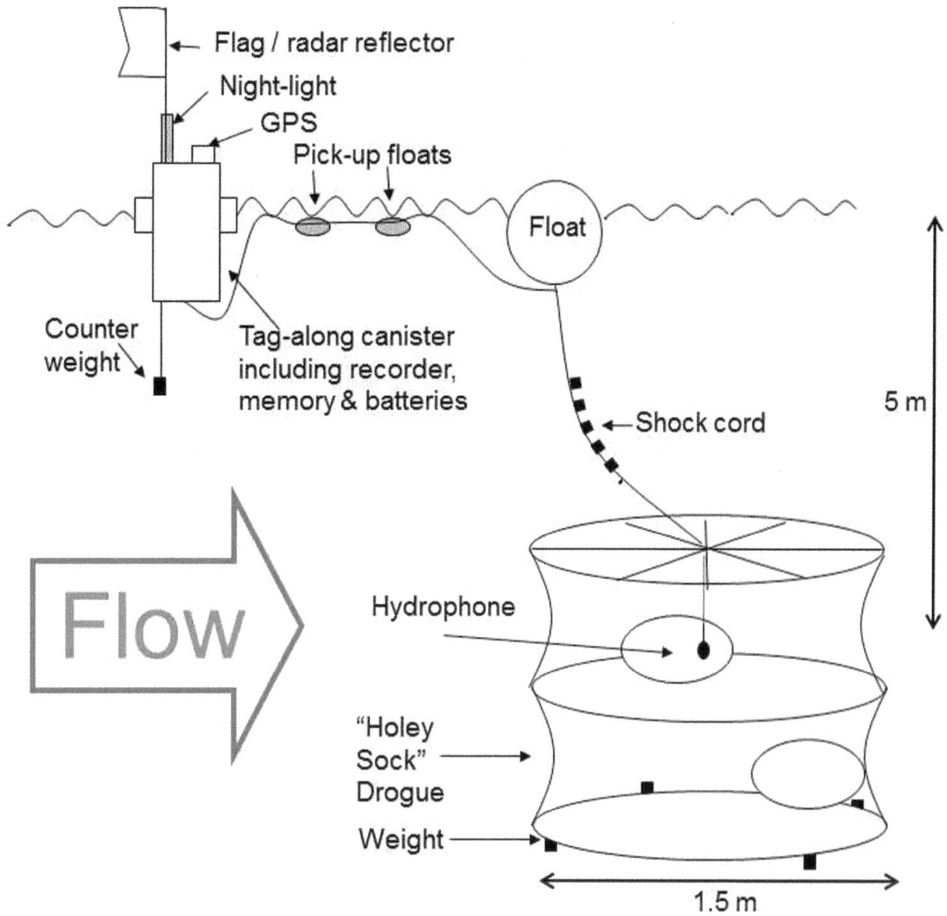

Fig. 9.1 Schematic of components of the drifting ears autonomous recording drifter developed for use in tidal streams

the flow noise, or (iii) deploying hydrophones from a boat allowed to drift with the current. The first two methods are attractive, but because the noise needing to be removed is unknown, its measurement, particularly background noise, will be biased to an unknown extent. The third method offers the simplest solution because it provides a mobile platform where water flow is effectively annulled, but it suffers from three drawbacks. First, even if the boat engine is not running and sonic electronics are extinguished, the boat will generate its own unwanted noise. Footfalls of the crew, loose material moving on deck or in the galley, slopping of the bilge, waves slapping the hull, etc, all require substantial attention to silence and/or the monitoring hydrophone(s) will have to be streamed a long distance from the craft. Second, the boat will not necessarily move in the precise direction or speed of the water in which the hydrophone is recording. Even a slight surface breeze or vertical stratification of water flow speeds will contribute to water flow around the hydrophone element or cable strumming, both of which will introduce spurious noise. The issue may be minor if the sound being monitored has a substantial signal-to-noise ratio (e.g. close to a particularly noisy operating turbine), but it becomes a major limitation for establishing background noise levels or recording turbine noise at distance. Finally, recording using over-the-side equipment from a disabled boat in aggressive tidal currents can only be undertaken with extreme care. The skipper needs to be confident that s/he can start the engines rapidly should there be a safety issue and there has to be a rehearsed procedure for retrieving (or jettisoning) the hydrophone equipment so that entanglement in the propeller(s) can be avoided during the rush of an emergency manoeuvre, should one be required. Given these three drawbacks, a different approach is needed to allow uncomplicated monitoring in tidal streams.

The Drifter-Hydrophone Concept

Drifting with a current offers significant advantages for recording ambient sound, having the recording equipment associated with a boat does not. We have therefore developed, and describe here, a drifting platform (termed "drifting ears") designed with a priority of keeping the hydrophone element fixed relative to the body of moving water while stripping the surface-platform down of everything except the essentials. To do this, the hydrophone is placed inside the bounds of a submerged underwater drogue (Fig. 9.1). The boat is dispensed with and replaced by a suspension buoy and a small floating tag-along case containing the ancillary electronics (batteries, recorder, etc.) and location/retrieval equipment. The entire unit is self-sufficient, dropped into the water upstream of the area of interest, then drifts passively with the current, recording as it goes for later retrieval downstream. The floating case includes a GPS recorder so that, on download, the drifter's precise timing and path is known. This approach has the dual benefit of providing recordings of ambient/device noise in a challenging environment and data gathering while moving so that sound intensities can be measured at a range of locations relative to the source(s) of interest. Moreover, one support boat can deploy multiple acoustic drifters to increase the spatial resolution of data recorded, allowing mapping, and making optimal use of weather windows.

The key feature of the drifter concept is the positioning of a hydrophone such that it is stationary relative to a body of water moving horizontally. The drogue (or sea-anchor, essentially an underwater parachute) must be capable of appropriately gripping the water while not introducing sound of its own. To explore this idea we tried two commercially available oceanographic drogue designs. The first was the "Microstar", a compact drogue intended for coastal waters, rivers and lakes (Pacific Gyre, CA) and consisting of an octahedron made from stiff plastic pipe and sail cloth. When not in use the units can be folded up umbrella-style. The hydrophone cable was mounted inside the body of the octahedron, with the hydrophone suspended below. The second design was a shortened version of the "holey-sock" drogue used for open-ocean current drifters (Pacific Gyre, CA). This consists of a series of plastic rings connected by tear-resistant fabric (Fig. 9.1), with holes cut in it to allow free passage of water. In that case, the monitoring hydrophone cable was mounted on suspension spokes at the top of the drogue such that the hydrophone itself hung within the free space in the middle of the sock.

Prior to deployment in Orkney, both configurations were field-tested at the Falls of Lora at the mouth of Loch Etive off western Scotland (56°27'N 05°23'W), a site with rapid tidal flows and areas of extreme turbulence such that the function of drogues as part of an acoustic recorder can be tested in all three dimensions. The Microstar units tracked the current well and were easy to handle from a small boat, but the horizontal fins were too efficient at capturing the vertical components of the flow so that the drogues were often either brought to the surface or the whole units dragged underwater despite the floatation and the weights attached. In addition, the support struts and sail cloth tended to introduce spurious creaking and rustling sounds when the units were flexed. The holey-sock drogues also tracked the current well but, being vertical cylinders with no horizontal resistance, were not assaulted by vertical currents, so maintained the hydrophone at a near-constant depth. They did, however, sometimes lean away from the vertical particularly in small but strong areas of upwelling. This could be corrected by reducing the number of hoops in the drogue from four to three and increasing both the size of the surface floatation buoy (from 30 to 40 cm diameter) and the mass of the bottom weights, to keep the unit vertically taut in the water. An even larger float was avoided because this would have introduced unnecessary windage. The drogue itself did not introduce its own sound, but occasionally the hydrophone element would bump or scrape the walls of the drogue. Such sound was clearly audible in the final sound files, however, and could be cut-out of the recording sequences; latterly, though, the hydrophone was suspended just below the drogue so that it was not able to contact the sides.

For the initial trials, commercially made hydrophones were used incorporating low internal noise, relatively high sensitivity (−185 dB re 1V μPa) broadband frequency response (0.02–44 kHz+2/−3 dB) and omnidirectional sensitivity (C54XRS, Cetacean Research Technologies, Seattle, USA). There is no reason, however, why models of other higher specification would not be suitable. The hydrophone cable was connected to the surface recording canister with sufficient excess to allow unrestrained stretch in the vertical and horizontal tether ropes.

The hydrophone cable was attached to a digital recorder (M-Audio Microtrack 24/96) housed with a floating surface canister. The resulting .wav files were stored on compact flash cards of 2 or 4 GB capacity, with cards swapped for each deployment. Again, the details of the precise recorder used are less important than the concept that the recorder be portable, have a sufficiently small current draw to function with a suitably sized battery pack, sample at a fast-enough rate and record without introducing excessive self-noise into the application. The canister also included appropriate high- and low-pass filters, batteries and switches. A logging GPS unit was also attached to the surface canister (Garmin eTrex) in a waterproof case (a mini-pelicase). The GPS was set to store a location every 3 or 15 s, depending on the rate of water flow and spatial resolution required. The physical separation of the GPS unit from the acoustic recording meant that they had to be time-aligned, an exercise performed simply by speaking out loud the satellite time into the hydrophone before and after its release into the sea. However, electronically joining the two data streams would be more convenient. The recorder and GPS units were powered by lead-acid gel or alkaline batteries with the recorder's own lithium ion battery being removed because of its limited power storage and seawater explosion risk.

To allow the unit to be tracked at sea, the floating canister was fitted with a pole, a high-visibility flag and a cylindrical radar reflector. The floatation buoy was a high-visibility spherical orange inflatable fishing float with additional 3M reflective tape added. For deployments in darkness, cyalume glow sticks were tested, but proved difficult to see far enough away to be useful. Instead, constant illumination lights (Ocean Safety, UK) were attached to the pole and were visible in darkness at ranges of up to 3.5 km.

Two lines are used to orientate the drogue, buoy and floating canister relative to each other. As thick and stiff a line as practicable (35 mm) was used to join the drogue and surface buoy; its principal function was to keep the various parts spatially aligned such that the hydrophone element would not be towed through the water by any motion at the surface. The vertical float-to-drogue line had sufficient rigidity that it would not strum. A length of shock cord was also added to it to arrest vertical-heave motion from waves impinging on the drogue. As there was less potential for strain between the buoy and floating canister, normal, floating 8 mm line was used, with two small fishing net floats threaded onto it to increase its visibility and ease of retrieval. The hydrophone cable was taped to the connecting ropes at intervals and with sufficient slack that strain on the ropes would not damage the cable.

The lines were arranged so that the hydrophone was suspended 5 m below the surface, a depth chosen because it kept the hydrophone clear of surface waves and guaranteed sufficient clearance between the bottom of the drogue and the likely clearance depth (10 m) of future submerged TSG devices. Depending on the sound frequencies of interest and the habitat being assessed, other depths may be more appropriate.

In our trials, four drifter-hydrophone units were constructed and tested in spring tides at the Falls of Lora, and following these trials, several minor modifications were made, primarily to reduce self-noise from the equipment. For example, clinking metal attachments (e.g. shackles to lugs) were eliminated or replaced with plastic or rope. Further, the surface marker, floats and tag-along canister were trimmed down to an acceptable minimum size to reduce their drag on the drogue-hydrophone arrangement.

Measurement Trials

Ambient noise recordings were then carried out over a swathe of the Fall of Warness (59°08'N 02°48'W) EMEC test site off Orkney in late January 2008 on both ebb and flood tides on the approach to full springs. During these periods, wind and sea state were suitable (Beaufort <2), with no precipitation. For each drift, the deployment boat (a 10 m fibreglass workboat) was stationed upstream of the tidal site and the four drifters were laid line abreast with the intention that they would drift through the site with the EMEC seabed cable ends (and future TSGs) at the near centre of the spread of drifters. Initially, it was a challenge to lay the drifters so they would travel in a well-spaced line, but after practice, good spatial coverage was achieved. Once the drifters were deployed the boat was moved upstream of the middle of the line and its engine and echosounder turned off so that it would drift passively several hundred metres behind the recorders. A listening hydrophone (Brüel and Kjær 8103) was hung over the side of the boat to monitor for stray acoustic activities such as over-the-horizon ferries or other unintended sound sources.

Immediately before each day's at-sea recordings, hydrophone sensitivities were checked using a test-tone and comparison with a reference hydrophone, which itself was calibrated at the National Physical Laboratory, Teddington, UK. The daily calibration took place in Kirkwall Harbour and was also used as an opportunity to soak both the hydrophones and the drogues so that they were hydrated at ambient temperature and sank quickly during the deployments immediately thereafter.

Subsequent to field recording, the sound files were downloaded and matched with the GPS tracks, using the verbal time stamps. As the path of each drifter was different and could not be predetermined, however, pre-prescription of sampling stations was not possible. Instead a series of lines running perpendicular to the path of the tide and drifters was calculated. For the EMEC example, these ran from northeast to southwest (65°T) and spaced every 463 m, i.e. every ¼ nautical mile. From the GPS tracks, it was then possible to calculate when a drifter passed each line, at which point a sound sample was taken from that location: this produced 15 lines for sampling across the EMEC study site. The sound characteristics at the point when each drifter passed a sampling line were then analysed for sound intensity and spectral properties. The most appropriate duration for this sound segment was considered by recording a 10-min sound file at the Fall of Warness site and analysing it five times at 1 kHz for seven different durations (2, 10, 20, 30, 60, 90 and 120 s). The results suggested that an optimum compromise between smoothing short-term variability and sampling in discrete locations at the site was 60 s (other durations may be more appropriate at other sites and flow rates), so this segment duration was used for analyses at that site to characterize the overall ambient noise floor at different states of tide and flow directions. It was also possible to use the data to map the spatial variation in ambient sound effectively over the site. An example map is shown in the right panel of Fig. 9.2. Note that one of the applications mentioned in EU Directive 2002/49/EC generates strategic noise maps, which are useful for spatial planning in relation to sound exposure. Similar sound maps for underwater sound are discussed by Ainslie et al. (2009) and can be valid provided the sound being mapped is relatively constant during the period of measurement. However, in addition to the two spatial dimensions basic to a map, versions based on different frequencies need to be made. Further, for tidal sites in particular, the direction, rate of flow and point in the spring-neap cycle at the time of measurement are key features to consider.

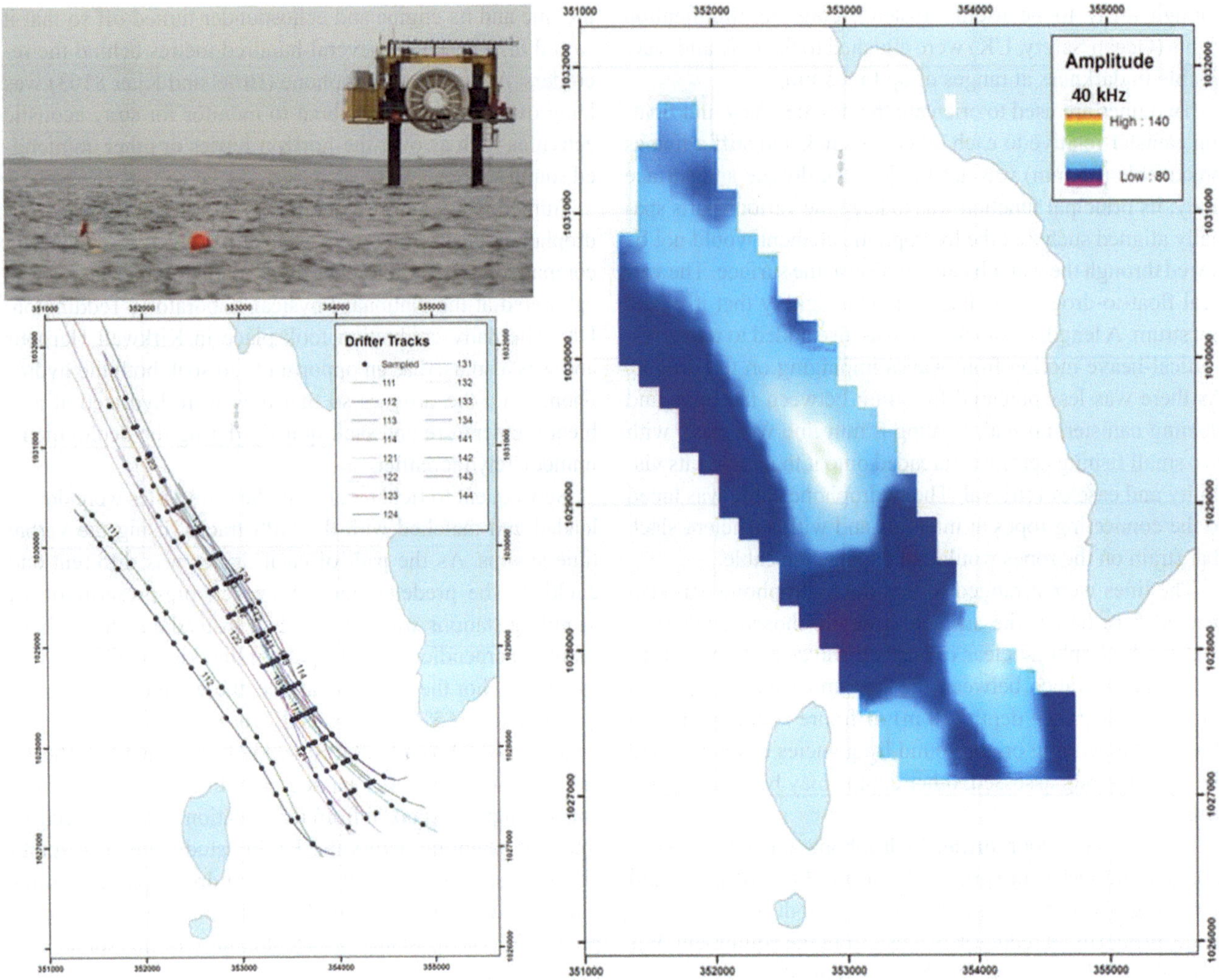

Fig. 9.2 Illustration of the use of multiple drifters and drifts to map ambient sound at the EMEC tidal test site at the Fall of Warness in Orkney. *Top left*: a drifter passing the OpenHydro test turbine parked above water. *Bottom left*: tracks of multiple drifts on an ebb tide on 21 January 2008, lines showing the paths of the drifters and dots showing the central locations of the 60-s sound samples used in the analyses. *Right*: an illustrative contour plot (dB re 1 μPa^2 Hz^{-1}) of the acoustic soundscape at an example frequency (40 kHz) from data interpolation from the sample locations in the bottom left plot with a 150 m grid cell size

Overall, the use of multiple drifters made it quick and relatively simple to sample sound over a wide area with a single boat. The system also demonstrated the potential to quantify transmission loss and hence the acoustic outputs of operating TSGs in a way equivalent to characterization of wind turbines/piling, as described earlier. This is because drifting over, past and around an operating turbine would be the equivalent of the sprint-stop-measure boat surveys conducted in less energetic environments. Moreover, because of their shape, it is unlikely that tidal turbines will emit submarine sound as symmetrically as wind turbines. Having the option of many sampling stations produced by several drifters planted across the flow allows emissions to be estimated on a variety of bearings.

In terms of results, the study quantified the ambient sound levels at the tidal site in both ebb and flood conditions during winter. Results from both tidal flow directions were similar, with the only differences being at low frequency. Discrete patches of high-frequency sound (visible in Fig. 9.2, right panel) were notable and raise interesting questions about the origin of the sounds detected at the site, but investigating them was beyond the scope of the present study. The issue is explored further by Carter (2013).

The drifter-hydrophone concept offers many solutions to the problems of recording in fast-moving water, but it does not solve all the issues. Its most significant drawback is that by aligning to a current that is itself moving over and past a site of interest (including TSGs), recording sessions are short-term events. Typical flow rates of 3–4 m s^{-1} mean that entire drifts are usually only 30–40 min duration. As a result, it is not possible to leave the equipment to record long-term temporal trends without continuously reseeding drifters upstream and hence having a continuous physical presence at the site. Looking at long-term trends such as diel or seasonal patterns in background noise or turbines in different states of operation/wear would therefore be problematic using drifter hydrophones. One solution might be specifically to bottom-mount a long-term monitoring hydrophone, akin to the "comparative" hydrophone used for trend analysis of offshore wind turbine/piling sound, as described earlier, but with the caveat that it would not be able to record events free of flow noise. Similarly, because the use of drifters requires having people in boats, it is not possible to record in all weather scenarios. Storm conditions not only render small-boat operation impractical, but they also render the drifter system inoperative because the extreme surface motion of the waves would introduce movement into the drogue setup. Again because the drifter concept uses water motion, the precise tracks that the drogues take cannot be pre-set, and instead the units have to be released upstream in the hope that they will pass over the areas of interest. However, the near-linear flows experienced at most tidal-stream energy sites (Fig. 9.2) tend to make the approach feasible, and when multiple drifters are seeded across the flow, one or even more are likely to cover the area required, particularly given that tidal sites and TSGs are themselves highly polarized to upstream–downstream orientation. Nevertheless, consideration of the appropriate number and distribution of drifts is needed to avoid selectively sampling at times or locations of particularly low or high noise, current speeds, back eddies, etc.

Although the above is the first published description of this new technique of recording, it will no doubt become more refined as the idea is taken forward. An obvious improvement would be to set multiple drifters at the same location but slightly staggered in time, giving the effect of replicating the sampling but allowing the spatially correlated time components of sound files to be disentangled. For example, brief events (such as distant boat engines starting up) would show up as transposed sound hotspots on neighbouring maps. Further, the tracking and retrieval methods described above are basic and rely on visual or radar tracking of the surface canister. The development of better communications with real-time remote tracking (VHF, satellite or mobile telephone) would allow more drifters to be deployed simultaneously over wider areas and for longer durations. Such tracking technology would almost certainly be required when surveying large areas in consideration for major tidal-stream developments such as the Pentland Firth off northern Scotland.

Measurement of Underwater Sound in Wave Energy Sites and from WECs

As with tidal-stream generators, the development and deployment of WECs are in their relative infancy relative to the longer established offshore wind sector. The growth of these technologies is coinciding with increasing requirements to understand the potential impacts of anthropogenic acoustic emissions on marine wildlife. Like tidal-stream systems, wave-energy generator systems offer some unique challenges. As described above, there are already several energy-conversion ideas and technologies in development, with different components making noise, but different layouts and configurations on or in the water column. Issues with such device diversity are clear when considering two of the most common classes of WEC.

(i) Floating surface distributed devices, for which an example is the Pelamis Wave Power device (www.pelamiswave.com) that relies on a series of horizontally distributed sections spread over > 100 m with most within a few metres of the surface. Further, the components can move with varying degrees of freedom to align with the dominant wave direction. Together, this adds up to a relatively mobile complex distributed system that is primarily but not entirely located at the surface.

(ii) Water-column-distributed devices such as the Oyster (www.aquamarinepower.com) may have rigid seabed foundations from which the system is hinged, so be relatively localized in the horizontal plane but with machinery distributed throughout the water column.

Unlike TSGs, understanding what acoustic outputs exist under different wave conditions is paramount for WECs as the devices become more energetic. Understanding or assessment of this is doubly important as the overall natural acoustic environment will also likely change as the wave regime and sea-states change. Compared with the predictability of tidal flows, wave conditions are harder to predict in advance, rendering boat-based surveys impractical because of the heavy costs of survey time to cover an appropriate range of wave conditions. In addition, the risk of increased boat noise increases with sea state, as do health and safety considerations. Likewise the use of drifting recorders is difficult because of the non-linear nature of drift direction that is common at wave sites. Importantly, however, hydrophones at wave-energy sites tend to suffer less from artificial noise from water flowing over the sensor owing to slower current speeds relative to tidal energy sites. This allows the use of fixed position, long-term acoustic dataloggers rather than the drifting concept discussed previously.

In a project supported by Scottish Natural Heritage, the Scottish Government commissioned work to develop a generalized measurement methodology for underwater noise generation from wave-energy systems deployed at the EMEC wave energy test sites in Orkney (EMEC, www.emec.org.uk). The proposed methodology includes determination of long-term temporal variation (e.g. changes in significant wave height) and device operational modes, consideration of the spatial properties of sound emanation (i.e. directionality of emissions) and radiated amplitudes over frequency bands relevant to potential receptors (Lepper et al. 2012).

Both types of WEC (floating or water-column distributed) involve physical mechanical movement or water motion, which is translated to electrical energy through a number of possible processes. These may generate noise through gearbox, generator motion, hydraulics motion, turbine blades, etc, and there may be additional noises peripheral to the energy generation method associated with mechanisms such as wave slap. It is likely that such noises will depend on sea state and tidal conditions (Lepper et al. 2012). Despite having some internal similarities to TSGs, WECs will be significantly unlike them because of the episodic and variable nature of wave events. Temporal variability of the sound signature of WECs is therefore likely to be much larger than that for TSGs.

Below, we outline methods to capture suitable and comparable datasets on underwater sound emissions from a diversity of WECs in their various operational modes and sea conditions. Because of these differences in generic device type, slightly different measurement methodologies were considered for the floating or water-column distributed systems. Further, the proposed methodologies were developed with current and developing UK and European guidelines on impact assessment in mind, to ensure that the data collected fall in line with current and possible future requirements of the consenting processes. The aim was to provide a robust reproducible generic methodology for assessing the acoustic output from WECs.

It is worth considering that in addition to measurement of received levels (acoustic levels some distance from a source), it is common practice in underwater acoustics to use the concept of a source level to characterize the acoustic output of an energy device. The source level can then be used as a source amplitude term with propagation loss models to predict sound-received levels at different ranges or in a new environment. Such predictive modelling is a common requirement of the EIA process both before and after systems are built. Traditionally, the source-level term refers to the acoustic output of a theoretical point source termed a monopole, which characterizes the source as infinitely small and independent of the environment. As described above, measurements of levels some distance (ideally in the farfield) from discrete sources using the drifting ear methodology proposed for individual TSGs depend too on the environment and will provide a series of measurements at different ranges as a consequence of the multiple recordings and the physical movement (drift) of the systems. These datasets combined with propagation-loss models provide potential for assessment of farfield source-level estimates of an equivalent point source at the TSG location independent of the environment.

As already stated, it is convenient for assessment and propagation modelling to summarize the acoustic output of a source down to a single point in space (a monopole). However, the wide variety of WEC device concepts poses a challenge, because WEC systems may be acoustically large (physically large relative to the radiated wavelength and relative to TSGs) and are often distributed throughout the entire water column or have a significant surface presence. In both cases, several individual sources of noise may exist, making characterization of the source in terms of a traditional source level difficult. For example, a surface attenuator may contain a series of subsystems located in separate floating sections spread over tens to hundreds of metres. Potential noise from individual components may arise from any of these sections at any time, making isolation of the noise source at a recorder some distance away problematic. At any single moment, a measurement of the soundfield made some distance from the source might contain contributions from several often incoherent (spatially and temporally) sources from within the device as a whole. The variety of distributed sources on the device potentially act as a form of complex spread-out array with different sources at different effective ranges from any single receiver location. As such, the array would likely have a highly complex nearfield. Only at greater ranges, where combined noise components appear to radiate from a single point can the monopole be documented and more easily used in propagation modelling.

Additionally, some sound energy from a potential source (e.g. a generator, a hinge, a hydraulic valve) in shallow water is likely to interact with the environment (non-free field) before arriving at the receiver system. For example, a distributed water-column device may radiate sound from several potential sources located at different positions in the water column with different sound-paths taken between source positions and the recorder system. This would result in constructive and destructive interference effects as soundwaves arrive from different routes. This problem of complex combinations of signals from a distributed source (the device) and multi-path arrivals (the environment) are also seen in noise assessment of many other systems, such as shipping (ANSI 2009) and marine dredging (Robinson et al. 2011), as well as marine piling, particularly in shallow water. An approach that has been adopted in these situations is to integrate the total received energy (from all the distributed noise source components) for a defined angular sector of travel either side of the closest point of approach (CPA; Lepper et al. 2012).

Using a similar approach for a WEC system recorder would require that recorders ideally be placed far enough away from the source to allow combined farfield arrival of complex (multiple sources distributed across the system), as well as environment-related (surface and seabed interactions) modifiers, for the signal to appear to radiate from a single point. The range at which both surface and seabed interactions are well developed and greater than the distributed array farfield distance is termed the source farfield and is related to the device geometry, distribution, water depth and bathymetry, and in most cases of these complex sources is practically impossible to separate from the ideal freefield condition. In such a case, greater measurement range reduces the relative effect of the spatial distribution of potential noise sources at the source and multipath arrivals. Numerical modelling approaches may then be used to estimate a theoretical source equivalent at an arbitrary position, i.e. a single point from which radiated noise appears to radiate when viewed from a distance.

As with measurement carried out on shipping and dredging, a wide variety of signal-processing methods can then be used (Lepper et al. 2012). They include integration of total energy between consecutive time-windows showing both short- and long-term temporal and spectral trends from various noise sources from the device, as well as detailed analysis of specific acoustic events and frequency-dependent source-level determination.

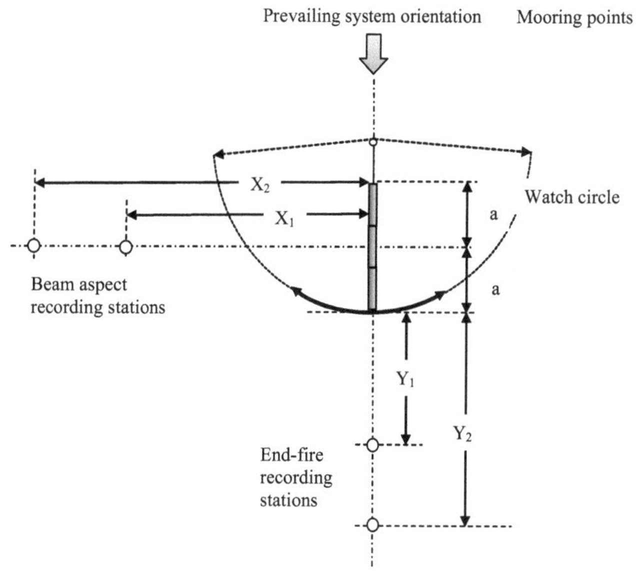

Fig. 9.3 Recorder configuration for a floating, surface-distributed WEC system. (Reproduced from Lepper et al. 2012)

Measurement of WECs

Because of the highly dynamic sea state likely at a WEC device location and the need for extended good temporal recording to encapsulate the inherent variability of wave regimes, we decided to pursue bottom-mounted recorder systems with receivers close to the seabed. Surface-buoyed systems or boat-based deployments tend to suffer from parasitic noise attributable to vertical and oscillatory wave motion, and in the case of boat-based deployments, increased platform self-noise, as discussed with tidal stream scenarios, as well as limited operational windows. Bottom-moored recorders strung vertically throughout the water column have also been used at wave test sites successfully; however, they may suffer from greater positional variation attributable to tidal flow or surge, resulting in increased uncertainty in precise WEC-to-receiver distances and consequent source-term estimates. Systems closer to the seabed also suffer less from unwanted flow noise (Lepper et al. 2012).

Floating surface-based energy devices may also have varying extents of positional freedom depending on mooring configuration, allowing them to align with dominant wave conditions. In the case of larger systems, this may result in considerable variation in horizontal position over time; potentially as much as hundreds of metres, particularly relative to a fixed position recorder station. This variation and the relative range to any recorder system have to be considered in post-analysis, so precise information on the WEC's position needs to be logged in parallel with the recording. WECs may also exhibit considerable directivity (i.e. radiate differently in different directions). To allow this variation to be captured, recorder units that can be left for days, weeks or longer in the field are needed. They also need to have suitable noise performance, bandwidth (for species of interest) and storage and/or use of duty-cycle and allow data capture during a variety of operational conditions across the spectral band of interest.

Figure 9.3 illustrates a plan view of a measurement methodology used for a floating WEC (in this example, a surface attenuator). Such devices often have considerable degrees of positional freedom, allowing movement within a watch circle. The device will, however, generally align to a preferred orientation determined by the prevailing wave direction and weather conditions. The acoustic recording methodology we developed used a number of seabed-mounted autonomous long-term recorders to capture the soundfield data from a range of position possibilities. The recorders were isolated from the surface to minimize the effects of surface interaction noise (cable strumming or tugging on surface buoys) and were placed outside the watch circle to avoid any danger of entanglement as the WEC moved.

The method used four recorders set to measure at the beam (orthogonal transect from the midpoint of the surface system) and end-fire aspect noise levels, with recorders ideally placed in the device farfield at ranges X_1, X_2, Y_1, Y_2 (Fig. 9.3). This, however, is often a compromise between being far enough into the farfield, not being contaminated by other noise sources, and obtaining sufficient signal-to-noise to be able to measure the WECs output. In order to understand source characteristics better, a determination of the local propagation conditions is required that can then be applied to propagation modelling. Ideally, as many recording systems as possible would be placed on a single transect

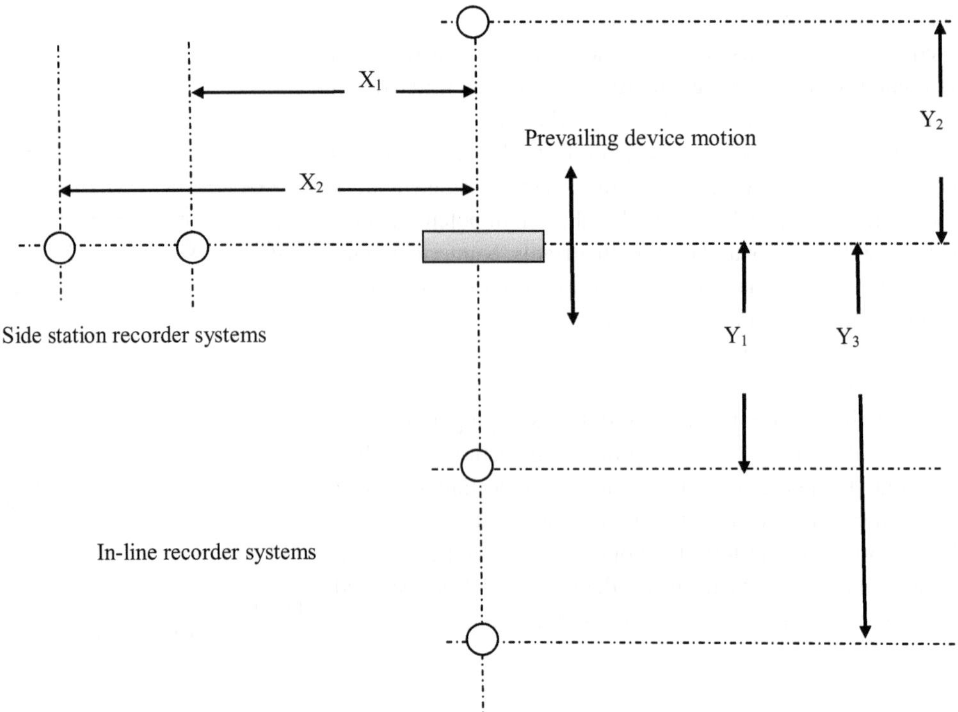

Fig. 9.4 Recorder configuration for a water-column-distributed WEC system. (Reproduced from Lepper et al. 2012)

from the source, so improving estimation of the propagation loss between in this case two (Fig. 9.3) on each orthogonal aspect, lessening the uncertainty associated with source-function estimates derived from propagation loss models. If the WEC is free to swing on its moorings, such movement could potentially allow the fine detail of the radiation pattern to be explored similar to the approach seen with drifting TSG recorders described above. However, to assess both positional and range variations between the WEC and the recorders, comparison was made of actual location data, usually provided by on-board (the WEC device) GPS systems provided by the WEC operator and the known position of the fixed recorder stations (also fixed with a GPS system at deployment). Further, data on the operational state of the WEC system during the measurement periods provided by the operators and/or ambient wave conditions allow comparison with device acoustic emissions under various wave scenarios.

Unlike surface-distributed WECs, most water-column-distributed devices are secured rigidly to the seabed and exhibit mechanical motion in fixed directions (horizontally or vertically), potentially leading to variation in the radiated soundfield in different azimuthal directions from the source. As with a surface-based WEC recording setup, a series of at least four bottom-mounted long-term recorders with two on each of two orthogonal transects around the energy device is needed, with additional systems added at varying ranges on the same transects to improve source-term estimates. At least one transect axis needs to be in line with the plane of any relative motion, as shown in Fig. 9.4.

As discussed above, recorder systems have to be placed sufficiently far into the source farfield to avoid nearfield effects. Factors such as water depth, device size configuration and inter-device separation would lead to a choice of offset distances X_1, X_2, Y_1, Y_2 and Y_3 (as in Fig. 9.4). In the case of an oscillating wave-surge energy-converter system, for example, potential varying noise sources exist throughout the water column, i.e. wave noise at the surface, hydraulic pumps at the base of the system, etc. These distributed sources and seabed and sea-surface multipaths will result in complex arrivals and hence measured received levels, similar to those that any marine receptor might hear. At shorter ranges (within the source nearfield) these fields may be highly variable for relatively short positional or range variations and with strong frequency dependence. At greater ranges, the combined effect of distributed source and multipath arrivals will show less spatial variability as the relative ratio of the source distribution and the range increases.

For both categories of WEC, the use of recorders on two orthogonal transects allows estimation of variation in radiated energy in these directions, and the use of secondary recorders on the same transect at different ranges can be used in conjunction with propagation loss modelling to improve source-term estimation. In the case of an oscillating wave-surge converter, there may be a difference too between the front and back relative to the wave direction. In that case, the use of recorders equally spaced either side of the system is suggested, because this would allow direct comparison of the front- and back-radiated energy under similar propagation loss conditions.

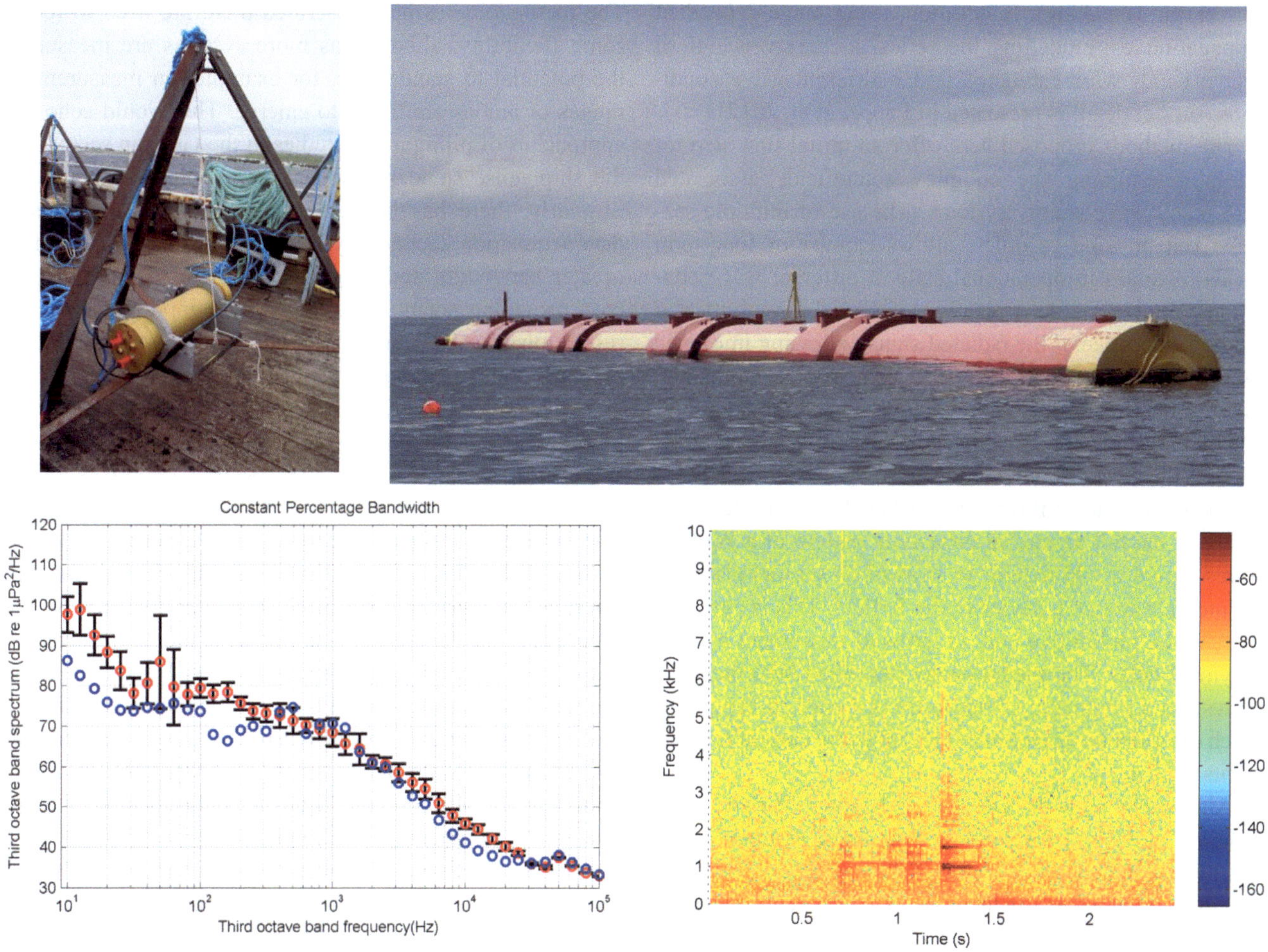

Fig. 9.5 Illustrations of the WEC recorder system trial. *Top right*: a bespoke autonomous recorder in a seabed frame ready for deployment. *Top left*: E.ON Pelamis (Pelamis wave Power) P2 system at the EMEC test site, May 2011. *Bottom left*: example sound levels derived from the trials, *red* circles denoting 10-min mean data from 13:01 to 13:16 on 12 May 2011 on the Inner South recorder with the Pelamis WEC present (bars ±1 s.d.), and *blue* circles being data from the northern recorder, 10:51–11:31 on 12 May 2011 when the WEC was not present (Beaufort scale 1–2). *Bottom right*: normalized spectral level plot of clanking noise, where the range from the recorder to the device midpoint was 333 m (reproduced from Lepper et al. 2012)

Measurement Trials

The proposed methodology was trialled on a full-scale floating Pelamis P2 surface attenuator system owned by E.ON at the EMEC Billia Croo wave test site in 2011 (see Annex A of Lepper et al. 2012) using a series of 4–5 seabed-mounted autonomous loggers deployed for a period of several days. Loggers were mounted in steel frames directly on the seabed with hydrophones in the freefield 75 cm above the seabed, two on a transect aligned with the end-fire position and the other two on an orthogonal transect from the device midpoint. Data acquisition was continuous at 16-bit resolution using bandwidths up to 48 kHz. Most of the measurements were made without support boats on site, so reducing additional noise contamination. The relatively recent improvement of commercially available autonomous acoustic loggers has made assessment of long-term noise trends possible for periods of weeks to months, allowing capture of a wide range of wave states at a wide range of bandwidths.

By correlation of data collected from acoustic loggers and wave state (waverider data from EMEC Ltd) and WEC data (GPS positioning at either end of the floating machine and operational status from Pelamis Wave Power Ltd, E.ON Ltd), it was possible to demonstrate device-noise variation under different sea states. Post-analysis allowed determination of noise trends for complex broadband signals and analysis of specific noise events temporally, spatially and in frequency. The bottom left panel in Fig. 9.5 shows the statistical distribution of third octave band analysis of device noise over a 10-min period under calm conditions (sea state 1–2) relative to an equivalent baseline measurement with the device not present. Similarly, the bottom right panel in Fig. 9.5 shows temporal and spectral analyses of a specific noise event (clanking). In addition, using range-dependent propagation loss modelling, theoretical source terms in third octave bands were estimated under different wave conditions (further detail is presented in Lepper et al. 2012).

The methods proposed here offer an initial first step towards determining the complex soundfields associated with many wave-energy systems. The use of multiple seabed fixed position recorders allows capture of long-term trends (changes in operational state at different wave conditions), and some determination of spatial directivity and range-dependence of the radiated signal, allowing improved source-term determination for future predictive modelling. These latter two properties are, however, particularly limited by the number and quality (noise performance, memory, battery life, etc.) of the recorders used vs. the associated instrument cost and deployment logistics. For example, to investigate the full azimuthal directivity pattern of a WEC to 5° resolution would require an array of 64 sensors placed at 5° intervals around a device in the farfield. This is currently likely to be prohibitive both in terms of instrument and deployment costs. The methodology described here therefore represents a pragmatic compromise designed to capture as much information as possible relevant to the current requirements of EIAs.

Alternative methods include an industry-wide fixed range standard for which all systems are measured akin to the marine piling noise assessments for offshore windfarm construction in German waters. In these studies, received levels were observed at a fixed distance of 750 m from the source, an approach with a number of advantages such as avoiding the need for complex environment-dependent modelling needing to be carried out. Mitigation procedures are then put in place to reduce risk for unknown levels at distances <750 m. Such a simplification reduces reliance on often hard-to interpret modelling results, but it precludes the development of a more rigorous understanding of differing scenarios required by UK regulations. In general terms, the wide variety of potential WEC design concepts, and therefore potential noise outputs and variation in deployment environments, makes definition of a fixed recorder range problematic. This is particularly so for surface WECs with a high degree of positional freedom. This drawback can be overcome, however, by the use of detailed positional data for both the WEC and the recorder systems. The ultimate choice of recorder deployment ranges, though, is constrained by WEC device noise and background sound levels, water depth and device size, potential degrees of freedom and contamination from other noise sources. The use of standard ranges for characterization where possible would of course be preferable.

Because of the relative infancy of measurement of noise from both wave and tidal energy systems, only a few TSG and WEC devices have been constructed at full scale and even fewer have had their acoustic properties measured. The methodologies here described therefore need to retain some flexibility, although as more systems are measured, the potential to standardize, for example, on measurement ranges or angles are likely to emerge. This would echo the approaches of piling noise studies in the German sector and the ship noise standards under international development. Similarly, there has been rapid improvement in acoustic data-acquisition technology, allowing better performance, greater bandwidth and longer deployment periods for less cost, providing potential for improved, temporal and spatial resolution. These advances can then be added to the general practical methods of sound acquisition proposed herein.

Discussion

The focus here has been on empirically measuring the sound output of existing tidal and wave energy devices *in situ*. There is an alternative approach, however, to focus on the specifics of the device engineering and to model the acoustic outputs of the actual parts, moving and fixed, and their interactions. Such a tactic would typically be undertaken during device design for engineering reasons, to minimize vibration and equipment wear. However, it is also possible to determine how the vibration and acoustic energy is projected out into the environment (Marmo and Carruthers 2010). Clearly, such an approach would look at the key elements of the device operation (gearbox meshing frequencies, rotor revolutions, etc.) and be a numerical prediction rather than a real measurement. Nevertheless, it would offer a significant advantage for this embryonic industry because it could predict the acoustic outputs of devices that have not been built yet or suggest options to tune devices towards particular acoustic outputs. When coupled with propagation modelling too, it could also forecast the potential soundscapes of multiple devices when they are eventually built into arrays. Being extremely powerful, such modelling approaches are ultimately predictions, and there is real value in coupling them with actual measurements in the field, as described herein. The field measurements will first validate or adjust the numerical predictions, second, provide ambient noise floor information upon which to superimpose any modelling, and finally quantify how less predictable features (such as device-breaking systems, cut in/out events) may actually sound. Additionally, with the marine renewables industry being in its infancy, there are no old (mechanically worn) devices in operation or devices with years of biological fouling attached. How the acoustic properties of ageing devices change will require repeat empirical measurement as the industry matures.

For there to be a long-term perspective on how device noise changes over time and for comparison between different machines and technologies in various locations, there needs to be some consistency in how device acoustic output is reported. Currently, there are no national or international standards for device characterization, unlike for wind turbines (IEC 2006). Consequently, comparisons between studies need to be approached carefully. However, given the rapid increase in scale-devices being placed in the sea, along with the regulatory need for information and the development of tailored methods to quantify device noise in these harsh environments, it is likely that reporting standards will begin to emerge soon. This will most likely be through vehicles such as the International Organization for Standardization (ISO Technical Committees TC8 and TC43, IEC TC114).

One of the positive by-products of trying to quantify the sound that propagates from operating marine renewable devices is the opportunity to discover what ambient noise already exists in the environment. As described above, the challenges of working in such dynamic habitats has dissuaded investigation of ambient sound in shallow vertically or horizontally moving coastal waters in the past. Doing so now, though, has demonstrated unexpected levels of spatial variability (Carter 2013) additional to better described variations with flow rate, sea state and rain (Urick 1983; Ma et al. 2005; Belleudy et al. 2010). The exact causes of the spatial variations, whether produced by the transport of bedload, bubbles or being of biological origin or a combination of all these issues remains unclear. However, the variable soundscape presents a significant and poorly understood acoustic topography against which the acoustic outputs of devices need to be considered (Carter 2013).

Other than simply measuring a source of anthropogenic noise pollution, any deliberation of device acoustic output in relation to ambient sound needs to focus on why the parameters are of interest. From an environmental perspective, it is unlikely that machines in their normal modes of operation would produce sound at levels capable of causing auditory harm to such vulnerable species as marine vertebrates (Southall et al. 2007). Instead, it is likely that animals will use the information for more subtle behavioural interactions, whether masking other signals, habitat avoidance/attraction or related to interactions with the devices themselves. As described earlier, injurious collisions with renewable devices are among the most pressing potential environmental impacts of the new technologies, so for animals to avoid collisions, they need to be aware spatially of the precise location of the devices, their moorings and other associated structures. Because of 24 h operation and low visibility in coastal waters, it is likely that most of the spatial interactions will be mediated by acoustic cues, so it is important to determine what an animal of interest can perceive on its approach to a TSG or a WEC. What it then does with this information (approach, avoid, ignore) is then up to the animal itself. Experimentation coupled with the close monitoring of existing devices will inform us of what behavioural responses animals show. However, key to any of these investigations is a precise understanding of the acoustic signals available for the animals to receive and act upon. Such an understanding will require the sort of basic and habitat relevant measurement methods outlined herein.

Acknowledgements We thank those involved in the development of these acoustic methods. The need to generate monitoring methods was initiated by the European Marine Energy Centre, particularly Jenny Norris, Matthew Finn, David Cowan and Dave Cousins, for whose guidance and facilitation we are grateful. Funding for the work came from Highlands and Islands Enterprise (HIE), the Scottish Government, Scottish Natural Heritage (SNH) and the Natural Environment Research Council (NERC) and the Department for Environment, Food and Rural Affairs (Defra) RESPONSE project NE/J004251/1. We also thank Ed Harland (Chickerell BioAcoustics), Gordon Hastie (SMRU) and Nicola Quick (SMRU Marine), Pelamis Wave Power and E.ON for their support. For help with fieldwork and analyses, we thank Kenny Black, Jim Elliott, Steve Gontarek and Steven Benjamins (SAMS), Steve Vile and his crew (Explorer Fast Sea Charters) and the Flamborough Light crew. Finally, we acknowledge the steadfast patience of guest editor Mark Shields throughout the drafting process and the valued suggestions for improvements from two anonymous referees.

References

Ainslie MA, De Jong CAF, Dol HS, Blacquière G, Marasini C (2009) Assessment of natural and anthropogenic sound sources and acoustic propagation in the North Sea. TNO report TNO-DV, C085

ANSI (American National Standards Institute) (2009) Quantities and procedures for description and measurement of underwater sound from ships. Part 1. General requirements. ANSI/ASA S12.64. American National Standards Institute, New York

Bailey H, Senior B, Simmons D, Rusin J, Picken G, Thompson PM (2010) Assessing underwater noise levels during pile-driving at an offshore windfarm and its potential effects on marine mammals. Mar Poll Bull 60:888–897

Belleudy P, Valette A, Graff B (2010) Passive hydrophone monitoring of bedload in river beds: first trials of signal spectral analysis. In: Gray JR, Laronne JB, Marr JDG (eds) Bedload-surrogate monitoring technologies. US Geological Survey, Scientific Investigations Report, 5091

Betke K, Schultz-von Glahn M, Matuschek R (2004) Underwater noise emissions from offshore wind turbines. Proceedings of the Joint Congress *CFA/DAGA* 2004—7ème Congrès Francais d'Acoustique and 30 Deutsche Jahrestagung für Akustik, Strasbourg

Boehlert GW, Gill AB (2010) Environmental and ecological effects of ocean renewable energy development: a current synthesis. Oceanography 23:68–81

Carter C (2013) Tidal energy underwater noise and marine mammals. PhD thesis, University of Aberdeen

Casper BM, Popper AN, Matthews F, Carlson TJ, Halvorsen MB (2012) Recovery of barotrauma injuries in Chinook salmon, *Oncorhynchus tshawytscha* from exposure to pile driving sound. PLoS ONE 7:e39593

De Jong CAF, Ainslie MA (2008) Underwater radiated noise due to piling for the Q7 Offshore Wind Park. Proceedings of the Acoustics08 (ECUA08) Conference, Paris, pp 117–122

Götz T, Janik VM (2010) Aversiveness of sound in phocid seals: psycho-physiological factors, learning processes and motivation. J Exp Biol 213:1536–1548

Harvey N, Clarke B (2012) Environmental impact assessment in practice. Oxford University Press, Melbourne, 320 pp

Huang M, Aggidis GA (2008) Developments, expectations of wave energy converters and mooring anchors in the UK. J Ocean Univ China 7:10–16

IEC (International Electrotechnical Commission) (2006) Wind turbine generator system. 2. Acoustic noise measurement techniques. IEC 61400-11. International Electrotechnical Commission Standard
Inger R, Attrill MJ, Bearhop S, Broderick AC, James GW, Hodgson DJ, Mills C et al (2009) Marine renewable energy: potential benefits to biodiversity? An urgent call for research. J Appl Ecol 46:1145–1153
IWC (International Whaling Commission) (2013) Report of the IWC Scientific Committee workshop on interactions between marine renewable projects and cetaceans worldwide. J Cet Res Manag (Suppl.) 14:393–415
Ketten DR (2012) Marine mammal auditory system noise impacts: evidence and incidence. In: Popper AN, Hawkins A (eds) The effects of noise on aquatic life. Springer, New York, pp 207–212
Lepper PA, Robinson SP (2008) Monitoring the temporal and spatial characteristics of the noise radiated from marine piling. J Acoust Soc Am 123:2987
Lepper PA, Robinson SP, Harland E, Theobald P, Hastie GD, Quick N (2012) Acoustic noise measurement methodology for the Billia Croo wave energy test site. EMEC report. http://www.emec.org.uk
Lucke K, Lepper PA, Hoeve B, Everaarts E, van Elk N, Siebert U (2007) Perception of low-frequency acoustic signals by a harbour porpoise (*Phocoena phocoena*) in the presence of simulated offshore wind turbine noise. Aquat Mamm 33:55–68
Ma BB, Nystuen JA, Lien R-C (2005) Prediction of underwater sound levels from rain and wind. J Acoust Soc Am 117:3555–3565
Madsen PT, Wahlberg M, Tougaard J, Lucke K, Tyack PL (2006) Wind turbine underwater noise and marine mammals: implications of current knowledge and data needs. Mar Ecol Prog Ser 309:279–295
Marmo BA, Carruthers BJ (2010) Modelling and analysis of acoustic emissions and structural vibration in a wind turbine. Proceedings of the COMSOL Conference, Paris
Miksis-Olds JL, Donagahy PL, Miller JH, Tyack PL, Reynolds JE (2007) Simulated vessel approaches elicit differential responses from manatees. Mar Mamm Sci 23:629–649
Nedwell JR, Langworthy J, Howell D (2003) Assessment of sub-sea acoustic noise and vibration from offshore wind turbines and its impact on marine wildlife; initial measurements of underwater noise during construction of offshore windfarms, and comparison with background noise. Subacoustech Report Reference 544 R 0424. Published by COWRIE
New LF, Harwood J, Thomas L, Donovan C, Clark JS, Hastie G, Thompson PM et al (2013) Modelling the biological significance of behavioural change in coastal bottlenose dolphins in response to disturbance. Funct Ecol 27:314–322
Norro AMJ, Rumes B, Degraer SJ (2013) Differentiating between underwater construction noise of monopile and jacket foundations for offshore windmills: a case study from the Belgian part of the North Sea. Sci World J 2013:7 (897624)
Nowacek DP, Thorne LH, Johnston DW, Tyack PL (2007) Responses of cetaceans to anthropogenic noise. Mamm Rev 37:81–115
NRC (National Research Council) (2005) Marine mammal populations and ocean noise: determining when noise causes biologically significant effects. National Academy Press, Washington, D. C., 142 pp
Piniak WED, Mann DA, Eckert SA, Harms CA (2012) Amphibious hearing in sea turtles. In: Popper AN, Hawkins A (eds) The effects of noise on aquatic life. Springer, New York, pp 83–87
Popper AN (2003) Effects of anthropogenic sounds on fishes. Fisheries 28:24–31
Richardson WJ, Greene CR, Malme CI, Thomsen DH (1998) Marine mammals and noise. Academic Press, San Diego, 576 pp
Robinson SP, Theobald PD, Hayman G, Wang LS, Lepper PA, Humphrey. V, Mumford S (2011) Measurement of noise arising from marine aggregate dredging operations. MALSF Report 09/P108. ISBN 978 0907545 57 6
Southall BL, Bowles AE, Ellison WT, Finneran JJ, Gentry RL, Greene CR, Kastak D et al (2007) Marine mammal noise exposure criteria: initial scientific recommendations. Aquat Mamm 33:411–521
Tougaard J, Henriksen OD, Miller LA (2009) Underwater noise from three types of offshore wind turbines: estimation of impact zones for harbor porpoises and harbor seals. J Acoust Soc Am 125:3766–3773
Tougaard J, Madsen PT, Wahlberg M (2008) Underwater noise from construction and operation of offshore wind farms. Bioacoustics 17:143–146
Urick RJ (1983) Principles of underwater sound. McGraw-Hill, New York, 423 pp
Van Parijs SM, Clark CW, Sousa-Lima RS, Parks SE, Rankin S, Risch D, van Opzeeland I (2009) Management and research applications of real-time and archival passive acoustic sensors over varying temporal and spatial scales. Mar Ecol Prog Ser 395:21–36
Wahlberg M, Westerberg H (2005) Hearing in fish and their reaction to sounds from offshore wind farms. Mar Ecol Prog Ser 288:295–309
Yelverton JT, Richmond DR (1981) Underwater explosion damage risk criteria for fish, birds, and mammals. J Acoust Soc Am 70:S84

Tracking Technologies for Quantifying Marine Mammal Interactions with Tidal Turbines: Pitfalls and Possibilities

10

Gordon D. Hastie, Douglas M. Gillespie, Jonathan C. D. Gordon, Jamie D. J. Macaulay, Bernie J. McConnell and Carol E. Sparling

Abstract

Currently, there is great uncertainty surrounding the environmental impacts of tidal turbines on marine mammals; one major concern derives from the potential for physical injury through direct contact with the moving structures of turbines. Collecting data to quantify these risks is challenging and methods for measuring movements underwater and interactions with turbines are limited. However, potential tools include a small number of cutting-edge technologies that are being used increasingly for research and monitoring; these include animal-borne telemetry, and active and passive acoustic tracking. Recent developments in these technologies are described along with their means of application in measuring fine-scale movements and avoidance or evasion responses by marine mammals around turbines. From a risk-characterization perspective, each technique can provide information to inform risk assessments or help parametrize collision risk models; however, each has its associated benefits and drawbacks and it is clear that, in isolation, none of them can provide all the data needed to address the problem. The three approaches appear highly complementary, with the strengths of one complementing the weaknesses in others; the solution to characterizing the risks posed by tidal turbines is likely to be a combination of such techniques.

Keywords

Collision risk · Marine mammals · Passive acoustics · Sonar · Telemetry · Tidal energy

Marine Mammals and Tidal Energy

Many countries have set ambitious targets for renewable energy, with energy from offshore sources anticipated to form an important part of this; for example, it is estimated that one-fifth of the UK's electrical supply could ultimately come from marine (wave and tidal stream) resources (Callaghan 2006). To achieve this aim, rapid progress needs to be made in understanding not only the latest energy technologies but also their likely impacts on the environment. Currently, there is great uncertainty surrounding the nature and extent of any environmental impacts of tidal stream energy devices (tidal turbines) on marine wildlife (particularly seals, whales and dolphins). This has the potential to curtail acceptance of new proposals, and can create barriers to commercial introduction of the technology.

One major environmental concern derives from the potential for physical injury to marine mammals through direct contact with moving structures of turbines. In light of these potential risks, regulators are faced with challenges when deciding whether to consent to developments. This clearly depends upon government policy, which often seeks to balance the desire to develop low carbon technologies with the need to ensure protection of the environment and natural heritage, and the specific requirements of wildlife legislation

G. D. (✉) · D. M. Gillespie · J. C. D. Gordon · J. D. J. Macaulay · B. J. McConnell
Sea Mammal Research Unit, Scottish Oceans Institute, School of Biology, University of St Andrews, St Andrews, Fife KY16 8LB, UK
e-mail: gdh10@st-andrews.ac.uk

G. D. Hastie · C. E. Sparling
SMRU Marine Ltd, New Technology Centre, North Haugh, University of St Andrews, St Andrews, Fife KY16 9SR, UK

M. A. Shields, A. I. L. Payne (eds), *Marine Renewable Energy Technology and Environmental Interactions,* Humanity and the Sea,
DOI 10.1007/978-94-017-8002-5_10,

in different countries. Faced with uncertainty about potential impacts on marine mammals, regulators may follow a precautionary approach requiring developers to implement costly monitoring and mitigation systems on all turbines. This potentially includes adaptive deploy-and-monitor approaches to turbine deployments where real-time mitigation measures such as turbine shut-down or slowing of rotors are initially required. The premise behind such risk-based approaches is that levels of risk informed by data from monitoring and operating conditions can be regularly reviewed and updated. These considerations demonstrate that research is urgently required to investigate the true level of the risk posed by tidal turbines. There is a clear need to learn whether marine mammals and tidal turbines can co-exist at the scales currently being envisaged for the industry.

For the purposes of this chapter, a simple framework for collision risk (loosely based on that proposed by Wilson et al. 2007) was considered. Collision risk depends on the natural densities of animals at the tidal sites and their dive behaviour, which in combination might be considered as providing a three-dimensional (3D) prediction of the likelihood of encounter in the absence of responsive movement. This information is useful for predicting what proportion of a given population might be considered to be at risk, a parameter that can be modified by responsive animal movement at two different scales. At a medium scale of hundreds of metres, animals might avoid the turbine site leading to a reduction in the rate of close encounters. At a finer scale of metres, individuals might respond directly to evade collision with turbine rotors. Detecting and measuring avoidance and evasion requires movement data with different levels of precision and accuracy. Throughout this chapter, these two terms are used to reflect these different types of responsive movement; however, it is important to stress that the distinction between avoidance and evasion is not absolute and that there is clearly a grey area of overlap between the two terms. Consideration of the measurement of natural densities or dive behaviour of animals at tidal sites under baseline conditions is not covered here.

Collecting such data is challenging, and available methods for measuring movements underwater and showing interactions of marine mammals with tidal turbines in high resolution are limited. Light does not transmit well through water, so underwater video technology has been used to a limited extent to image marine mammals and to record their behaviour underwater (e.g. Similä and Ugarte 1993; Herzing 1996; Ridoux et al. 1997; Davis et al. 1999). Such research has also generally been carried out at relatively short range (a few metres) and only during daylight in waters with good visibility. In most tidal areas around the UK, poor visibility attributable to suspended sediment or relatively long periods of darkness are likely to seriously constrain the use of video.

Potential alternative tools for measuring marine mammal 3D movements in high resolution include a small number of cutting-edge technologies that are being used increasingly in research and monitoring. For example, animal-borne instrumentation is a technology that is widely used to track individuals underwater and can provide data on 3D movements at very high resolution (Madsen et al. 2002; Tyack et al. 2011), passive acoustic techniques using hydrophone arrays have been used to locate and track cetaceans underwater (Watkins and Schevill 1972; Clark et al. 1985; Leaper et al. 1992; Freitag and Tyack 1993; Jensen and Miller 1999; Janik et al. 2000; Hastie et al. 2006), and accelerated development of active sonar systems for the defence sector for sub-sea monitoring of potential security threats and for fisheries research and management may provide a basis for tracking animals.

This chapter describes recent developments in each of these technologies and how they might be applied to assessing encounter rates and measuring fine-scale movements and avoidance or evasion responses by marine mammals around tidal turbines. Further, the way in which these techniques aim to provide an understanding of the potential risks posed by tidal stream energy to marine mammals is discussed.

Animal-Borne Instrumentation

For most of their lives, marine mammals are difficult to observe directly; some travel many hundreds of miles out to sea and all spend most of their time underwater. The development of telemetry systems has revolutionized our ability to observe and understand how marine mammals behave at sea (McConnell et al. 2010). Telemetry generally consists of tags that are attached to individual animals allowing the collection of data on their movements and behaviour. These data are either transmitted to a receiver or downloaded directly from the tag after recovery. Several innovative technological solutions have been developed over the past few decades to elucidate where animals go and how they interact with their prey, oceanographic conditions and conspecifics in a range of different settings and circumstances. However, when the question is specifically how animals behave in relation to a fixed structure such as a tidal turbine, there are a number of specific challenges that push technology and innovation to their limits.

The primary data required to quantify marine mammal interactions with tidal turbines are two-dimensional (2D) locations at the surface and the corresponding depth at a fine spatial and temporal resolution. The D-tag (Johnson and Tyack 2003) collects such data at the required resolution by means of dead-reckoning, which uses data from movement sensors (acceleration, attitude and speed through water) on the animal to interpolate the position of a tagged animal between estimated locations at the surface. However, absolute surface locations are not estimated by D-tags and this uncertainty can result in inaccurate geo-referenced reconstructions of an underwater track. Also, because large quantities of raw data are stored in memory the device needs to be recovered to retrieve the information, limiting its use to circumstances and species where tag recovery is possible. Although automatically timed tag-release mechanisms are a possibility, it is frequently logistically difficult to recover detached tags from animals that roam widely.

The Argos satellite system (Argos 2008) has provided both a means to relay data and to estimate approximate locations for many species of marine mammal. Its primary advantage is that coverage is global and data (including stored dive depths) can be sent immediately on surfacing. For these reasons it has been used to track seals (McConnell et al. 1999), and large (Mate et al. 2007) and small cetaceans (Sveegaard et al. 2011). However, the location estimates it provides are sparse (perhaps 1–6 per day) and of low precision (errors of >1 km are common). The data are therefore not of sufficient quantity or quality for the purpose of investigating fine-scale movements around marine renewable devices.

The use of the Global Positioning System (GPS) might appear to be an attractive option to increase precision, but the duration of surfacing intervals when animals breathe are usually too short (or interrupted by wave wash) for a conventional GPS to obtain a fix from a cold start. This issue was resolved by the Fastloc innovation; Fastloc obtains a snapshot (<0.2 s) of GPS satellite transmission when the animal surfaces. This is then processed and condensed into about 32 bytes of pseudo-range data that are time-stamped and stored for subsequent transmission. Once these data are received ashore, the pseudo-range data are post-processed to provide a series of accurate GPS fixes. Fastloc data can be relayed using Argos, but that system imposes severe restrictions on the amount of data (fix data) that can be relayed. Note that Lonergan et al. (2009) emphasized that accurate re-creation of an animal's track depends not just on fix precision, but also on obtaining a sufficiently high fix rate.

Since 2004, tags deployed on seals have used the mobile (cell) phone network (GSM) to relay data ashore. In these GPS/GSM tags, data (including stored Fastloc and depth data) are collected routinely over periods of 6 months or more. Every time a seal swims within suitable network coverage, the stored data are sent ashore using 2.5 GSM channels (GPRS). This allows high rates of data flow to be achieved at low energy cost, albeit with potential high data latency (i.e. not real time). Although this technique is currently suitable for pinnipeds (and potentially large cetaceans), the GSM registration period (often 20 s or more) prohibits GSM as a method of data transfer for short-surfacing species such as harbour porpoises (*Phocoena phocoena*).

Fig. 10.1 A harbour seal with a GPS/GSM tag glued to its fur at Strangford Lough, Northern Ireland. The tag will detach during the annual moult in August

As the tidal energy industry is in its infancy, there have been no studies yet that have measured fine-scale marine mammal behavioural interactions (evasion) with tidal turbines. However, a recent study used GSM/GPS tags to study the impact the SeaGen turbine may have had on the avoidance behaviour of harbour seals (*Phoca vitulina*). The SeaGen tidal turbine is a 1.2 MW tidal energy convertor located in the narrow entrance to Strangford Lough, Northern Ireland; for more detail on the environmental monitoring associated with the turbine, see Chap. 12. SeaGen was installed in the narrow entrance to Strangford Lough (Strangford Narrows) in 2008 with the aim of testing whether the introduction of a turbine would change the behaviour of seals passing through the narrow entrance to the Lough. Whereas fine-scale interaction (evasion) was not the primary focus of the telemetry work, the results thus far do illustrate the types of data that can be obtained, and highlight both the limitations of the data and the potential of what might be obtained in future. Data were derived primarily from three deployments each of GPS/GSM telemetry tags on 12 harbour seals in the vicinity of the Strangford Lough tidal turbine site: prior to the installation of the turbine (2006), during the installation of the turbine (2008) and after the turbine became fully operational (2010). A harbour seal with a tag glued to its fur following net capture is shown in Fig. 10.1; the tag detaches during the annual moult in August. These tags incorporated pressure sensors to measure depth, and the information was summarized into dive records. Dive records started when the animal was at least 1 m below the surface for 10 s and ended when it was above that depth. For each individual dive, the tag stored a depth at nine points equally spaced throughout the duration of the dive.

The study accrued a total of 2772 seal-days of track data, and from these data, the position at which each seal crossed a line across the narrows at the location of the turbine was estimated by linear interpolation between the closest fixes before and after the crossing. With a mean interval of approximately 30 min between fixes, there was error in the lateral crossing position of the order of tens of metres (Lonergan et al. 2011). Despite this poor precision, the results do indicate some avoidance of the turbine in general and a small reduction in transit rates past the turbine when the turbine was operating. As position underwater was inferred by interpolation between surface location fixes and there were few surfacing events close to the turbine, there remained uncertainty about the exact location and timing of the transits past the turbine, and that led also to uncertainty about the depth of animals close to the turbine. Typical dives lasted around 4 min and, because each dive was time-stamped, the depth of each seal as it crossed the turbine line could be estimated, potentially allowing an assessment of whether seals crossed at the same depth as the turbine. However, inaccuracies in crossing interpolation resulted in errors in crossing time that were too large to estimate crossing depth with accuracy.

A concern prior to that study was that the seals tagged in the vicinity of Strangford Narrows might use the area of interest only occasionally, reducing the ability to obtain a large enough sample to be able to say anything meaningful about animal behaviour around the turbine. This is a general concern when using telemetry to investigate an influence that applies to only a small part of an animal's typical home range. This study had an advantage, though, in that the particular geography of the Lough meant that animals tagged within the Lough would have to move past the turbine location to access offshore areas. As a result, they did pass the turbine frequently. However, in less confined waters this is clearly not the case, and the reliance on individual animals effectively to dictate the study area remains a key limitation of telemetry to understand avoidance or evasion behaviour around tidal turbines.

An important result of the study was that there was generally a high degree of consistency in the patterns of harbour seal behaviour within individual animals, but great variability between animals. This individual effect accompanied by small sample sizes (generally because of the financial cost of tags and challenges in catching animals) reduces the statistical power to detect changes at the level of a population. However, because of the intermittent nature of turbine operation (with the turbine stationary for period of hours to days), it was possible to use the same tagged individuals to investigate avoidance or behaviour around the turbine. The Strangford Lough deployment was the first ever deployment of GPS/GSM tags; since then, tag resolution has improved significantly. For example, in a recent study on harbour seals in Denmark (McConnell et al. 2012), tags were programmed to provide a location fix rate of 90 per day (albeit at reduced tag longevity). Doing this has the potential to reduce the errors in calculating the proximity of a seal to a tidal turbine. Programming the tags to attempt (albeit not always successfully) to get a GPS fix every surfacing period would also reduce the temporal fix resolution to a maximum of about 5 min, reducing the error in estimating interaction, but still only down to a scale of a few hundred metres. For example, harbour seals often spend periods of approximately 3–6 min underwater; if seals were travelling at speeds of ~1.5–2 m s^{-1}, they could easily travel several hundred metres between locations, possibly more if travelling with a current. A clear improvement that needs to be considered for applications such as this would be the combination of dead-reckoning (as used in D-tags described above) with the GPS/GSM technology to obtain frequent fixes at the surface and to relay data ashore. In theory, an accurate 3D interpolation between GPS surface fixes could then be obtained; in practice, there remain issues of error magnification in regions of strong tidal current (Shiomi et al. 2008, 2010). Indeed, the limiting factor is to be able to model and predict the current with sufficient accuracy and resolution within several hundred metres of the turbine. Another telemetry technique would be to attach ultrasonic acoustic devices (e.g. pingers) to marine mammals; their fine-scale trajectories can be calculated from data provided by a static array of hydrophones in the vicinity of the turbines. This technique is described further in the section below on passive acoustic techniques.

Although not yet able to provide the location and movement data at sufficient resolution to determine evasion behaviour by marine mammals to tidal turbines that are needed to populate collision risk models, telemetry techniques do provide background information about how target animals behave when not in the vicinity of a turbine. They also have the potential to measure overt behavioural responses at greater ranges, such as when animals avoid sites.

Active Acoustic Techniques

In recent years, there has been accelerated development of target-tracking using active sonar systems in the defence sector for sub-sea monitoring of potential security threats, and for fisheries research and management. Recent research has shown that a new generation of sonar systems has the capacity to produce acoustic images of marine mammals, and may provide a basis for monitoring avoidance or evasion behaviour by marine mammals around tidal turbines. For example, Nøttestad et al. (2002) used a 95 kHz Simrad SA 950 multibeam sonar to measure the behaviour of fin whales (*Balaenoptera physalus*) foraging on herring schools, and Benoit-Bird and Au (2003a) used a Kongsberg SM2000 to locate and track spinner dolphins (*Stenella longirostris*) in the water column in Hawaii. Further, Benoit-Bird and Au (2003b) used a Tournament Master Fishfinder NCC 5300 to integrate the behaviour of spinner and dusky dolphins

(*Lagenorhynchus obscurus*) and their prey (Benoit-Bird et al. 2004). More recently, West Indian manatee (*Trichechus manatus*) behaviour was measured in waters with poor visibility (caused by turbidity and sediment load) using a range of sidescan sonar systems (Gonzalez-Socoloske et al. 2009; Gonzalez-Socoloske and Olivera-Gomez 2012).

The fundamentals of all active sonar systems are essentially the same; pulses of sound (pings) are produced electronically using a sonar projector, then the system listens for echoes of these pulses as they reflect off objects, using a series of hydrophones. With knowledge of the speed of sound in water and the time for the ping to travel to the target and back, the range between the sonar and the underwater object can be calculated. To calculate the bearing, several hydrophones are used to determine the relative arrival time at each, or with a receiver array of hydrophones, by measuring the relative amplitude in beams formed through a process called beam-forming (Thorner 1990). Sonar efficiency can be affected by variations in the speed of sound, particularly in the vertical plane, which arise as a result of temperature, dissolved impurities (usually salinity) and pressure changes. Further, scattering caused by small objects in the sea, from the seabed and the surface can be a major source of interference. Together, these effects potentially render the use of active sonar to detect and track marine mammals in energetic tidal areas particularly challenging.

There are many commercially available sonar systems. A recent review collated an inventory detailing more than 200 systems from 39 sonar manufacturers (Hastie 2012). These are designed for a wide range of use including swathe bathymetry, underwater navigation, fisheries research and seabed profiling; fundamental transmission frequencies range from 12 to 2250 kHz. Source levels were also provided by manufacturers in 99 of the systems and ranged from ~187 to 237 dB re 1 μPa at 1 m. Of these systems, 24 incorporated automated target detection and tracking software, but most were designed for vessel or port security rather than for marine wildlife tracking.

To be able to measure the behaviour of marine mammals around tidal turbines, a sonar system needs to meet a number of key specifications. For example, for research or monitoring applications it has to provide appropriate spatial coverage (both horizontally and vertically), effectively determining the volume of water that can be monitored around the turbine. It also needs to have suitable temporal (ping rate), angular (degrees) and range (cm) resolution to allow marine mammals to be detected, classified and tracked effectively. A suitable system also needs to have an efficient detection capability; this depends heavily on the proportion of sound that is reflected by the animal back to the receiver array. This is often termed the target strength and is usually expressed in decibels (dB). There is little empirical information on the target strength (*TS*) of marine mammals; it is a challenging parameter to measure accurately and most information has been obtained opportunistically with large species of whale. For example, in the 1960s, Dunn (1969) reported for sperm whales (*Physeter macrocephalus*) a *TS* measurement of −8 dB at 1 kHz, and Love (1973) reported a *TS* measurement of humpback whales (*Megaptera novaeangliae*) of +8 dB at broadside for a whale 15 m long ensonified at 20 kHz. For smaller marine mammals, there are few data available, but Au (1996) reported mean broadside aspect *TS* measurements of a stationary bottlenose dolphin (*Tursiops truncatus*) under controlled conditions ranging from −11 to −24 dB depending on transmission frequency. Most acoustic energy was reflected from the area between the dorsal and pectoral fins, corresponding to the location of the dolphin's lungs. Similarly, Doksæter et al. (2009) measured the *TS* of 22 marine mammals (assumed to be dolphins or small whales) from a seabed-mounted Simrad EK60 (38 kHz); mean *TS* ranged from −5 to −35 dB, with an overall mean of −20 dB. Measurements of *TS* such as these provide an indication of the effective range at which a marine mammal could be detected by sonar and also potentially provide one metric to help discriminate marine mammals from other targets (e.g. fish or debris).

A critical factor rarely considered when using active sonar to monitor behaviour is that most marine mammals rely heavily on sound as a means of navigation and for detecting prey, and that the hearing and vocal ranges of many species (Richardson et al. 1991) overlap with the transmission frequencies of many commercially available sonar systems (~12–150 kHz). Therefore, there is clear potential that the acoustic signals produced by sonar systems could cause a range of negative impacts from interference with communication (Fristrup et al. 2003) or changes in behaviour (for a review, see Richardson et al. 1991) to auditory injury (Southall et al. 2007). Although research on the impacts of sonar has focused on relatively low frequency military systems with fundamental transmission frequencies within the hearing ranges of marine mammals (Tyack et al. 2011), low frequency components of the signals from sonar systems with higher fundamental transmission frequencies could still be audible to animals and elicit the negative reactions described above. This is particularly important when using sonar as a monitoring tool designed to measure behavioural responses, because it is important that any observed behavioural response be attributed if appropriate to the tidal turbine rather than to the sonar being used to measure animal behaviour.

When considering whether an animal may respond to a sound, it is important to note that hearing ability varies markedly with the frequency of a sound. For example, the harbour porpoise hearing threshold at 500 Hz is about 90 dB re 1 μPa, whereas its hearing threshold at 50 kHz is about 35 dB re 1 μPa (Kastelein et al. 2002). This would mean that a sound with a pressure level of 100 dB re 1 μPa and a frequency of 500 Hz would be barely audible to the porpoise, but the same sound pressure level at a frequency of 50 kHz would be perceived as relatively loud. It is also relevant that different species of marine mammal differ markedly in the frequency ranges that they can hear effectively. Although the primary frequencies of most systems are well above the hearing ranges of marine mammals, the characteristics of many sonar signals (short acoustic pulses with rapid rise times) can lead to the introduction of significant low frequency components, making them audible to animals and potentially eliciting a behavioural reaction (Hastie et al., in press). This effectively limits the sonar systems suitable for measuring behaviour to those operating at higher frequencies, where detection ranges may be limited to tens of metres, suitable for visualizing evasion behaviour. Larger range-avoidance behaviour by marine mammals could still be measured using lower frequency active sonar systems, but the trade-offs with potential responses to the sonar need to be considered too. Therefore, deployment of any sonar systems to monitor the behaviour of marine mammals should only be carried out after a thorough review of sonar sound characteristics and the hearing sensitivities of the species of interest.

Although several studies have used active sonar to measure the behaviour of marine mammals (Nøttestad et al. 2002; Benoit-Bird and Au 2003a, b; Benoit-Bird et al. 2004; Doksæter et al. 2009; Gonzalez-Socoloske et al. 2009; Gonzalez-Socoloske and Olivera-Gomez 2012), none have measured fine-scale marine mammal interactions with tidal turbines. In a recent study of marine mammal interactions with a tidal turbine, two manually scanning sonar systems (Tritech Super SeaKing: 375 kHz) were deployed on the SeaGen tidal turbine. Strangford Lough hosts a number of large marine species that have the potential to interact with the tidal turbine. The primary aim of that study was to evaluate the efficiency and reliability of sonar as a basis for monitoring and mitigation tool for marine mammals on an operational tidal turbine. Further aims were to evaluate the frequency of close range interactions between marine mammals and tidal turbines and to compare movement metrics of marine mammals and other mobile targets as a basis for automated classification of marine mammals. Sonar images were monitored visually by a user and times when targets were detected were noted. In addition, an observer located on top of the turbine control room simultaneously monitored marine mammals upstream of the turbine (on both flood and ebb tides). Each time a marine mammal was sighted, that observer noted its species, number of animals, location and time of the sighting.

In all, 135 h of real-time monitoring using a combination of visual and sonar techniques was carried out. To assess the efficiency of the sonar at detecting marine mammals, the timing and locations of all observer sightings were compared with target detections made using the sonar. If a target detection made by the sonar was within 30 s of a visual sighting and was close to the sighting location, the target was tentatively confirmed as a marine mammal. The number of targets confirmed as marine mammals by spatial and temporal data from the observer, and the number of other targets (those detected using the sonar but not correlated with a visual sighting of a marine mammal) were compared. The tracks of all targets were plotted in X–Y coordinates around the turbine to assess the frequency and proximity of marine mammals to the turbine. Mean target speed was estimated for each track as the mean of the direct path distance between two consecutive target locations divided by the time between each location (taking into account current speed). These allowed assessment of the variation in the tracks of likely marine mammals and other targets to be compared. In all, 72 marine mammals were sighted close to the turbine and 159 mobile targets were detected using the active sonar. Comparison of the sonar targets with the spatial and temporal information on sightings made by the observer suggested that a number of the sonar targets (22 targets; 14 % of all targets) were marine mammals. They included harbour seals, harbour porpoises and grey seals (*Halichoerus grypus*). The overall rate of target detection was 1.18 targets per hour and the rate for confirmed marine mammal targets was 0.16 h^{-1} when the tide was running. Although marine mammals were detected on both flood and ebb tides, further analyses of presence vs. tidal pattern were precluded by the small sample sizes. When sightings of marine mammals within the area covered (up to 50 m) by the sonar were compared with sonar targets, the percentage of sightings that could be matched with sonar targets was 46.7 %. Comparison of the track metrics suggested that the speed of targets confirmed as marine mammals (2.3 m s^{-1}) was significantly faster than unconfirmed targets (1.6 m s^{-1}); the marine mammals also moved straighter past the turbine than the unconfirmed targets. This finding supports the indication that confirmed marine mammals actively swim in the water column and suggests that most unconfirmed targets move passively in the water column and become susceptible to lateral movements because of the turbulence upstream of the turbine.

The results of that study illustrate that small marine mammals (and other mobile targets) can be detected in a tidally turbulent water column in real time using sonar, with an effective range of several tens of metres. The combination of visual observations and active sonar allowed an assessment of the reliability of the sonar at detecting marine mammals when they are close to the surface. Although a number of the detected sonar targets could be matched with sightings at the surface, the overall percentage was relatively low at 46.7 %. This low rate of detection is potentially attributable to the inherent problems associated with active acoustics in tidal environments and targets close to the water surface. It is known, for instance, that the highly heterogeneous water characteristics near the surface or wind-generated clutter are likely to have significant impacts on the imaging capabilities of sonar. For example, wind-generated whitecaps on the surface are an excellent acoustic reflector and the surface return clutter from the whitecaps can corrupt the quality of the acoustic data to such an extent that they become unreliable for small target detection (Kozak 2006). Moreover, density variations in the water column can cause the path of the sound to follow a distorted or curved path. Thermoclines are the most frequent cause of such distortion, but the effect can be observed too wherever water masses of differing salinity occur together (Kozak 2006). It is possible, therefore, that animals close to the surface will be effectively masked by the acoustic clutter associated with the surface. Although the detection rates appear relatively low compared with sightings, this may partly be a function of animals being close to the surface. This is highlighted by the fact that most detections (90 %) were made when the sea state was calm (Beaufort sea state ≤ 1) and surface clutter was likely to be at a minimum. It should be noted, however, that many other mobile targets (0.64 h^{-1}) were detected in the absence of sightings at the surface, so it is possible that a portion of these were marine mammals.

Results from the target tracks confirmed that marine mammals (and other targets) frequently move past the turbine in relatively close proximity (within 10 m). On several occasions, the tracks indicate that a marine mammal moved between the turbine pile and the end of the crossbeam supporting the hub of the turbine blades. However, because the sonar targets have no associated vertical information, the depth at which the targets were when they passed the turbine and whether they would be within the area of the rotors is unknown. Further, it was generally impossible to detect and track targets using the sonar immediately downstream of the turbine because of the turbulence produced by the turbine leg and crossbeam.

It is clear from the above that active sonar can be used to detect and track marine mammals in the vicinity of fixed structures such as tidal turbines. However, few off-the-shelf systems have the spatial and temporal resolution, range and 3D detection capabilities required to track marine mammals. Moreover, it is critical to consider carefully the acoustic characteristics of the system and the hearing ranges of the species of interest in order to avoid accidentally conflating responses attributable to the turbine with responses attributable to the sonar. It is also clear that most sonar systems are currently very user-intensive so that for them to become an efficient monitoring tool, a number of key developments are required. Most important is that it is critical that marine mammals can be differentiated automatically from other underwater targets, e.g. marine debris. Results of the Strangford Lough study suggest that, given sufficient resolution and range of imaging capabilities, features of the target (*TS*, speed, depth changes, number of targets, movement in relation to tidal direction etc.) can potentially be used to identify marine mammals and to distinguish them from other targets (including other wildlife species). Work is currently underway by a number of sonar manufacturers to develop new multibeam systems (see Fig. 10.2) to help address this issue and recent trials have been encouraging; these new systems have greater spatial and temporal resolution than earlier ones and provide data that potentially allow marine mammals to be effectively differentiated from debris using automated classification algorithms based on target size and shape and their movement characteristics (Hastie 2012). It is therefore likely that active sonar could become a powerful tool for monitoring fine-scale interactions between marine mammals and tidal turbines over the next few years.

Passive Acoustic Techniques

Marine mammals use both passive and active acoustic detection as their principal means of sensing their environment; dolphins and porpoises in particular produce echolocation clicks for navigation and finding prey, and these potentially provide a means of locating and tracking individual animals in 3D. For example, it should be possible to use arrays of sensors (hydrophones) mounted on tidal energy devices to locate the clicks of cetaceans (and therefore the animal) swimming around the devices. Based on their abundance and widespread distribution, the species of cetacean most likely to interact with tidal turbines in European coastal temperate waters is the harbour porpoise. It produces trains of characteristic narrow band ultrasonic clicks (peak frequency 140 kHz), which are projected forward in a narrow beam (3 dB beamwidth of 16°) and have an on-beam source level of 178–205 dB re 1 μPa p–p (Villadsgaard et al. 2007). The primary function of these clicks is echolocation and click rate varies with behaviour and the echolocation task being undertaken. Click rates typically vary between 5 and 35 but can reach > 1000 clicks per second (Clausen et al. 2010). It is believed that in the wild, porpoises vocalize frequently with 90 % of the intervals between clicks being < 20 s (Akamatsu et al. 2007). Several species of dolphin are also found in inshore waters and are likely to interact with tidal energy devices. They produce communication whistle vocalizations as well as echolocation clicks. Their clicks are louder and have a broader bandwidth than those of porpoises, and their rate of click production may be more variable.

Passive acoustic systems have been used extensively for detecting vocalizing animals but their use to locate and track animals is less well developed. The location of a vocalizing animal can be calculated by determining the time difference between the sound arriving at two or more hydrophones. Although there are a number of different ways to use this information, the most common technique for localizing the

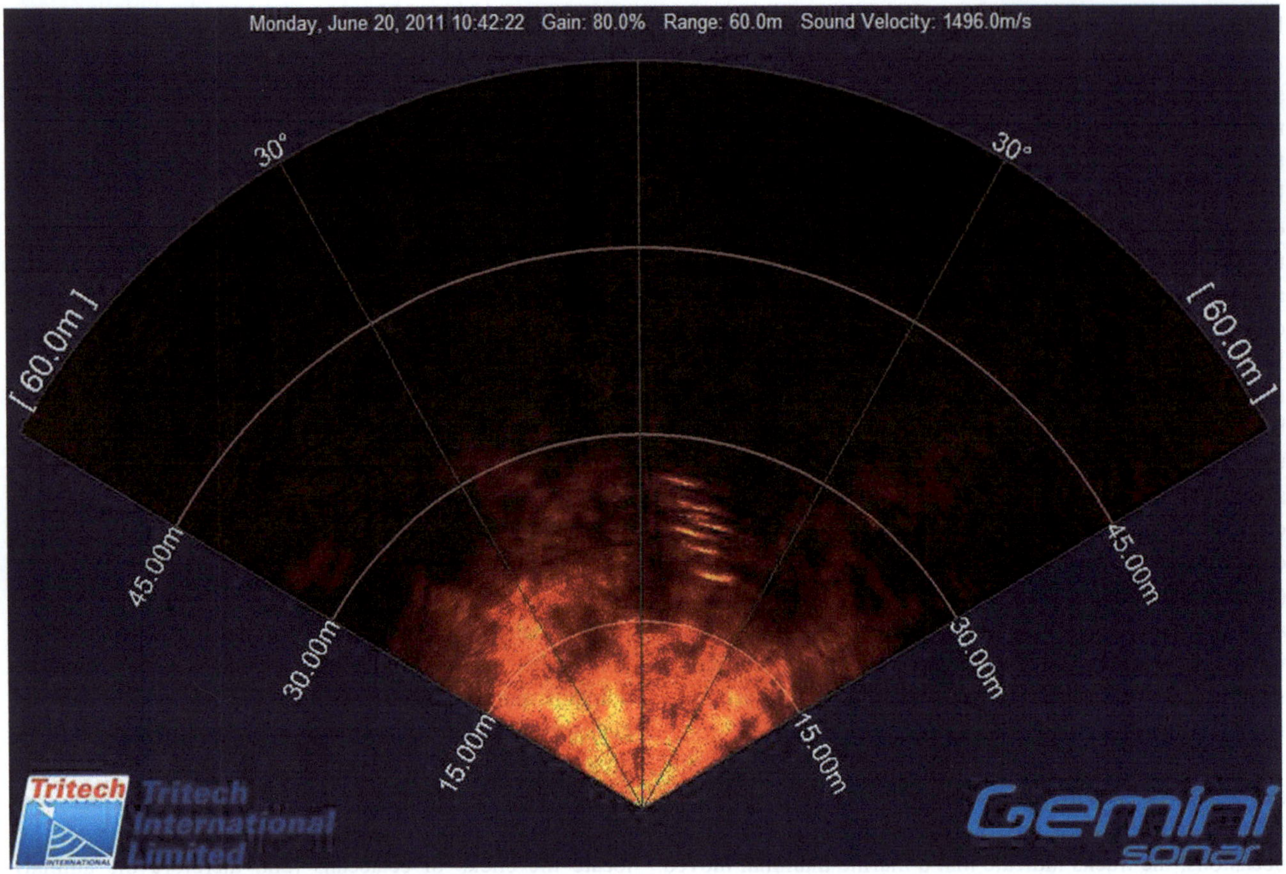

Fig. 10.2 Multibeam sonar image of a group of seven bottlenose dolphins in the Firth of Tay, Scotland. The image is from a forward-looking multibeam system (720 kHz) and shows a plan view out to 60 m with the sonar transducer located at the lower apex. The dolphins are visible as distinct targets between 0 and 30° at a range of 15–30 m from the sonar

sound source is based on measuring time of arrival differences (TOADs) of the sound at each hydrophone. A single time-delay measurement between a pair of hydrophones allows the location of the vocalizing marine mammal to be determined along a hyperbolic surface of infinite length.

Simple two-hydrophone towed arrays are now used routinely to carry out passive acoustic line-transect surveys for marine mammals. Each detected vocalization restricts the vocalizing marine mammal to a unique hyperbolic surface. However, because this type of array is moving, a sequence of detections results in a series of surfaces all of which cross at a common location. This allows a range to the vocalizing animal to be calculated, key information for line-transect survey. This methodology has been used extensively with the largest of the toothed whales, the sperm whale (Leaper et al. 1992; Lewis et al. 2007) and the smallest, the harbour porpoise (Chappell et al. 1996; Gillespie and Chappell 2002). Recently, Gordon et al. (2011) used stereo-hydrophone towed array surveys to determine porpoise densities and their patterns of spatial distribution in high tidal current areas off the coast of Wales. Freely available software such as PAMGUARD (Gillespie et al. 2008; www.pamguard.org) and relatively affordable towed hydrophone equipment renders this methodology accessible for surveys at tidal sites.

Arrays with a larger number of hydrophones provide a greater number of TOADs; for an array with N hydrophones there are $N(N-1)/2$ TOADs, each of which will provide a hyperbolic surface; however, only $(N-1)$ of these will be fully independent. Theoretically, the location of the vocalizing animal is restricted to the point(s) at which all these surfaces cross, so at least two independent time delays are needed to calculate a source location in 2D and three time delays (requiring a four-hydrophone array) to calculate one in 3D. The accuracy of these locations depends on a range of factors including the physical environment, the array design and the acoustic behaviour of the marine mammals being studied.

Standard acoustic analytical techniques assume that sound travels at a constant speed and in straight lines. In reality, the speed of sound in seawater is affected by temperature, pressure and salinity, all of which can be measured easily and/or reliably predicted. More problematic is the possibility of there being changing sound speed profiles within water masses, which would result in acoustic refraction and lead to curved sound paths and a number of concomitant errors.

However, these effects may be less of a problem in strong tidal current areas where waters are well mixed. Acoustic reverberation, background noise and the directional nature of cetacean vocalizations can all result in variable signal waveforms at different hydrophones within an array, often introducing timing errors, especially for species with spectrally pure vocalizations, such as the harbour porpoises. These effects result in errors within the measured TOADs. Additionally, for arrays in which the sensors are not rigidly fixed, error in the location of the hydrophones (through for example the effects of waves and tidal currents) is another substantial potential source of uncertainty. All of these issues contribute to the overall error in the localized position of the marine mammal and the estimation of the magnitude of each, and how they feed through to uncertainty in the final source location is an important aspect of acoustic localization.

Generally, the effect of these errors is determined by their magnitude in proportion to the TOADs. Therefore, larger arrays, which will have larger TOADs, will tend to provide more reliable locations than smaller ones. As a rule of thumb, it is not realistic to expect good localization at ranges more than ten times the dimensions of an array. However, large array dimensions can bring their own problems. Some come from the practical considerations of deploying and maintaining a large rigid array in an extremely energetic marine environment, whereas others relate to the nature of the signals themselves. For example, cetacean echolocation clicks are highly directional, so as two hydrophones are moved farther apart, the received waveforms on each will become increasingly dissimilar, resulting in increasing timing errors. With large separations and low received sound levels, it is likely that some clicks will be detected on only a subset of hydrophones, leading to difficulties in tracking cetacean movements using widely separate hydrophones. This issue was explored empirically by deploying a large 3D array (dimensions ~20 m) on a fish farm (a cost-effective means of deploying a large floating array in porpoise habitat). With this it was possible to detect coherent clicks from porpoises within a range of 150–200 m. It was also possible to localize and track animals, although the fact that the hydrophones were not rigidly fixed in this array compromised location accuracy. More importantly, the exercise served to demonstrate that, in practice, sufficiently coherent clicks are detected at multiple hydrophones to allow tracking.

A second issue with larger arrays is a problem that can be termed "aliasing". To calculate a coherent series of TOADs it is important that the same signal be identified on all hydrophones. When clicks are received at a fast rate, with an inter-click interval less than the travel time of sound between the hydrophones, the possibility for calculating a TOAD between the "wrong" signals becomes a problem; this risk increases both with the rate at which clicks are received (because of the high rates of vocalization and with many animals vocalizing at the same time) and array size. Methodology for dealing with this issue is currently being developed and one promising technique is simply to localize all possible combinations of clicks. Many of the resulting localizations can be discarded because they are in impossible locations (e.g. above the sea surface or below the seabed) or at too great a range. Of the balance, the location with the lowest error or the highest fit probability can be selected.

Although the distance between hydrophones is a significant consideration, the distribution of hydrophones within an array also has an important bearing on the type and dimensionality of location information to be calculated. For consistency in all dimensions, hydrophones need to be evenly distributed in 3D space. However, achieving high positional accuracy may be more important in some dimensions or locations than others (accuracy close to turbine rotors would be prioritized, for example) and there may be times when practical considerations dictate a particular configuration. As an example of this, recent work has been carried out using a linear array (a string of hydrophones distributed on a single straight line) to measure the depth to which porpoises dive and the proportion of time spent at each depth in tidal rapid areas (Gordon et al. 2011). Information so obtained has obvious relevance for assessing the risk of collision with tidal turbines placed in those areas. In that case, a linear vertical array was chosen because a line of hydrophones mounted on a heavily weighted cable suspended from a drifting vessel was a practical configuration to maintain even in strong tidal currents. As the hydrophones were deployed in only one dimension (on a straight line), the array could only locate animals on a horizontal circle centred on the array's axis, providing depth and range information but not an unambiguous $X-Y$ location. In this study, depth was the key information required, and the trade-off between reduced information and enhanced practicality is warranted. It has been possible to use vertical arrays in a number of high tidal current areas, providing an effective and pragmatic tool for preconstruction surveys that can provide valuable data on how animals use tidal rapids and their expected encounter probabilities if turbines are deployed within them (e.g. Fig. 10.3).

A system that would allow fine-scale 3D tracking of animals within a few hundred metres of operating turbines has the potential to provide information essential for understanding how small cetaceans interact with tidal turbines, the real extent of collision risk and (if necessary) a mitigation system to reduce the risks. To achieve this conventionally would require a rigid 3D array with dimensions of tens of metres. It would be challenging and expensive to provide a dedicated construction to hold sensors in such an array in tidal areas, but many turbine designs include substantial support structures that could provide a cost-effective rigid support for an array of appropriate size. By calculating measurement errors, where possible using data collected during field trials,

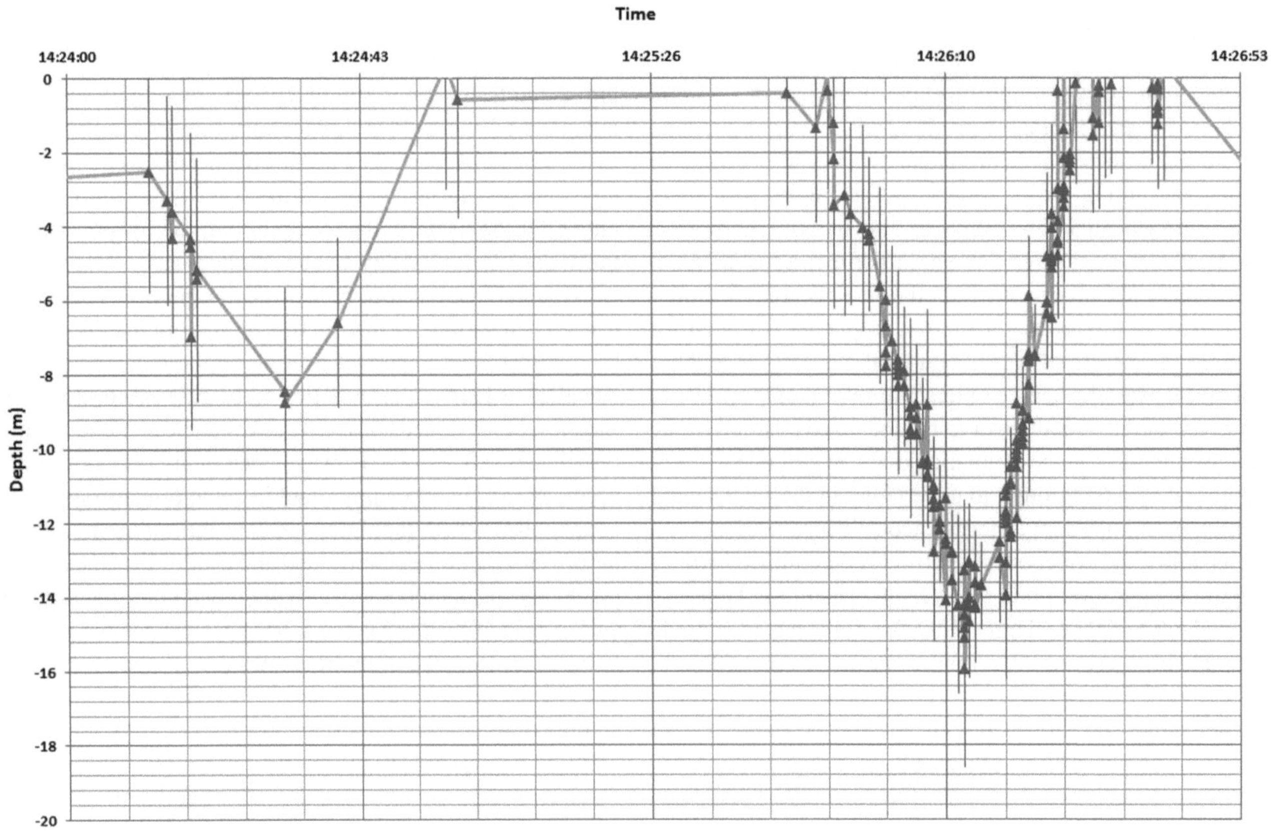

Fig. 10.3 Plot of the depth of localized harbour porpoise clicks using a vertical linear array of hydrophones. The graph shows time (hh:mm:ss) along the *x*-axis and depth on the *y*-axis with each localized click represented by a triangle

it is possible to determine the localization accuracy achievable with any potential hydrophone array, making it feasible to explore the performance of different hydrophone arrays deployed on future commercial scale tidal turbines. Using data collected in the field and making the assumption that the location of hydrophones on any fixed support structure will be known precisely, it is possible that a properly designed array, deployed on a future commercial scale tidal turbine, could provide sub-metre accuracy in 3D out to ranges of several tens of metres. This would allow precise tracking, especially at ranges close to the turbine, allowing detailed studies of avoidance and evasion behaviour around tidal turbines.

To measure TOADs with sufficient precision, signals need to be digitized by precisely synchronized acquisition devices. Where turbines have large support structures and hydrophone arrays are planned to be incorporated at an early stage, it should be feasible to deploy systems with multiple rigidly fixed hydrophones hard-wired to a single digital acquisition device, with digital data being streamed ashore for detailed analysis. However, this may not always be feasible, and data collection would only be possible once the turbine has been installed. A potential solution to this would be to deploy clusters of hydrophones in small arrays of the order of a metre or so with waveforms and/or click detections being recorded autonomously within each cluster. TOAD analysis of the synchronized signals within clusters would provide accurate and unambiguous bearing and azimuth data for vocalizing animals, and "crossing" such 3D bearings from multiple clusters deployed around a tidal turbine should provide locations and tracks for vocalizing animals. Although these locations might be less accurate than those that could be provided from a larger rigid array of synchronized hydrophones, they should still provide data of value for management application.

Thus far, only acoustic localization derived from the animals' own vocalizations have been considered, with clear limitations especially for seals that vocalize infrequently in these waters. For such less vocal species, however, a potential solution (as mentioned above) is to use a combination of telemetry and passive acoustic localization approaches by tagging animals with ultrasonic acoustic tags and using the array to localize them when they approach within acoustic range. This would be a particularly useful approach if tags could be produced and deployed on animals inexpensively so that large numbers of animals could be tagged. Once hydrophone arrays are established to track one species group, it will be become increasingly attractive to tag other groups too, to maximize benefit from such installations. Large

Table 10.1 A summary of the attributes of each technology to detect and track marine mammals in the vicinity of a turbine; the values are largely illustrative and the actual values depend upon the location of the study, technology configuration and the local density of marine mammals

Technology	Species suitability			Individual identification	Resolution		Plausible sample size		Data latency	Range
	Seals	Odontocetes	Mysticetes		Spatial	Temporal	Individuals	Duration		
GPS/GSM tags with dead reckoning	Y	N	N	Y	10–500 m	10 s	30 % of tagged	6 months	≤2 d/4 h/4 min with direct UHF link	Global
Passive acoustic monitoring	N	Y	N	N	5 m	Continuous	Species density	Indefinite	Logged/0	200 m
Passive acoustic monitoring and acoustic tags	Y	N	N	Y	5 m	5 s	30 % of tagged	Indefinite	Logged/0	200 m
Active sonar	Y	Y	Y	N	1 m	Continuous	Species density	Indefinite	Logged/0	50 m

numbers of fish are routinely tracked using this approach. For seals, long-term low-cost tags attached to their flippers might be used, but practical and ethical difficulties with applying long-term tags to cetaceans is likely to rule out the approach with that group.

Pitfalls and Possibilities: The Way Forward

Until studies can provide data demonstrating that marine mammals can take appropriate avoidance and evasion around operational turbines, or more efficient, cost-effective mitigation solutions are developed, uncertainty about the impacts of tidal energy on marine mammals will continue to provide a barrier to the progression of the tidal energy industry. Measuring baseline conditions will only go so far in terms of informing a prediction of interactions or risk of collision. A clear pathway to understanding the true impacts is to deploy turbines in areas used by marine mammals alongside an appropriate behavioural research scheme that allows for any impacts to be detected quickly. The practical application of the techniques described above is potentially critical in reducing uncertainty.

From a risk-characterization perspective, each technique can provide some information to inform risk assessments or help parametrize formal collision risk models for some species. However, each has its associated benefits and drawbacks (Table 10.1). Telemetry provides a comprehensive dataset for individual animals, but only a few animals (potentially an unrepresentative subset of the population) are likely to be tagged. Further, tagged animals range freely and the areas close to tidal turbines are likely to represent a very small proportion of the animal's home range; few of the data collected might then be directly relevant. The spatial resolution of currently available systems is on the order of hundreds of metres, appropriate for investigating broad-scale avoidance of sites but not fine-scale collision evasion. Telemetry is currently practical with seals and large cetaceans, but not with small cetaceans; moreover, in most cases, deployment duration for cetaceans is markedly shorter than for seals, being limited to hours or days.

Active acoustics can provide fine-scale movement data at short ranges suitable for measuring evasion behaviour, and the data for all marine mammals can be from precisely the area of interest. However, spatial coverage is limited and several systems may be required to cover the aperture of a single turbine rotor, for example. Further, with current systems it is challenging to identify targets to species level and much operator input is required. Passive acoustic systems can provide data at both fine and medium spatial scales. However, their use is limited to species that vocalize frequently or animals that have been specifically tagged with acoustic transmitters.

It is therefore clear that none of these approaches alone can provide all the data to address the issue satisfactorily. To a large extent, however, the three approaches are complementary, with the strengths of one complementing the weaknesses in others, and the way forward will probably be to deploy more than one system concurrently. For example, a turbine which incorporated both active and passive acoustic systems might be considered. To allow for passive acoustic monitoring of non-vocalizing species such as seals, a relatively large sample of animals from local populations might be fitted with low-cost acoustic transmitters. Active acoustic data should provide a good measure of the number of likely mammal targets approaching the turbine rotors from upstream while operational along with the extent of any responsive movements. Passive acoustic monitoring could complement this by providing additional information on species identity and echolocation behaviour as well as providing tracking information that would extend outside the relatively narrow beam of the active sonar and out to greater ranges. Further, other methods not specifically discussed here, such as shore-based visual observation, can provide valuable partner monitoring techniques to help measure animal movements.

It should be noted here that none of these methods can be relied upon to provide a reliable indication that a collision has taken place or of its consequences to either device or animal involved. The development and testing of additional partner technologies, including video monitoring, strain gauges and accelerometers to facilitate this is a priority. In parallel, work needs to be carried out that will further understanding of the consequences for animals should they be involved in a collision; currently the precautionary approach is to assume mortality, regardless of rotor speed or location of the impact. Modelling techniques that combine biomechanics and morphology/anatomy to estimate the potential effects of strikes on bone and tissue need to be developed (Carlson et al. 2012).

The level of monitoring suggested here will inevitably be expensive. Although it is not proposed that this should be carried out at every device or that it should form the basis for a continuous real-time mitigation scheme (see the Strangford Lough chapter), where it is appropriate is for detailed studies of collision risk in early deployments. An understanding of this is urgently needed if the wider development of tidal power is to proceed sustainably and without onerous regulation.

References

Akamatsu T, Teilmann J, Miller LA, Tougaard J, Dietz R, Wang D, Wang K et al (2007) Comparison of echolocation behaviour between coastal and riverine porpoises. Deep Sea Res II: Top Stud Oceanog 54:290–297

Argos (2008) Argos user's manual. CLS Argos, Toulouse

Au WWL (1996) Acoustic reflectivity of a dolphin. J Acoust Soc Am 99:3844–3848

Benoit-Bird KJ, Au WL (2003a) Hawaiian spinner dolphins aggregate midwater food resources through cooperative foraging. J Acoust Soc Am 114:2300

Benoit-Bird KJ, Au WL (2003b) Prey dynamics affect foraging by a pelagic predator (*Stenella longirostris*) over a range of spatial and temporal scales. Behav Ecol Sociobiol 53:364–373

Benoit-Bird KJ, Wursig B, McFadden CJ (2004) Dusky dolphin (*Lagenorhynchus obscurus*) foraging in two different habitats: active acoustic detection of dolphins and their prey. Mar Mamm Sci 20:215–231

Callaghan J (2006) Future marine energy results of the Marine Energy Challenge: cost competitiveness and growth of wave and tidal stream energy. Carbon Trust Report. pp 40. http://www.oceanrenewable.com/wp-content/uploads/2007/03/futuremarineenergy.pdf

Carlson TJ, Watson BE, Elster JL, Copping AE, Jones ME, Watkins M, Jepsen R et al (2012) Assessment of strike of adult killer whales by an OpenHydro tidal turbine blade. Report to the US Department of Energy under Contract DE-AC05–76RL01830

Chappell O, Leaper R, Gordon J (1996) Development and performance of an automated harbour porpoise click detector. Rep Int Whal Comm 46:587–594

Clark CW, Ellison WT, Beeman K (1985) Progress report on the analysis of the spring 1985 acoustic data regarding migrating bowhead whales, *Balaena mysticetus*, near Point Barrow, Alaska. Rep Int Whal Comm 36:587–597

Clausen KT, Wahlberg M, Beedholm K, Deruiter S, Madsen PT (2010) Click communication in harbour porpoises *Phocoena phocoena*. Bioacoustics 20:1–28

Davis RW, Fuiman LA, Williams TM, Collier SO, Hagey WP, Kanatous SB, Kohin S et al (1999) Hunting behaviour of a marine mammal beneath the Antarctic fast ice. Science 283:993–996

Doksæter L, Godø OR, Olsen E, Nøttestad L, Patel R (2009) Ecological studies of marine mammals using a seabed-mounted echosounder. ICES J Mar Sci 66:1029–1036

Dunn JL (1969) Airborne measurements of the acoustic characteristics of a sperm whale. J Acoust Soc Am 46:1052–1054

Freitag LE, Tyack PL (1993) Passive acoustic localisation of the Atlantic bottlenose dolphin using whistles and echolocation clicks. J Acoust Soc Am 93:2197–2205

Fristrup KM, Hatch LT, Clark CW (2003) Variation in humpback whale song (*Megaptera novaeangliae*) song length in relation to low-frequency sound broadcasts. J Acoust Soc Am 113:3411–3424

Gillespie D, Chappell O (2002) An automatic system for detecting and classifying the vocalisations of harbour porpoises. Bioacoustics 13:37–61

Gillespie D, Gordon J, McHugh R, McLaren D, Mellinger DK, Redmond P, Thode A et al (2008) PAMGUARD: semiautomated open source software for real-time acoustic detection and localisation of cetaceans. Proc Inst Acoust 30:67–75.

Gonzalez-Socoloske D, Olivera-Gomez LD (2012) Gentle giants in dark waters: using side-scan sonar for manatee research. Open Remote Sens J 5:1–14

Gonzalez-Socoloske D, Olivera-Gomez LD, Ford RE (2009) Detection of free-ranging West Indian manatees *Trichechus manatus* using side-scan sonar. Endang Species Res 8:249–257

Gordon J, Thompson D, Leaper R, Gillespie D, Pierpoint C, Calderan S, Macauley J et al (2011) Assessment of risk to marine mammals from underwater marine renewable devices in Welsh waters. Phase 2. Studies of marine mammals in Welsh highly tidal waters. Report produced for the Welsh Assembly Government, Document JER3688R100707JG, by EcologicUK. 126 pp

Hastie GD (2012) Tracking marine mammals around marine renewable energy devices using active sonar. SMRU Ltd, St Andrews. Report to the Department of Energy and Climate Change. Report SMRUL-DEC-2012–002

Hastie GD, Donovan C, Götz T, Janik VM (in press) Behavioural responses by grey seals (*Halichoerusgrypus*) to high frequency sonar. Mar Poll Bull

Hastie GD, Wilson B, Thompson PM (2006) Diving deep in a foraging hotspot: acoustic insights into bottlenose dolphin dive depths and feeding behaviour. Mar Biol 148:1181–1188

Herzing DL (1996) Vocalizations and associated underwater behavior of free-ranging Atlantic spotted dolphins, *Stenella frontalis* and bottlenose dolphins, *Tursiops truncatus*. Aquat Mamm 22:61–79

Janik VM, Van Parijs SM, Thompson PM (2000) A two-dimensional acoustic localization system for marine mammals. Mar Mamm Sci 16:437–447

Jensen ME, Miller LM (1999) Echolocation signals of the bat *Eptesicus serotinus* recorded using a vertical microphone array: effect of flight altitude on searching signals. Behav Ecol Sociobiol 47:60–69

Johnson MP, Tyack PL (2003) A digital acoustic recording tag for measuring the response of wild marine mammals to sound. J Ocean Eng 28:3–12

Kastelein RA, Bunskoek P, Hagedoorn M, Au WWL, de Haan D (2002) Audiogram of a harbor porpoise (*Phocoena phocoena*) measured with narrow-band frequency-modulated signals. J Acoust Soc Am 112:334–344

Kozak G (2006) Side scan sonar target comparative techniques for port security and MCM Q-Route requirements. Report to L-3 Communications Klein Associates, Inc

Leaper R, Chappell O, Gordon JCD (1992) The development of practical techniques for surveying sperm whale populations acoustically. Rep Int Whal Comm 42:549–560

Lewis T, Gillespie D, Matthews LJ, Danbolt M, Leaper R, McLanaghan R, Moscrop A (2007) Sperm whale abundance estimates from acoustic surveys of the Ionian Sea and Straits of Sicily in 2003. J Mar Biol Assoc UK 87:353–357

Lonergan ME, Duck CD, Thompson D, Moss S (2011) British grey seal (*Halichoerus grypus*) numbers in 2008; an assessment based on using electronic tags to scale up from the results of aerial surveys. ICES J Mar Sci 68:2201–2209

Lonergan ME, Fedak M, McConnell B (2009) The effects of interpolation error and location quality on animal track reconstruction. Mar Mamm Sci 25:275–282

Love RH (1973) Target strengths of humpback whales *Megaptera novaeangliae*. J Acoust Soc Am 54:1312–1315

Madsen PT, Payne R, Kristiansen NU, Wahlberg M, Kerr I, Mohl B (2002) Sperm whale sound production studied with ultrasound time/depth-recording tags. J Exp Biol 205:1899–1906

Mate B, Mesecar R, Lagerquist B (2007) The evolution of satellite-monitored radio tags for large whales: one laboratory's experience. Deep Sea Res II: Top Stud Oceanog 54:224–247

McConnell BJ, Fedak M, Hooker SK, Patterson T (2010) Telemetry. In: Boyd IL, Bowen WD, Iverson SJ (eds) Marine mammal ecology and conservation. Oxford University Press, Oxford, pp 222–262

McConnell BJ, Fedak MA, Lovell P, Hammond PS (1999) Movements and foraging areas of grey seals in the North Sea. J Appl Ecol 36:573–590

McConnell BJ, Lonergan ME, Dietz R (2012) Interactions between seals and offshore wind farms. The Crown Estates Report 41

Nøttestad L, Ferno A, Mackinson S, Pitcher TJ, Misund OA (2002) How whales influence herring school dynamics in a cold-front area of the Norwegian Sea. ICES J Mar Sci 59:393–400

Richardson WJ, Greene CR, Malme CI, Thompson DH, Moore SE, Wursig B (1991) Effects of noise on marine mammals. LGL Ecological Research Association Inc., TX, for the US Department of the Interior

Ridoux V, Guinet C, Liret C, Creton P, Steenstrup R, Beauplet G (1997) A video sonar as a new tool to study marine mammals in the wild: measurements of dolphin swimming speed. Mar Mamm Sci 13:196–206

Shiomi K, Narazaki T, Sato K, Shimatani K, Arai N, Ponganis PJ, Miyazaki N (2010) Data-processing artefacts in three-dimensional dive path reconstruction from geomagnetic and acceleration data. Aquat Biol 8:299–304

Shiomi K, Sato K, Mitamura H, Arai N, Naito Y, Ponganis PJ (2008) Effect of ocean current on the dead-reckoning estimation of 3-D dive paths of emperor penguins. Aquat Biol 3:265–270

Similä T, Ugarte F (1993) Surface and underwater observations of cooperatively feeding killer whales in northern Norway. Can J Zool 71:1494–1499

Southall BL, Bowles AE, Ellison WT, Finneran JJ, Gentry RG, Greene CH, Kastak D et al (2007) Marine mammal noise exposure criteria: initial scientific recommendations. Aquat Mamm 33:411–521

Sveegaard S, Teilmann J, Tougaard J, Dietz R, Mouritsen KN, Desportes G, Siebert U (2011) High-density areas for harbor porpoises (*Phocoena phocoena*) identified by satellite tracking. Mar Mamm Sci 27:230–246

Thorner JE (1990) Approaches to sonar beamforming. In Southern Tier Technical Conference; Proceedings of the 1990 IEEE, pp 69–78

Tyack PL, Zimmer WMX, Moretti D, Southall BL, Claridge DE, Durban JW, Clark CW et al (2011) Beaked whales respond to simulated and actual Navy sonar. PLoS ONE 6:e17009. doi:17010.11371/journal.pone.0017009.

Villadsgaard A, Wahlberg M, Tougaard J (2007) Echolocation signals of wild harbour porpoises *Phocoena phocoena*. J Exp Biol 210:56–64

Watkins WA, Schevill WE (1972) Sound source location by arrival-times on a non-rigid three-dimensional hydrophone array. Deep-Sea Res 19:691–706

Wilson B, Batty RS, Daunt F, Carter C (2007) Collision risks between marine renewable energy devices and mammals, fish, and diving birds. Report to the Scottish Executive Scottish Association for Marine Science, Oban, Scotland PA37 1QA

Strategic Sectoral Planning for Offshore Renewable Energy in Scotland

11

Ian M. Davies and David Pratt

Abstract

The development-plan-based approach being taken to the establishment of the offshore wind, wave and tidal sectors in Scotland is outlined. The strategic planning framework and processes leading to the development of Draft Plans are explained, along with the relationships between these and the sustainability appraisal process. Detail is also given of the spatial planning methods used to identify development opportunities for one sector in Scottish waters.

Keywords

Marine · Planning · Renewable energy · Scotland

Introduction

The Scottish Government has set a range of challenging targets for energy and climate change. They recognize the potential to take advantage of the extensive offshore renewable energy resources (wind, wave and tidal power) available in Scottish waters and include meeting at least 30% of total energy demand from renewable sources by 2020. In addition, the Climate Change (Scotland) Act 2009 sets statutory targets of at least 42% carbon emissions cuts by 2020, and at least 80% by 2050.

To assist in meeting these targets, the Scottish Government has adopted a non-statutory process of strategic sectoral marine planning to establish a spatial strategy for commercial-scale offshore renewable energy developments in its waters. The process incorporates key strategic environmental, social and economic considerations as well as seeking to capture the views of communities in order to facilitate sustainable development of the offshore renewable energy industry in and around Scotland.

In Scotland, spatial marine planning is being undertaken in the context of international, European, UK and Scottish marine legislation, policy and guidance. At the international level, the United Nations Convention on the Law of the Sea establishes the right of coastal nations to set laws and regulate the use of the marine area out to 12 nautical miles from the coast. That Convention also establishes exclusive economic zones from 12 to 200 nautical miles from the coast. Within these zones, the coastal nation has sole rights over all natural resources. The UN has also produced guidance (Ehler and Douvere 2009) on marine spatial planning at national and regional levels and stated the need for it to take place within a broader system for ecosystem-based management.

At the EU level, the Integrated Maritime Policy seeks to provide a more coherent approach to maritime issues, with increased coordination between different policy areas including blue growth, marine data and knowledge, maritime spatial planning, integrated maritime surveillance, and sea basin strategies. In terms of legislative drivers, the Marine Strategy Framework Directive (MSFD) requires Member States to achieve good environmental status in regional seas by 2020, to apply an ecosystem approach to marine management, and to ensure that pressure from human activities

I. M. Davies (✉)
Marine Laboratory, Marine Scotland, 375 Victoria Road,
Aberdeen AB11 9DB, UK
e-mail: ian.davies@scotland.gsi.gov.uk

D. Pratt
Marine Scotland, Victoria Quay, Leith,
Edinburgh EH6 6QQ, UK
david.pratt@scotland.gsi.gov.uk

M. A. Shields, A. I. L. Payne (eds), *Marine Renewable Energy Technology and Environmental Interactions,* Humanity and the Sea,
DOI 10.1007/978-94-017-8002-5_11, © Springer Science+Business Media Dordrecht 2014

is compatible with good environmental status. The Directive is transposed into UK legislation by the Marine Strategy Regulations 2010. Member States are required to report on the status of their seas and monitoring arrangements by 2012. Although maritime activities are not regulated directly through the MSFD, they will influence marine environmental quality, so their impact will be taken into account in the assessment of good environmental status. In addition to the MSFD, the Water Framework Directive, the Habitats and Birds Directives and the Common Fisheries Policy contain provisions for the use and protection of the marine environment.

At the UK level, the Marine and Coastal Access Act 2009 requires that marine plans are prepared for the UK marine area (0–200 nautical miles). The devolved administrations (the Scottish Government, the Welsh Assembly Government and the Northern Ireland Executive) have jurisdiction over marine planning matters out to 12 nautical miles from the coast. For the purposes of marine planning, the marine area from 12 to 200 nautical miles offshore is also executively devolved to Scottish Ministers. In accord with the 2009 Act, the UK Government and devolved administrations are required to prepare a joint Marine Policy Statement (MPS). This Statement was published in 2011 and provides the framework for preparing Marine Plans and decision-making in relation to the marine environment, and establishes policies and objectives for specific sectors and activities. The MPS builds on the UK vision for clean, healthy, safe, productive and biologically diverse oceans and seas (Defra 2002), and the High Level Objectives for the marine environment agreed among the four UK administrations (HM Government 2008) to fulfill this vision.

In Scotland, the new legislative and management framework for the marine environment is established by the Marine (Scotland) Act 2010. As noted above, the Scottish Government has jurisdiction over marine planning matters out to 12 nautical miles from the coast, and for the purposes of marine planning; the marine area from 12 to 200 nautical miles offshore is executively devolved to Scottish Ministers. As a result, the development of a National Marine Plan for Scotland will reflect the legislative provisions outlined in both the Marine (Scotland) Act 2010 and the Marine and Coastal Access Act 2009.

The Marine (Scotland) Act 2010 also allows for a system of regional marine planning to be developed for areas within Scottish waters. The Regional Plans will be directed by the objectives and policies of the Scottish National Marine Plan and will draw on existing work undertaken as part of the Scottish Sustainable Marine Environment Initiative (SSMEI 2012). This initiative consists of a series of local marine planning pilot projects, undertaken with the aim of gaining a greater understanding of the nature, value and management needs of Scotland's marine environment.

Spatial Marine Planning and Offshore Wind Energy

The Scottish Government is responsible for marine planning and also for the licensing of most marine developments (including renewable energy projects). Although the strategic and legislative framework for marine spatial planning continues to emerge, there is a growing requirement and ambition to fulfill future energy needs by harnessing the power of the marine environment.

To address these issues, the Scottish Government is applying a strategic sectoral marine planning process to offshore wind, wave and tidal stream energy resources. This process is principally driven by the Strategic Environmental Assessment (SEA) Directive (2001/42/EC) and the UK Marine Policy Statement requirement to assess marine plans against social, economic and environmental objectives. The objectives of MSFD are considered through the requirement to set environmental objectives within the SEA process and are also applied because the spatial outputs of the sectoral planning process are considered within the context of the national marine planning regime. What follows concentrates on the applied process for offshore wind energy, but comparable processes have been undertaken for wave and tidal energy too.

In 2009, The Crown Estate (TCE) undertook the first stage of lease bidding in Scottish territorial waters. TCE acts as owner of the seabed, on behalf of the UK Government, and is responsible for leasing areas of the seabed for a range of marine activities, including renewable energy. Exclusivity Agreements (the first step towards securing a commercial lease) were awarded for ten commercial development sites, in Solway Firth, Wigtown Bay, Kintyre, Islay, Argyll Array, Beatrice, Inch Cape, Neart na Gaoithe, Forth Array, and Bell Rock. The identification and progression of these sites was considered to be a programme for development under Section 4 of the Environmental Assessment (Scotland) Act 2005. Therefore, in response to the TCE leasing round and to support the sustainable delivery of the potential for offshore wind around Scotland, the Scottish Government made a commitment to undertake a SEA of the potential for offshore wind development in Scottish Territorial Waters (STW), to include the ten site options. A draft Plan was developed to accompany the SEA Environment Report, and thereby to ensure that those reviewing the assessment findings during statutory consultation were clear about the emerging proposals.

Around this time, TCE also announced the third round of leasing for offshore wind energy in UK Waters (excluding Scottish Territorial Waters). In response to this, the UK Department for Energy and Climate Change (DECC) published and consulted upon the SEA for UK Offshore Energy (DECC 2009). As with the SEA for the leasing round in STW, this exercise focused on identifying, assessing and, where possible, mitigating the potential impacts of offshore renewable energy developments on key environmental receptors, as listed in EU Directive 2001/42/EC Annex 1(f). A key difference was in the output of these processes, with the Scottish SEA providing an assessment of the ten sites and the UK SEA identifying strategic areas suitable for development. A key similarity was that neither of these processes addressed the potential social and economic effects that may arise as a result of offshore wind developments.

The Scottish Government received 856 responses during the statutory consultation on the SEA and draft Plan for STW, and 483 expressed objections to the potential development sites located to the southwest of the Scottish mainland. The grounds for objection emphasized the potential for adverse environmental effects, and significant levels of public concern were raised too regarding the potential for adverse socio-economic impacts. In response, the Scottish Government commissioned an economic impact assessment of the ten short-term sites considered in the Plan. The outcome of the this exercise, together with the SEA, led to the publication in March 2011 of the document "Blue Seas Green Energy—A Sectoral Marine Plan for Offshore Wind Energy in Scottish Territorial Waters" (Scottish Government 2011). In finalizing the Plan, Scottish Ministers decided that six short-term sites would be suitable to progress the development of offshore wind energy, and that three others, including the two situated within the southwest region, were removed from the Plan because of issues previously outlined. A site in eastern Scotland had previously been removed by TCE because of issues around aviation.

In addition to the short-term sites identified by TCE, the Scottish Government undertook constraint and opportunity mapping in order to identify further options within which there might be further potential for development. The Crown Estate marine spatial modelling tool, Marine Resource System (MaRS), was used to identify options by mapping environmental and technical constraints along with resource opportunities. In total, 30 medium-term options (areas of search) were identified initially, and these were subject to SEA. Based on the results of that assessment, five options were ruled out and 25 medium-term options (areas of search) were taken forward in the Sectoral Marine Plan. These are now being reviewed over a 2-year period in order to determine further areas for the development of offshore wind energy.

In summary, the Scottish Government (2011) document provides a strategic planning framework and adopted spatial policy in which six offshore wind energy sites from the 2009 Crown Estate Leasing Round might progress through to project licensing. However, the assessment and consultation process applied to the original ten commercial development sites resulted in the removal of three from the Plan as a consequence of the resultant evidence base. For each of these three sites, prospective commercial developers had exclusivity agreements which indicated that a process that seeks to identify sustainable locations for the development of offshore renewable energy prior to the involvement of developer companies would be more appropriate for the other 25 areas of search. The next iteration of the "Blue Seas Green Energy" document will therefore involve a plan-led approach in seeking to identify future opportunities for renewable energy development in Scottish Waters.

The Sectoral Marine Planning Process for Offshore Renewable Energy

The Sectoral Marine Plan for Offshore Wind Energy concluded that as additional data and monitoring information and improved data handling procedures become available, they should be incorporated into the emerging iterative marine planning process, as applied to the further areas of search for windfarm development in Scottish Territorial Waters, and to opportunities farther offshore.

As sectoral marine planning has developed in Scotland, it has become clearer that its primary purpose is to improve the efficiency of the licensing/consenting process. An integrated process (Fig. 11.1) has been developed that meets the requirements of relevant EU and national legislation, progressively increasing the clarity of the definition of areas of search within a sectoral plan, with open public consultation. The process is iterative, and the sectoral Plans derived from the process are subject to periodic review and updating, as new information or policy requirements emerge.

The full sectoral planning process has been applied to offshore wind in STW, leading to the Scottish Government (2011) document. Work is currently in progress to undertake similar exercises for wave and tidal stream energy, building on experience gained in the Saltire Prize process (Harrald et al. 2010). Scoping studies (Davies et al. 2012a, b) and Regional Locational Guidance (RLG; Scottish Government 2012a, b) documents for wave and tidal energy have been published and sustainability appraisal will follow in 2013.

Scoping for Areas of Search

The initial stage of the sectoral planning process is the scoping study leading to areas of search, and the provision of RLG, which leads (after consultation) to Initial Plan Frameworks with Plan Option areas. The scoping process will be described in some detail, because it represents the first stage in the sectoral marine planning process.

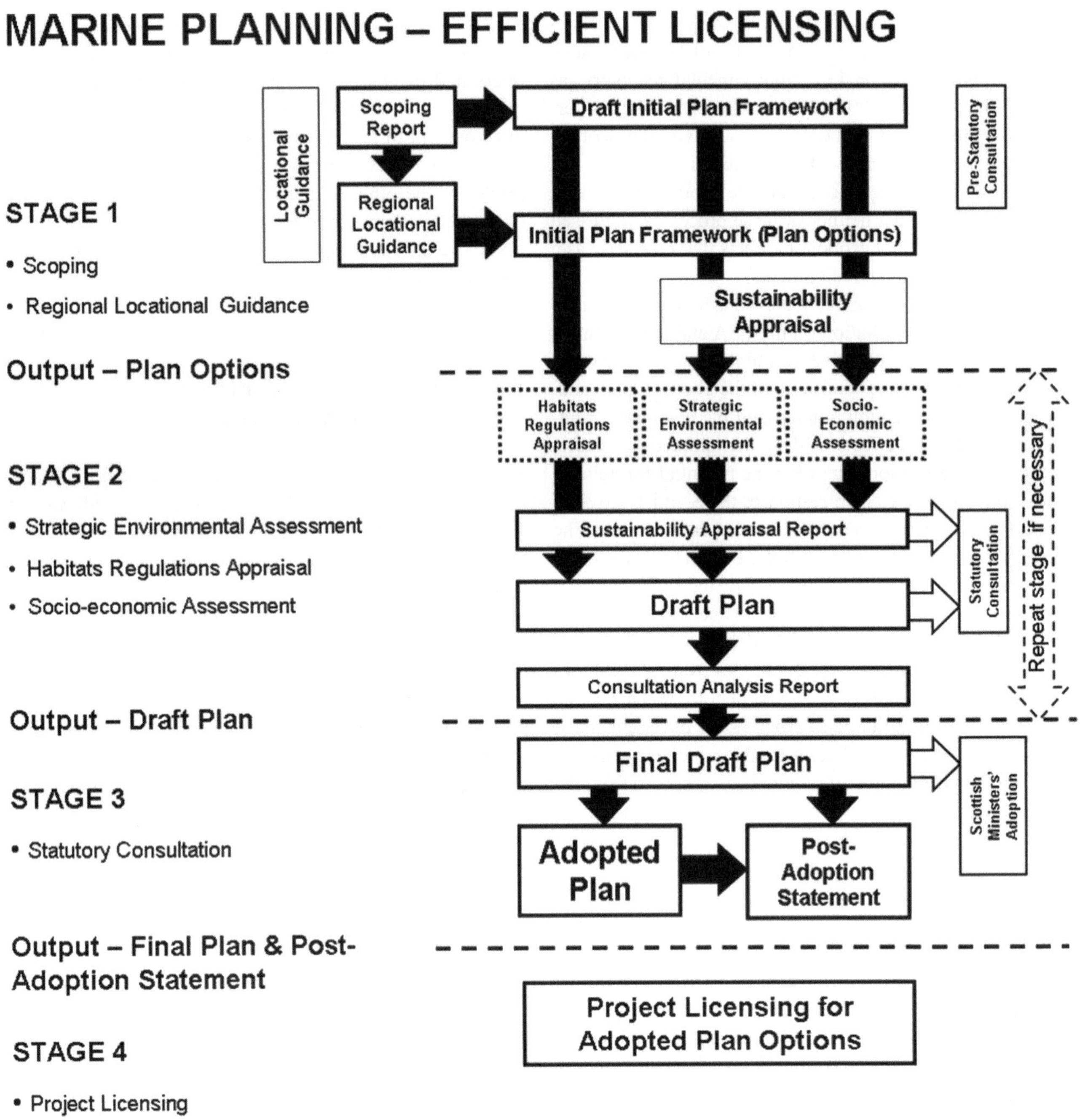

Fig. 11.1 Structure of sectoral marine planning for renewable energy in Scottish marine waters

Marine Scotland has undertaken scoping studies for marine renewable energy using TCE MaRS spatial modelling (GIS) system. That system is used to map zones of broad environmental sensitivity and technical opportunities as well as the constraints relevant to marine renewables developments. It is a powerful tool for handling and integrating a wide range of spatial data referring to the environmental and technical factors that can influence the development of tidal stream, wave and offshore wind energy projects (and other activities). Individual layers of data are held in a supporting geodatabase, and can be selected for use in the spatial modelling. They are derived from a wide variety of sources, including published data, developed by TCE or provided by users such as Marine Scotland, to bring in information of importance in individual applications. The layers of data are brought together in the modelling system to develop integrated layers, presented as spatial models, that map the opportunities and constraints applying in potential development areas.

In order to apply the MaRS tool, it is necessary for the user to make decisions on the data to be included in the models and the way in which the data are to be handled. These decisions consider matters such as: (i) the factors that require consideration when locating marine renewable energy developments and the availability of spatial data that can be included in the models, including terrestrial factors such as the location of protected areas and monuments, and potential for landscape and visual impacts; (ii) whether particular activities or uses should be considered as incompatible with particular forms of renewable energy developments, or whether activities or uses should be considered as presenting gradations of limitation to development potential; (iii) the relative importance (weighting and scoring) that should be applied to the different layers of data in the final integrated model; and (iv) the relative quality and reliability of data layers. A system of scoring and weighting of information held in MaRS is used to produce graduated maps of the least to the greatest technical, and subsequently environmental, sensitivity. From these outputs, broad areas of technical opportunity and relatively low constraint on development can be identified and explored in more detail through RLG.

The first step in the MaRS analysis is to identify broad availability of resource (wind, wave, tidal stream and technical constraints (e.g. distance from shore). Tidal stream resource that currently is considered to have potential for commercial-scale exploitation is confined to a number of distinct areas around Scotland, often in sounds or around headlands, where mean spring tidal current speeds exceed 1.5 m s^{-1}. By contrast, wave resource of potential commercial value is widely distributed west and north of Scotland, and in the northern part of the North Sea, and wind resource is even more widely distributed, although significantly stronger in western and northern areas.

The next step in the analysis is to identify current uses of the sea that are incompatible with renewable energy developments. Generally, such uses include existing infrastructure, e.g. cables, pipelines and other aspects of the offshore oil and gas industry, areas leased for other purposes such as aquaculture, navigation aids and defined shipping zones, and protected wrecks. The areas covered by these uses are combined to create an overall spatial model of areas from which particular forms of marine renewables should be excluded at this point in time. Features that have been treated as exclusions include: all offshore cables inside UK waters; all pipelines in UK waters; anchorage areas; aquaculture leases (current and pending); open (i.e. operational) waste disposal sites; International Maritime Organization routing, excluding Areas To Be Avoided (ATBAs); munitions dumps; navigation aids; offshore shipping zones; offshore windfarm demonstration sites; land areas; UK offshore wind activity; shipping density exclusion areas; live tidal energy leases; UK DEAL oil and gas safety zones, and surface and subsurface features; UK continental shelf exclusion buffers around oil and gas infrastructure, i.e. 500 m; live wave energy leases; operational anemometers in UK waters; protected wreck exclusion buffers.

In addition to factors that can be considered to be incompatible with one or more forms of renewable energy development, there are many that act as partial constraints on development and increase the complexity and scale of risk in the consenting process. Partial constraints include importance to commercial fisheries, abundance of protected species, potential for archaeological remains, and many others. The third step in the scoping process is to collate or develop layers of data representing relevant constraints, using sources such as published data and data derived in-house by TCE or Marine Scotland. The constraint layers are allocated to one of three thematic restriction models, covering constraints arising from industrial activity, environmental factors and socio-cultural interests.

The industry model includes features related to oil and gas activity, shipping, fisheries and military practice/exercise areas. The environment model includes areas with international and national designations for conservation purposes, fish spawning and nursery areas and areas of importance to seabirds and marine mammals. The socio-cultural layer is broad in its scope, covering visual and recreational factors as well as historical heritage and archaeological potential. More complete listings of the factors typically included in the three themes are listed in Table 11.1 together with information on the scores and weights applied to the layers. The weights and scores are adjusted between wind, wave and tidal energy scoping studies to reflect the relative importance of the various factors to different types of energy project. For example, commercial fishing is unlikely to be possible within the footprint of wave or tidal energy development, but some forms (e.g. static gear fisheries) may be feasible with windfarms. Static gear fisheries would therefore be given less weight than mobile gear fisheries in wind energy planning. Presentation of the information by theme reduces the difficulties inherent in developing relative weightings for diverse types of data (e.g. the relative weighting of seabird colonies, wrecks, fish landings and sightings of basking sharks).

The overlaying of the various layers within each theme, plus application of the scoring and weighting schemes and integration with the exclusion model, results in individual integrated maps of relative levels of constraint within each theme.

The output from the industry restriction model is dominated by current "industrial" activity in the coastal zone, shipping routes and military exercise areas. For example, aquaculture is currently entirely limited to waters within a short distance of the shore, as is much of the shipping activity (ferries and vessels on passage around Scotland), and some of the most valuable fishing grounds are in the shel-

Table 11.1 Typical layers of data included in the three thematic constraint models, along with indications of how weighting and scoring can be used to manage the relative influence of each layer in the constraint model and theme

Layer	Weight	Maximum score	Potential relative influence
	Socio-cultural restriction model		
Landscape	1,000	182	182,000
Royal Yachting Association cruising routes	500	50	25,000
Royal Yachting Association racing areas	500	50	25,000
Royal Yachting Association sailing areas	500	50	25,000
Scheduled Ancient Monuments	800	80	64,000
Surfing beaches	700	100	70,000
World Heritage sites	1,000	100	100,000
Wrecks	700	70	49,000
Protected wrecks	700	70	49,000
Potential for marine archaeological remains	700	70	49,000
	Environment restriction model		
Bird reserves	800	80	64,000
Important bird areas	500	50	25,000
Local nature reserves	800	80	64,000
Special Areas of Conservation (SACs)	1,000	100	100,000
Special Protection Areas (SPAs)	1,000	100	100,000
Sites of Special Scientific Interest (SSSIs)	900	100	90,000
Offshore candidate SACs and SPAs	1,000	100	100,000
Offshore draft SACs and SPAs	1 000	100	100,000
Offshore possible SACs and SPAs	1 000	100	100,000
Ramsar sites	1,000	100	100,000
Possible seal haul-out sites	600	60	36,000
Areas of importance to basking sharks	400	73	29,200
Nursery areas for commercial fish	300	55	16,500
Spawning areas for commercial fish	300	55	16,500
Areas of search for potential Marine Protected Areas (MPAs)	600	60	36,000
Areas of search for seabird aggregations	400	40	16,000
Areas of importance to breeding seabirds	400	73	29,200
Areas of importance to seabirds in winter	400	73	29,200
Areas of importance to marine mammals	800	145	116,000
	Industry restriction model		
Offshore cables in UK waters (not active)	500	100	50,000
Pipelines in UK waters (not active)	500	100	50,000
Potential gas and CO_2 storage sites	800	80	64,000
Carbon capture and gas storage infrastructure	800	80	64,000
Current licensed areas for hydrocarbons	700	70	49,000
Closed waste disposal sites	700	70	49,000
Military practice and exercise areas	1,000	180	180,000
Shipping density	800	145	116,000
Ferry routes	1,000	100	100,000
Commercial inshore and offshore fisheries, mobile and static gear landings from mobile gear in inshore waters	1,000	182	182,000
Dredging	1,000	100	100,000

tered internal waters of the Minches, to the northwest of Scotland. The output from the environment restriction model again indicates greater levels of constraint in inshore waters. Relatively high levels of constraint in the North and South Minches are influenced by their importance to seabirds and marine mammals. Important areas include designated sites around Rhum and St Kilda, and in the inner Moray Firth, as well as the important seabird areas off the east coast between Peterhead and Berwick. The socio-cultural restriction model is strongly influenced in the case of offshore wind by national and local landscape designations (as expressions of sensitivity to visual impacts). Other factors include recreational uses and the potential for underwater archaeological remains.

An expression of the overall level of constraint on development of wind, wave or tidal energy projects in Scottish waters needs to take account of environmental, indus-

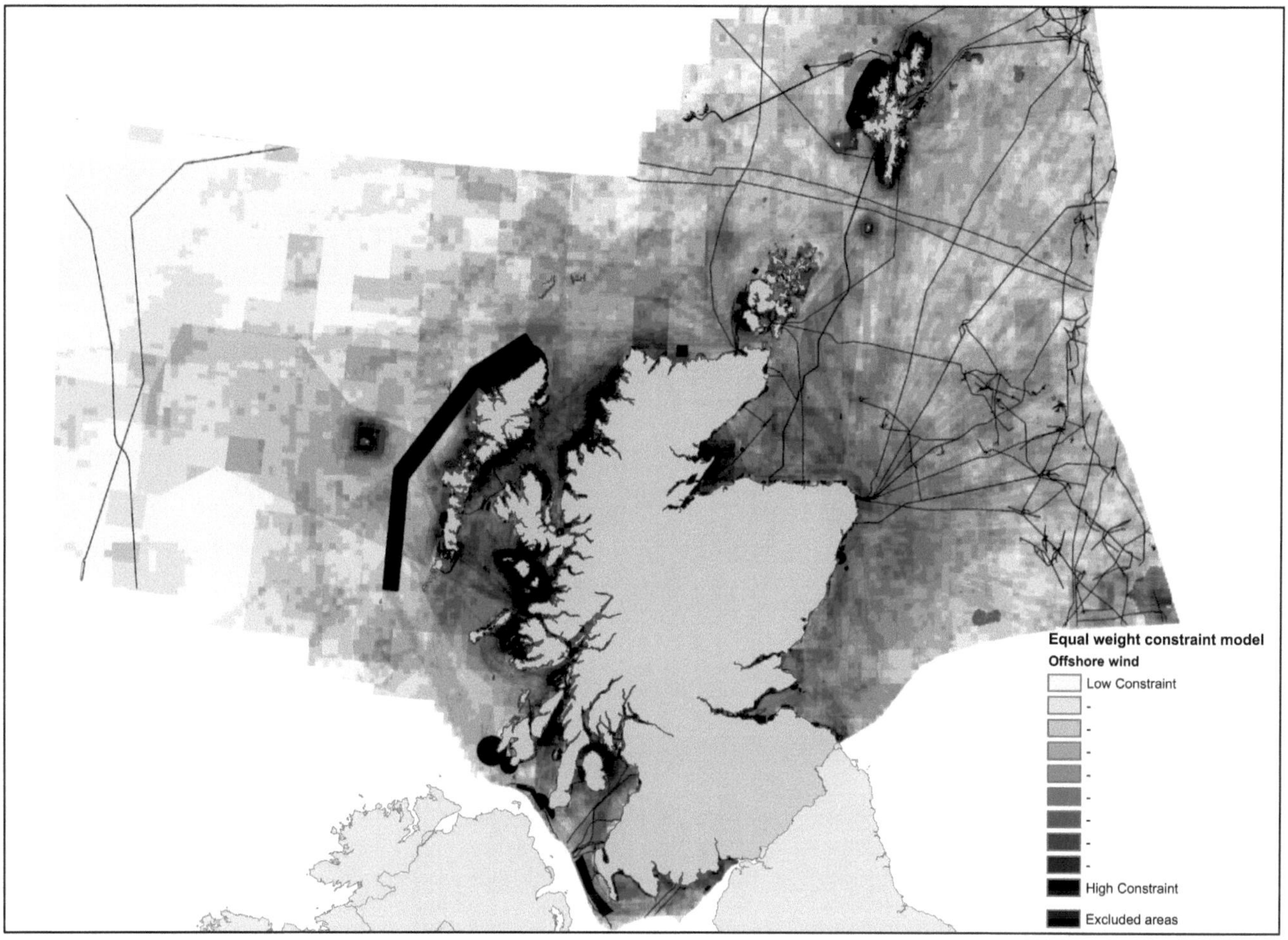

Fig. 11.2 Combined constraint model (equal weighting of industry, environment and socio-cultural themes) showing the relative levels of constraint on windfarm development in Scottish waters

Table 11.2 Composition of the combined models, showing the differences in weightings between the four

Theme	Environment	Industry	Socio-cultural
Model			
Equal weighting	100	100	100
Environmentally focused	200	100	100
Industry focused	100	200	100
Socio-culturally focused	100	100	200

trial and socio-cultural constraints simultaneously. Having grouped the data and developed thematic restriction models, it is possible to combine the models within MaRS and to assess the sensitivity of the outputs to variation in the overall weighting between themes. Typically, four combined models have been created at this point in time, and the relative weightings of the themes have varied. In an equal weighting model (Fig. 11.2), the three themes are weighted equally, but three other models have been used wherein each of the themes has been assigned a weighting equal to the sum of the weightings for the other two themes (Table 11.2).

There are generally broad similarities between the outputs of the four combined models, i.e. features that are not particularly sensitive to the relative weightings of the three themes:

1. Constraint is generally a coastal phenomenon. Most activities in the sea, in terms of all three themes, are concentrated in coastal waters.
2. On the east coast of Scotland, the most constrained areas are in the inner parts of the major firths, Moray, Forth and Tay. The degree of constraint decreases seawards.
3. North and South Minch are generally moderately to strongly constrained, but there are areas on the west coast farther south, west and southwest of the Inner Hebrides, where the degree of constraint is much less.
4. The degree of constraint off the east coast of Scotland is less than in the Minch, but moderate constraint is present over much of the Moray Firth and persists for 30 miles or more offshore of most of the east coast.
5. The model emphasizing socio-cultural interests is dominated by seascape and visual impact issues in the inshore waters west of Scotland and around Orkney and Shetland.

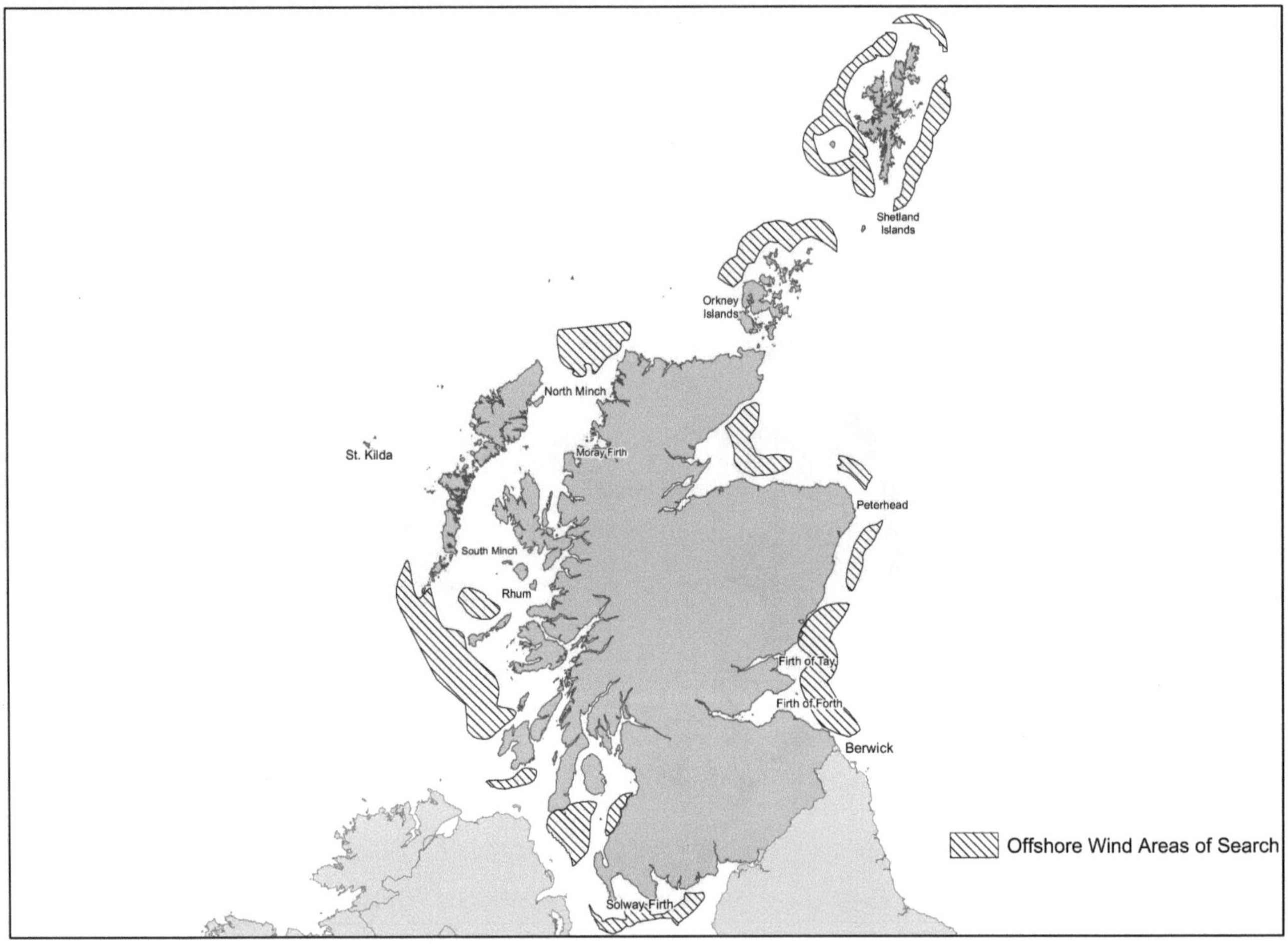

Fig. 11.3 Areas of search for offshore wind plan options within Scottish territorial waters

Visual impact is generally considered to be a less significant issue for wave or tidal energy projects than for wind-power projects,

6. Generally, the levels of constraint outside STW are much less than those within. They also decrease with distance offshore outside STW, such that at 30–40 miles offshore, the levels of constraint are generally very low. There will be some sensitive areas, such as those associated with the oil and gas industry, where development may not be appropriate.

The observation of similarities between outputs from the four combined models acts as a sensitivity check and lends some confidence that the outputs from that stage in the sectoral plan development process are robust and not overly sensitive to the decisions made on relative weightings of factors. Consequently, the equal weight combined model has generally been used to identify areas of search for renewable energy plan option areas in Scottish waters. The area considered is limited in each case to the area identified as having potentially adequate power resource, and emphasis is given to those areas that generally show relatively low overall levels of constraint. Further filters can be applied as appropriate, e.g. a filter by water depth can be used to distinguish areas suitable for particular combinations of technologies. Regional outputs (e.g. for wind power in STW off southwest Scotland) can be combined into a national scale representation of areas of search for renewable energy plan option areas (e.g. Fig. 11.3 for offshore wind).

RLG and Initial Plan Frameworks

Building on the outcomes of the scoping study, the second step of the sectoral marine planning process is the development of RLG. The guidance provides a detailed account of the characteristics and current uses of sea areas that are relevant considerations in the assessment of development prospects at a regional level. It gives consideration to more-detailed environmental, technical and socio-economic issues, including interactions with commercial fishing, aquaculture developments and other related uses of the seas, showing the information used in the MaRS models in a disaggregated form. It also considers issues relating to the buildability of offshore wind developments, including access to and the potential provision of grid infrastructure, and the relevant planning policy considerations when potential developments will interact with the terrestrial environment, for example through the construction of electrical substations and other landward activities. The information contained within the RLG is used to refine the search areas identified at the scoping stage to plan options for consideration within the development of the sectoral marine plan. An Initial Plan Framework is then produced.

In accord with Article 5 (1) of the SEA Directive, the Framework details the Plan Options and reasonable alternatives to be considered in the plan development process. It also provides an outline of the sustainability appraisal process, including consultation methods for developing the plan. The sustainability appraisal process used in sectoral marine planning consists of distinct components addressing strategic environmental assessment, socio-economic assessment, habitats regulations appraisal and subsequent consultation analysis.

Strategic Environmental Assessment

To date, the development of Sectoral Marine Plans has fallen under Section 5(4) of the Environmental Assessment (Scotland) Act 2005. As they are generally considered to have the potential to give rise to significant environmental effects at the screening stage of the process, a full strategic environmental assessment is required for each plan. Where the potential geographic scope of a sectoral marine plan's impacts covers both STW and the wider Scottish marine area, SEAs are undertaken in accord with both the requirements of the Environmental Assessment (Scotland) Act 2005 and The Environmental Assessment of Plans and Programmes Regulations 2004.

On determining that a full strategic environmental assessment is required, the next stage is scoping. At this point, the scope and content of the assessment is established along with the proposed duration of the consultation period for the SEA Environmental Report and Draft Sectoral Marine Plan. Although a traditional SEA, such as that carried out Scottish Government (2007) for wave and tidal energy, would involve an environmental assessment of the spatial content of the related plan, the Sectoral Marine Planning process allows the key environmental baseline data and potential environmental receptors to be integrated into the development of RLG. As the potential plan options and alternatives are derived from the RLG, it allows strategic environmental considerations to be integrated into the process at an early stage.

Socio-Economic Assessment

Once the options and alternatives within the Initial Plan Framework have been determined, a social and economic impact assessment is undertaken, seeking to ask the following questions:

1. What marine activities might be affected by the plan options contained within the Initial Plan Frameworks for offshore wind and wave and tidal energy, and in what parts of Scotland's waters might these impacts arise?
2. What are the potential costs associated with the impacts of the plan options on other marine activities and relevant downstream sectors?
3. What are the potential benefits associated with the impacts of the plan options on other marine activities and relevant downstream sectors?
4. What are the potential social impacts, positive and negative, associated with the plan options?
5. What is the potential distribution of costs and benefits between marine activities, and between offshore renewable energy regions?

The outcomes, conclusions and evidence gathered as a result of this social and economic impact assessment are integrated in the planning process and inform the development of the Draft Sectoral Marine Plans.

Habitats Regulations Appraisal

In fulfilment of obligations under Habitats Regulations and Offshore Habitats Regulations, and to support the development of Sectoral Marine Plans, there is also the need to undertake Habitats Regulations Appraisal (HRA) and, if necessary, to produce an appropriate assessment. The Regulations implement the EC Habitats and Birds Directives in Scottish Waters and require that an appropriate assessment be undertaken where a plan is not directly connected with or necessary for the management of designated European Sites or Offshore European Sites, including Special Areas of Conservation (SACs) and Special Protected Areas (SPAs), and where the possibility of there being a significant effect from the implementation of the plan cannot be excluded. In Scotland, and elsewhere the UK, these requirements are extended to consideration of effects on Ramsar sites and on sites that are proposed for designation, such as potential SPAs, candidate SACs, and Sites of Community Importance (SCIs).

The process for the developing sectoral marine plans seeks to incorporate the consideration of designated sites and protected species at an early stage. In the event that areas of planned development for renewable energy interact with these features, the HRA process allows measures for mitigation to be advanced, where appropriate, or alteration of the plan to ensure that there are no adverse effects on the integrity of designated European sites as a result of plan implementation.

Sectoral Marine Plan Stage and Consultation Analysis

Once the sustainability appraisal has been undertaken, the outcomes inform the development of a draft plan for wind, wave or tidal energy. As well as containing potential areas for the development of offshore wind energy, the draft plan for offshore wind (for example) also contains the key strategic issues captured during the respective environmental and socio-economic assessments. This allows regulators, potential developers and other stakeholders to be aware of likely significant strategic issues that might arise when developments are being considered in specific regions of Scotland.

The draft plan and sustainability appraisal report are subject to consultation with both statutory consultees and the public, with consultation undertaken on national, regional and sectoral bases. Stakeholders include the fishing, shipping, tourism and recreation, defense and aviation sectors along with central and local government authorities, agencies, nongovernmental organizations and the communities of Scotland. Emphasis is placed on those likely to be affected by any proposed area for development contained within the draft Plans.

Following the consultation, a consultation analysis report is produced that documents all consultation responses as well as providing an analysis of the key issues arising. The issues and responses arising from the consultation on both the plan and sustainability appraisal report are then used to inform and refine the Final Sectoral Marine Plan, which is put before Scottish Ministers for adoption. If the plan is adopted, a Post-Adoption Statement is produced to provide an account of its development process and an audit of the consultation exercise.

The adopted plans for wind, wave and tidal energy will form a spatial basis upon which future leasing for commercial-scale development areas can take place. Leasing or development proposals within the areas identified within the plans need to be consistent with the Scottish Government's development plans or a justification provided at the project consenting stage as to why this is not the case.

The Integration of Sectoral and National Marine Planning Processes

As outlined above, the Scottish Government is required to develop an integrated National Marine Plan for Scotland reflecting the legislative provisions outlined in both the Marine (Scotland) Act 2010 and the Marine and Coastal Access Act 2009. The non-statutory nature of sectoral marine plans may yield questions on how this regime will integrate into the framework for both national and regional marine planning. In practice, however, the process is likely to be relatively straightforward. It is ultimately a tool for spatially identifying strategic development areas for commercial-scale renewable energy in line with the principles of sustainable development required for marine spatial planning. The National Marine Plan will provide the framework in which any areas identified will be taken forward and assessed again at the licensing stage, reflecting the requirements of existing EU and UK legislative and policy drivers.

The outputs of sectoral marine plans, i.e. the representation of spatially defined strategic locations for offshore renewable energy developments, will be placed in the broader context of all existing activities within Scottish marine areas through the National Marine Plan. It is then the role of the National Marine Plan to determine if and how these areas are adopted into the National Marine Plan.

With adoption in the National Marine Plan comes an assumed presumption in favour of commercial-scale developments in areas identified within sectoral marine plans, when such developments are assessed against national marine planning policy at the project consenting stage. It is still necessary for developers wishing to use sites within the areas identified in the plans to follow the full process of application for the necessary permits for development, including the Marine Licence. It is also still possible to seek consent for development outside Plan areas, for example as small demonstration projects, or to deploy new types of device with requirements not well covered by the Plans.

As with the terrestrial planning system, the absence of adoption in development planning does not necessarily mean that development cannot take place in these areas. Should a development area be adopted within a sectoral marine plan, it becomes the policy of Scottish Ministers, and this would also be a material consideration at project consenting stage.

The Evolution of Sectoral Marine Planning

Building on the identification and adoption of areas for the development of offshore renewable energy within sectoral marine plans, the broader discipline of sectoral marine planning can play a role in many stages of project development.

Unlike the National Marine Plan, sectoral plans are not confined by geographic scope and could consider a multitude of strategic issues associated with the development of offshore renewable energy, some of which may exist within the scope of the terrestrial planning regime, such as the provision of strategic grid solutions. Additionally, through extensive consultation with statutory consultees, other sectors, key stakeholders and communities, the sectoral process allows for key issues and potential sectoral conflicts to be advanced at an early stage in the development process. This allows the implementation of initiatives such as sectoral engagement strategies, which can identify areas of common benefit and analyse conflicts, seeking to identify measures to mitigate or minimize potential impacts that might arise as a result of development.

As an area is adopted within a plan, RLG needs to be refined to present key regional and strategic information that a potential developer may wish to consider during the planning phase of a project within that area. Strategic information in relation to sectors, including aviation, fishing and shipping, can be presented to regulators and potential developers, along with broader information on considerations such as protected areas and the use of sea areas by protected species. This can inform the early stages of

Table 11.3 Comparison of the structure of the IOC/UNESCO Guide approach to marine spatial planning with the sectoral planning process used in Scotland

Step	Activities in UNESCO step	Parallel activity in sectoral planning for marine renewable energy in Scotland
1	Identifying need and establishing authority	Growth in marine renewables is necessary to meet Scottish Government targets for energy. Marine Scotland is designated in legislation as the responsible authority for marine planning.
2	Obtaining financial support	Obtained through normal government resourcing channels.
3	Organizing the process through pre-planning	The project team grew from policy and science divisions in Marine Scotland. The goals are to improve the efficiency of renewables licensing and to deliver energy targets on time.
4	Organizing stakeholder participation	Stakeholder participation occurs as broad statutory and non-statutory activities in the process, and also through regular meetings with smaller stakeholder groupings.
5	Defining and analysing existing conditions	Extensive existing information is collated in the MaRS spatial modelling tool, and specific new work is carried out to strengthen areas of weakness, where necessary.
6	Defining and analysing future conditions	Preferred development areas with limited interactions with other activities are identified as areas of search. Estimates of capacity for growth are research targets.
7	Preparing and approving the spatial management plan	Plans are completed after public consultation, published and adopted by Ministers.
8	Implementing and enforcing the spatial management plan	Plans are implemented through leasing of the seabed, and through licensing/consenting activities by the regulator (Marine Scotland).
9	Monitoring and evaluating performance	The operation and the performance of the plans are monitored through tracking of project progress through the licensing process. Work is in progress to modify marine monitoring and assessment processes to inform future iterations of the SEA process.
10	Adapting the spatial management proces	Sectoral planning is a cyclical process. Between cycles, data will be updated, new data collected, new analytical methods developed and research needs addressed.

selection of sites and the development of environmental impact assessments and any other relevant consenting requirements. Further, in line with the requirements of strategic environment assessment, the sectoral marine planning process provides a framework in which to monitor, review evidence baselines and undertake data-gap analyses in a targeted fashion. This in turn can inform a strategic environmental research strategy that targets key data gaps and seeks to inform better policy- and decision-making. As new information becomes available, it would inform sectoral plan reviews and ultimately enhance the framework for project consenting, to ensure that decisions are as far as possible both robust and defensible.

The Scottish approach to sectoral marine planning has been developed in response to strong policy drivers from the Scottish Government. These have been applied and assessed in the framework of both European Directives (e.g. SEA Directive, Habitats Directive) and national legislation (e.g. Habitats Regulations). The outcome is a structured approach (Fig. 11.1) based on marine spatial modelling within an iterative planning framework.

Marine spatial planning is a relatively new discipline, but it is being used with increasing frequency to address development or ecosystem objectives in the face of complex demands for marine space and associated goods and services. In most countries, there has been insufficient time for a standardized, tried and tested, approach to development. In this respect, the role of the International Oceanographic Commission (IOC) and United Nations Educational, Scientific and Cultural Organization (UNESCO) in publishing a stepwise approach to marine planning (Ehler and Douvere 2009) has been helpful, particularly in providing a structured framework suitable for tailoring to meet local needs and conditions.

The IOC/UNESCO guide breaks the marine planning process down into ten steps (Table 11.3), from identifying the need for planning to the use of adaptive management in applying the plan. It is possible to make close parallels between these ten steps and activities within the Scottish sectoral planning process for marine renewables. Some aspects of the Scottish process have moved ahead more rapidly than others. For example, the use of spatial planning tools at the scoping stage and in the presentation of RLG is well established.

All sectors of the marine renewables industry in Scotland are moving ahead strongly, with the support of a strategic sectoral planning process. The success of the new industries raises new questions that need to be addressed through a combination of targeted research and spatial analysis. For example, how can best use be made of the energy resources available in Scottish waters, and what factors will limit the growth of marine renewables?

References

Davies IM, Gubbins M, Watret R (2012a) Scoping study for tidal stream energy development in Scottish waters. Scottish Mar Freshwat Sci 3(1):42

Davies IM, Gubbins M, Watret R (2012b) Scoping study for offshore wave energy development in Scottish waters. Scottish Mar Freshwat Sci 3(2):41

DECC (2009) Strategic environmental assessment for UK offshore energy. http://www.offshore-sea.org.uk/site/index.php

Defra (2002) Safeguarding our seas: a strategy for the conservation and sustainable development of our marine environment. http://archive.defra.gov.uk/environment/marine/legislation/strategy.htm

Ehler C, Douvere F (2009) Marine spatial planning: a step-by-step approach toward ecosystem-based management. Intergovernmental Oceanographic Commission and Man and the Biosphere Programme. IOC Manual and Guides, 53. ICAM Dossier, 6. UNESCO, Paris

Harrald M, Aires C, Davies I (2010) Further Scottish leasing round (Saltire prize projects): regional locational guidance. Scottish Mar Freshwat Sci 1(18):168

HM Government (2008) Our seas—a shared resource (Defra) http://www.scotland.gov.uk/Topics/marine/seamanagement/marineact/Ourseas

Scottish Government (2007) Scottish marine renewables. Strategic environmental assessment. http://www.scotland.gov.uk/Topics/marine/marineenergy/wave/WaveTidalSEA

Scottish Government (2011) Blue seas green energy—a sectoral marine plan for offshore wind energy in scottish territorial waters. http://www.scotland.gov.uk/Publications/2011/03/18141232/0

Scottish Government (2012a) Draft regional locational guidance—tidal energy. http://www.scotland.gov.uk/Topics/marine/marineenergy/Planning/tidalrlg

Scottish Government (2012b) Draft regional locational guidance—wave energy. http://www.scotland.gov.uk/Topics/marine/marineenergy/Planning/waverlg

SSMEI (2012) Local planning pilots (SSMEI). http://www.scotland.gov.uk/Topics/Environment/Wildlife-Habitats/protectedareas/SSMEI). Accessed Sept 2012.

Strangford Lough and the SeaGen Tidal Turbine

12

Graham Savidge, David Ainsworth, Stuart Bearhop, Nadja Christen, Bjoern Elsaesser, Frank Fortune, Rich Inger, Robert Kennedy, Angus McRobert, Kate E. Plummer, Daniel W. Pritchard, Carol E. Sparling and Trevor J. T. Whittaker

Abstract

The background to and outcomes of the Environmental Monitoring Programme (EMP) required by statutory regulators for the deployment of the SeaGen tidal turbine in Strangford Lough, Northern Ireland, an area with many conservation designations, are described. The EMP, which was set within the context of an adaptive management approach, considered possible effects of the device on local populations of seals and harbour porpoises, representative seabirds and benthic communities. The studies on seals were carried out on both local and regional scales. The ecological studies were complemented by detailed field and hydrodynamic modelling investigations together with a programme of mitigation measures designed to reduce collisions between seals and turbine rotors. In general only minor statistically significant changes in abundance, distribution and animal behaviour patterns were recorded, principally associated with small distributional shifts close to the turbine structure and with the likelihood that these changes were ecologically of little significance. The seal–rotor collision mitigation studies provided a base for the establishment of acceptable collision risk strategies. The EMP highlighted observational, methodological and statistical challenges in assessing the environmental consequences of marine energy devices. A brief review of related studies in Strangford Lough is included.

Keywords

Benthos · Current measurements · Environmental monitoring programme · Marine mammals · Mitigation measures · Seabirds · SeaGen tidal turbine

The names of the contributing authors to each section are identified against each.

G. Savidge (✉)
Marine Laboratory, Queen's University Belfast, Portaferry, Co. Down BT22 1PF, UK
e-mail: g.savidge@qub.ac.uk

D. Ainsworth
Marine Current Turbines, Bristol and Bath Science Park, Dirac Crescent, Emersons Green, Bristol BS16 7FR, UK

S. Bearhop · N. Christen · K. E. Plummer
Centre for Ecology and Conservation, College of Life and Environmental Sciences, University of Exeter, Cornwall Campus, Tremough, Penryn, Cornwall TR10 9EZ, UK

Introduction [GS, DA]

Strangford Lough, located on the east coast of Northern Ireland, is one of the largest marine embayments in the United Kingdom. The overall morphology of the Lough, which is ~30 km long and 8 km wide, reflects strongly the glaciation associated with the last Ice Age, and is dominated by the presence of boulder reefs, or pladdies, and islands formed from partially drowned drumlins. The range of tides within the Lough averages 3.3 m, and there is a standard diel cycle, with the restricted fetch constraining wave activity.

One of the Lough's principal features is the Narrows (Fig. 12.1), a restricted channel some 8 km long, minimum width 1 km and depth varying between 30 and 60 m,

M. A. Shields, A. I. L. Payne (eds), *Marine Renewable Energy Technology and Environmental Interactions,* Humanity and the Sea,
DOI 10.1007/978-94-017-8002-5_12, © Springer Science+Business Media Dordrecht 2014

that connects it to the Irish Sea. It has been estimated that ~350×10^6 m^3 of water enters and leaves the Lough through the Narrows on each tide (Brown 1990). The constriction of the Narrows creates two secondary tide-related effects: first, there is an approximately 2 h offset in the times of high and low water inside and outside the Lough and, second, the flow in the Narrows is characterized by very strong currents peaking in excess of 4.5 m s^{-1} on spring tides (Boake et al. 2012). It is this latter feature that offers clear potential for the exploitation of the Narrows as a tidal energy resource.

The history of the use of Strangford Lough as a tidal energy resource dates back to pre-medieval times. The existence of a tidal mill, dated to 532 AD, depended on enclosure of a local bay and was constructed as part of a monastic settlement (McErlean and Crothers 2007). Such tidal mills were common around the Lough up to the 18th and 19th centuries. In the 1970s and 1980s, considerable interest was shown in the possible construction of a tidal barrage across the Narrows to allow energy to be extracted by a system of tidal turbines similar to those employed in the La Rance development in Brittany. However, the scheme was not progressed because of the relatively small tidal range of the Lough and the possibility of severe environmental impacts arising from a reduction in physical mixing and associated increase in stratification and consequent deterioration of water quality (Newbould and Carter 1984).

The environmental status of the Lough has long been recognized, starting with the first major ecological surveys initiated in the 1950s (Gotto 1951; Williams 1954). In 1998 the Lough was designated as a Special Protection Area (SPA) and a Ramsar site, then in 2008 was confirmed as a Marine Nature Reserve and subsequently as an EU-listed Special Area of Conservation (SAC). Interest in the potential of the Strangford Narrows as a site for the exploitation of tidal energy on a commercial scale came in 2004 following an Environmental Scoping Report (Royal Haskoning 2004) for Marine Current Turbines Ltd (MCT). The Report confirmed the suitability of the main channel of the Strangford Narrows (Fig. 12.1) for the deployment of a twin-rotor tidal device, to be known as SeaGen. Deployment of SeaGen was aimed to assess its design features and efficiency, a follow-up to earlier successful trials of a single-rotor device (SeaFlow) in the Bristol Channel. The Lough test site provided a sheltered location with low wave activity to allow access to SeaGen for servicing, as well as a readily accessible established grid connection.

A successful application was made by MCT in 2005 for a Food and Environmental Protection Act (FEPA) Licence to the Northern Ireland Environment and Heritage Service, the predecessor to the Northern Ireland Environment Agency (NIEA), to allow the deployment of a SeaGen demonstrator tidal device in Strangford Narrows for an initial period of five years. The application was supported by an Environmental Statement (ES) dated the same year. A key condition of the FEPA Licence was the development of a comprehensive Environmental Monitoring Plan (EMP) covering the pre-installation, installation, operation and decommissioning phases of the project. The EMP defined not only a comprehensive monitoring programme but also the application of an adaptive management approach to the interpretation of data, entailing the continuous review of observational data and identification of any negative ecological issues that may have developed. Detailed reviews of the data were carried out at set intervals by a scientific panel and also in summary at meetings open to all stakeholders. A condition of the FEPA licence was the development of an Environmental Action and Safety Management Plan (EASMP) in 2011. The EASMP was based on the EMP and included details of an extensive programme of ecological observations to be made in support of the FEPA Licence, but also, importantly, incorporated a series of environmental mitigation measures to be undertaken in order to address concerns raised by the Regulator, particularly associated with avoiding seal–turbine rotor collisions. Observations were conducted between 2006 and 2011 and covered the two-year pre-installation period for SeaGen, the four-month installation phase and the subsequent commissioning and operation of the device. A series of interim reports was prepared over the course of the project, culminating in the submission of the final report based on the terms of the EMP (Royal Haskoning 2012).

The SeaGen EASMP in Strangford Lough is the first such programme relating to the environmental impact of a full-scale, grid-connected, commercial tidal device. It is

B. Elsaesser · D. W. Pritchard · T. J. T. Whittaker
School of Planning, Architecture and Civil Engineering, Queen's University Belfast, University Road, Belfast BT7 1NN, UK

F. Fortune
Royal Haskoning DHV, 10 Bernard Street, Leith, Edinburgh EH6 6PP, UK

R. Inger
Environment and Sustainability Institute, University of Exeter, Penryn Campus, Treliever Road, Penryn, Cornwall TR10 9EZ, UK

R. Kennedy
Marine Ecosystem Research Laboratory, Martin Ryan Building, Ryan Institute, School of Natural Sciences, National University of Ireland Galway, Galway, Ireland

A. McRobert
Water Management Unit, Northern Ireland Environment Agency, 17 Antrim Road, Lisburn, Co. Antrim BT28 3AL, UK

K. E. Plummer
British Trust for Ornithology, The Nunnery, Thetford, Norfolk IP24 2PU, UK

C. E. Sparling
SMRU Marine Ltd, St Andrews New Technology Centre, North Haugh, St Andrews, Fife KY16 9SR, UK

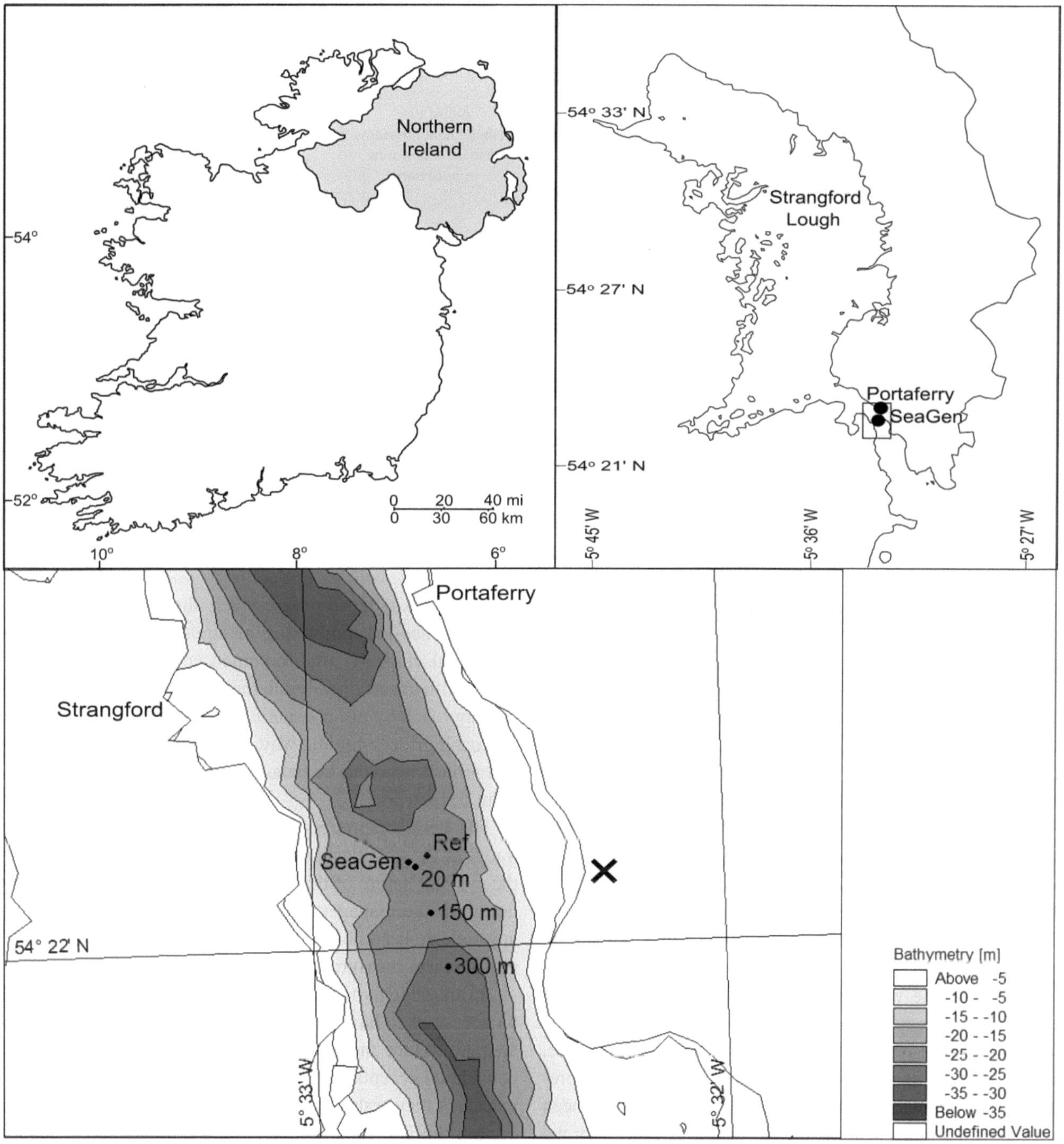

Fig. 12.1 Location of the SeaGen tidal turbine in Strangford Narrows, Northern Ireland. In the lower diagram, the locations "Ref", "20 m", "150 m" and "300 m" refer to the sampling sites surveyed annually for the benthic study. The large cross east of the Narrows shows the location of the observation point for shore-based monitoring of seals and porpoises and the first round of seabird monitoring. Full details of the background to the figure are provided in Kregting and Elsaesser (in review)

appreciated that the results relate to a single device, but it is considered that many of the data obtained will help inform consenting of future array developments or as input to predictive models for the ecological influence of tidal arrays. In addition, the SeaGen programme has provided a major incentive for the development of further studies into the ecological consequences of the deployment of marine renewable energy devices.

This chapter provides an overview of research into the environmental interactions of SeaGen in the Strangford Narrows at the entrance to Strangford Lough, carried out to fulfil the requirements of the EMP, itself a major component of the

EASMP. The initial section provides a brief general background to adaptive management, followed by an outline of how the approach was applied to the SeaGen EASMP. The following three sections highlight the outcomes of the ecological monitoring components of the EMP, components that were defined primarily by the EU Habitats Directive and include observations of common (or harbour; *Phoca vitulina*) and grey seal (*Halichoerus grypus*) populations, cetaceans, diving and other seabirds and the benthic communities of the Narrows. Because of specific concerns about the status of common seals in Atlantic waters, emphasis was placed on the local population of that species, but with behavioural pattern observations of both local populations and populations farther afield. The attention given to seabirds reflected the importance of Strangford Lough as an overwintering site for several migrant species, as a summer nesting area for internationally important numbers of common (*Sterna hirundo*), Arctic (*S. paradisaea*) and sandwich terns (*S. sandvicensis*) and also as a feeding area, flight route and breeding and roosting area for several resident species. The benthic component of the EMP recognized the ecological importance of the rich biodiversity of benthic communities in fast-flowing tidal current zones (Connor et al. 2004) and their relatively restricted occurrence in UK coastal waters. The sections on ecological consequences of the deployment of SeaGen in Strangford Narrows are followed by a description of the modification of the flow regime by SeaGen. Arguably, changes in the flow and turbulence regimes resulting from the deployment of tidal stream devices potentially represent one of the most fundamental and significant impacts of this technology on the environment. Finally we highlight some of the ongoing developments in marine energy research within or adjacent to the Strangford Narrows and the Lough. These developments have received considerable momentum from the SeaGen project.

Owing to the varied time-scales in the development of the various aspects of the environmental studies associated with the overall project, more or less detail, as appropriate, is provided. Certain components of the data have been presented before, whereas other data have not previously been presented, so more detail is provided; this is especially the case for the section on current measurements in the vicinity of SeaGen.

Adaptive Management and the SeaGen Project in Strangford Lough [FF, AM]

Adaptive management in the context of environmental management has been the subject of much study in recent decades, notably by Allan and Stankey (2009) and Williams et al. (2009). It is an iterative process in which uncertainty about the environmental effects of a project (developments

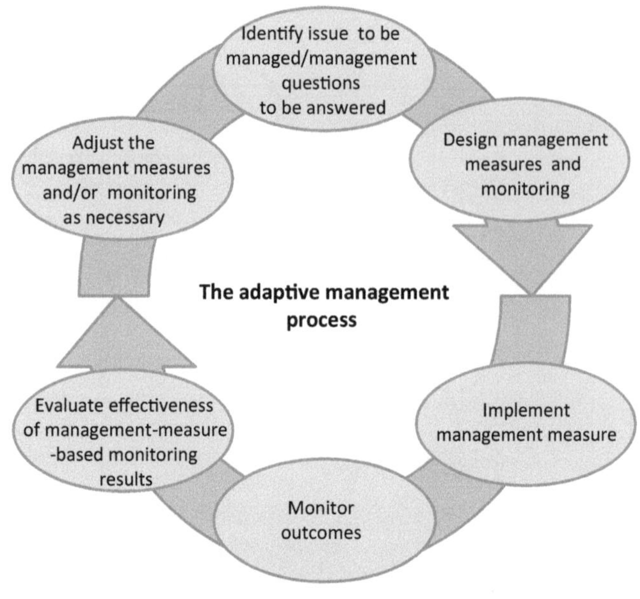

Fig. 12.2 The adaptive management process (based on Williams et al. 2009)

or activity) is reduced progressively through carefully managed, science-led monitoring of agreed indicators of environmental impacts after commencement of the activity. Faced with uncertainty, regulators tend to favour a conservative approach by not consenting to a project, so an adaptive approach offers a middle way, allowing environmental risks and project needs to be balanced within an agreed management framework.

The iterative process of adaptive management is illustrated in its most basic concept in Fig. 12.2, demonstrating the cyclical nature of this approach to management. Careful design of the management measures established to address any concern is combined with monitoring of the outcomes of management measures.

In areas of particular environmental sensitivity, it may be necessary to have in place precautionary mitigation measures to reduce the potential for negative effects to a level considered acceptable by regulators and stakeholders. Such measures may restrict project operations as data are collected, but the aim for a project would normally be to reduce or remove them where new data on environmental effects that become available over the duration of the project indicate that doing so would be appropriate. Of course, the opposite outcome is also a possibility, where the monitoring undertaken indicates an impact whose significance is such that increased mitigation or cessation of the project may be appropriate.

An adaptive management approach to project development, mitigation and monitoring may be particularly appropriate where one or more of the following criteria apply:

- the project involves new or novel technologies or applications of technology;

- initial environmental impact assessment (EIA) indicates that impacts should not be significant, but that data on the potential effects on important features (physical or biological) are either uncertain or unavailable;
- the construction and operation phases can be undertaken in an iterative manner, whereby in the early stages of a project the magnitude of potential effects can be minimized, either by reduced operation during the commissioning phase, or through a modular approach to installation, allowing scaling of parallel mitigation and monitoring and allowing additional capacity to be added throughout the project's life;
- there is a desire among policy-makers and regulators to support the purpose of the project, with a balancing requirement for environmental protection, combined with innovative approaches with the potential for achieving both objectives.

Where adaptive management is the agreed approach, then a number of elements may be required to ensure that it works well and is widely accepted. These include:

1. actively engaging stakeholders in all phases of a project, facilitating mutual learning and reinforcing the commitment to learning-based management (Williams et al. 2009);
2. mechanisms being in place for the active communication of findings and management decisions to stakeholders, a prerequisite for gaining the support of stakeholders for the decisions made by managers;
3. clear distinctions being drawn between environmental monitoring and parallel mitigation measures;
4. measures of the effectiveness of mitigation measures being in place;
5. clearly defined monitoring questions being set to define what the monitoring measures are targeted to answer;
6. a monitoring plan being designed to ensure that all data being collected are sufficient in terms of quality, timeframe and quantity, to answer the monitoring questions;
7. a management group (or groups) being set up, capable of bringing together the key regulator and stakeholder groups along with technical and scientific experts, to ensure that management decisions are based on sound science and expert judgement informed by appropriate scientific knowledge and understanding.

Licensing of the Device

The application to install SeaGen in the Narrows of Strangford Lough presented the regulator, the NIEA, and stakeholders with a series of unique challenges and opportunities. Although the site has the necessary significant tidal resources and is close to habitation, research facilities, grid connection and manufacturing centres, it is also designated for its nature conservation features through both UK domestic legislation and internationally through European legislation.

An EIA was undertaken by the environmental consultants Royal Haskoning to accompany the licence application. It identified a number of potential environmental receptors, for which the nature and extent of effect from the device, and possible negative impacts, was unknown. In particular, effects on marine mammals (resident harbour and grey seals, as well as harbour porpoises, *Phocoena phocoena*) and benthic rocky reef ecology were identified as requiring further study. Of special concern to regulators was the potential impact on the populations of seals, particularly harbour seals, using Strangford Lough.

A licence for installation and operation of SeaGen was granted by the regulator in 2005 (updated subsequently in 2006 and 2007, to accommodate design and installation changes), with installation taking place in 2008. The licence was granted subject to the agreement of mitigation measures to reduce the potential for collision impacts on marine mammals from the tidal device and the establishment of a wider EMP for the SeaGen project encompassing both monitoring to inform managers as to the effectiveness of the mitigation measures in safeguarding seals and the wider environmental monitoring of other receptors. The EMP was required to run for three years from the installation of SeaGen with the key adaptive management element of the EMP focused on management of the potential impacts on seals.

Adaptive Management, Monitoring and Mitigation

MCT, together with NIEA, agreed to establish two nested working groups to oversee the EMP. The first, a Science Group (SG), was set up to oversee, scientifically review and advise the management of the EMP. The SG included the developer MCT, the regulator NIEA, the UK's Sea Mammal Research Unit (SMRU), Queen's University Belfast, the UK Joint Nature Conservation Committee (JNCC), and Royal Haskoning. A second group, a Liaison Group (LG), with much wider membership, was open to all interested parties, with key members including the Royal Society for the Protection of Birds (RSPB), Ulster Wildlife Trust, National Trust (NT), the Maritime and Coastguard Agency (MCA) and local nature conservation groups. An independent joint chairman for both groups was appointed who had been involved in the management of Strangford Lough for more than 30 years.

For each of the receptors monitored (marine mammals, seabirds, benthic ecology, hydrodynamics) a series of key questions was formulated, which a number of modules within the EMP were designed to answer. Monitoring of receptors is discussed in more detail in other parts of this chapter.

The key mitigation measure employed to reduce the perceived collision risk between marine mammals and SeaGen rotors, allowing SeaGen to begin operation, was the use of marine mammal observers (MMOs) stationed on SeaGen for watching an agreed area upstream of SeaGen for marine mammals. Following a protocol agreed with the SG, the MMOs had the ability to shut down SeaGen for a short period of time, until a marine mammal had moved away or past an agreed action area. The movement of an animal into an action area where it was possible for the animal to interact with the turbine was key to triggering a shutdown. The action area was defined on the basis of an assessment of the potential behaviour of marine mammals and the physical characteristics of the tidal stream. As MMOs could only operate during the day, operation of the device was limited to daylight hours.

In addition to the deployment of MMOs, an experimental active sonar system was trialled as a potential mitigation measure, with its effectiveness as a tool for mitigation measured against the use of MMOs over several months. The technological aspects of the active sonar system are described more fully by Hastie et al. (2014, Chap. 10). The active sonar assisted in the detection of marine mammals approaching SeaGen, allowing time for SeaGen to be shut down when required and mitigating the collision risk. Monitoring outcomes demonstrated that active sonar effectively mitigated against collision risk in a manner comparable with MMOs, and the SG agreed to adjust monitoring by replacing MMOs with active sonar. Unlike MMOs, active sonar is not daylight-dependent, so switching to a solely active sonar mitigation method allowed SeaGen to change to 24 h operation.

As knowledge of the operating characteristics of both SeaGen and the active sonar became better understood, it was agreed that the action distance at which a SeaGen shutdown was enacted when a marine mammal was near SeaGen could be reduced. At the start of the EMP, an animal up to 250 m upstream of SeaGen would trigger a SeaGen shutdown. Continually reviewing the data gathered provided the SG with evidence to support progressively reducing the shutdown action distance to 30 m over a number of years, so reducing the number of shutdowns.

All decisions to alter mitigation measures were made within the wider monitoring framework, which allowed the SG to be comfortable that no significant adverse impacts were affecting marine mammals using the Narrows. The SG and the PG met regularly to discuss results and the effectiveness of monitoring and mitigation measures. Where appropriate, changes to mitigation were recommended and adopted. Results of the EMP were reported twice annually, with a final report produced in January 2012 (Royal Haskoning 2012).

Outcomes and Lessons Learned

MCT successfully installed and operated a grid-connected tidal device for more than three years, justifying confidence in the technology and supporting the development of future projects elsewhere. Without an EMP and an adaptive management approach, it is doubtful whether a licence would have been granted; it is stressed that throughout the process, the regulator gained confidence from the mitigation methods adopted. Notably, continuous review of the results of the monitoring programme and the mitigation measures by the SG, together with the success of those measures, provided an essential structure to the project, contributing greatly to its success. The LG played a vital role in the transmission of information about the EMP to stakeholders.

The key lesson learned was the potential of adaptive management as a tool to support responsible development in areas of high conservation value for safeguarding conservation interests while allowing data collection on the impacts of a development. At a project-specific level, the effectiveness of active sonar to detect marine mammals near a tidal turbine was demonstrated. Further, the results from the adaptive sonar monitoring programme are helping now to develop active sonar systems for use in Strangford Lough and other locations suitable for both tidal and wave-energy installations.

SeaGen EMP: Marine Mammals [CES]

The harbour seal population of Strangford Lough is a qualifying feature of the Strangford Lough SAC. The conservation objectives of the SAC pertaining to harbour seals are the maintenance of the population at 200 animals within the Lough. Although not listed as qualifying species, grey seals and harbour porpoises are also present in the Lough. The latter are listed on Annex I of the EU Habitats Directive and as such are afforded protection against death, injury and disturbance. Consequently, the EMP considered potential impacts primarily on harbour seals but also considered whether there was any impact on the populations of harbour porpoises and grey seals.

Several objectives were defined based on the EMP (Table 12.1) and led to the development and implementation of a number of separate marine mammal monitoring studies involving a range of spatial and temporal scales. In some cases, the studies were continuations of studies carried out during the baseline characterization period prior to the deployment of SeaGen; in others, bespoke methodology was designed and implemented. Detailed accounts of the methodology and primary data will be published in the primary literature in due course, but a brief overview of the methods and a summary of the main findings are presented here.

Table 12.1 Objectives of the EMP relating to marine mammals

Objective	Monitoring put in place
Ensuring no mortality of marine mammals as a consequence of physical interactions with the turbine rotors	A system of active acoustic monitoring that detects marine mammals within 30 m of the rotors and allows precautionary shut-down of the turbine Carcass surveys and *post mortem* evaluation of all strandings
Ensuring that the turbine does not present a barrier to the free passage of marine mammals through Strangford Narrows	Pile-based marine mammal observations (July 2008–August 2009) Seal telemetry studies, tracking individual harbour seals using GPS phone tags Acoustic monitoring of harbour porpoise activity in the Narrows and Lough using TPODs
Ensuring that the relative abundance of marine mammals in Strangford Narrows is not modified significantly by the operation of the turbine	Shore-based visual observation of marine mammals in the Narrows around the turbine site Acoustic monitoring of harbour porpoise activity in the Narrows using TPODs
Ensuring that the subsurface noise generated by the turbine does not cause a level of disturbance to marine mammals sufficient to displace them from areas important for their foraging and social activities	Measurement of operational noise, modelling how this noise travels through water and predicting any likely impacts on marine mammals Sightings of marine mammals in close proximity to the turbine during operation from shore-based visual observation, pile-based observation and seal telemetry
Ensuring that the number of harbour and grey seal adults and pups present within the Strangford Lough SAC does not decrease significantly as a result of the installation and operation of the SeaGen turbine	Aerial survey of population size and distribution (set within the context of historical data) Number of harbour seals using the Lough from NIEA/NT boat counts supplements the data
Ensuring that the SeaGen turbine does not cause a significant change in the use of important harbour or grey seal haul-out sites within the Strangford Lough SAC	Aerial survey of population size and distribution (set within the context of historical data) Number of harbour seals using the Lough from NIEA/NT boat counts supplements the data
Ensuring that the SeaGen turbine does not displace harbour porpoises from Strangford Narrows and the adjacent Strangford Lough SAC	Acoustic monitoring of harbour porpoise activity in the Narrows and Lough Sightings data from shore- and pile-based observers

Monitoring

Shore-based Visual Surveys

A fixed point watch station was established on the eastern shore of the Strangford Narrows (Fig. 12.1), chosen at the baseline assessment stage of the project, for an optimum view of the proposed location of the SeaGen device in the Narrows. Data on the activities of harbour and grey seals and harbour porpoises were collected at the watch station by Queen's University Belfast using systematic surveys over a 200 ha sector surrounding the SeaGen site, recording the position and activities of target species monthly between May 2005 and December 2012. Concurrent data on selected avian taxa were also obtained (see below). The mammal data provided a measure of the relative usage by seals and porpoises of the area of the Narrows visible from the observation point, but did not provide an absolute number of these animals in the Narrows; rather, the data provided an index of relative abundance that was used to examine temporal and spatial trends in their pattern of use of the area.

Monthly fluctuations in average relative abundance of harbour seals across the year were apparent, with numbers greatest over summer and lowest during winter. Although there were annual fluctuations in relative abundance, the patterns indicated that the fluctuations were the result of natural variability rather than being related to the presence of SeaGen. There was evidence for a redistribution of harbour seals in relation to turbine operation, however, with some parts of the survey area closest to the turbine exhibiting significant decreases in relative abundance when the turbine was operating and others exhibiting increases. Notwithstanding, there was no statistical evidence for a change in the relative abundance of harbour seals overall while the turbine was operating.

The analysis of grey seal sightings also revealed significant relationships between sighting rates and the state of the tide, year, time of day, time of year and spatial location. There was no statistically significant change in the relative numbers of grey seals seen, or in their distribution during turbine operation, nor was there any evidence for an underlying change in numbers or distribution on days when the turbine was operating. The analysis of harbour porpoise sightings revealed no evidence of any effect of turbine operation on sighting rates, but simulations revealed that the power to detect significant changes in sighting rates was particularly low for harbour porpoises and grey seals.

Passive Acoustic Monitoring of Harbour Porpoise Activity

TPODs (Chelonia Ltd), which operate by logging the start and end of echolocation clicks of porpoises and dolphins, were employed to monitor harbour porpoise acoustic activity. It is important here to stress that TPODs detect porpoise presence in a given area but do not provide a direct indication of the number of porpoises present. The data obtained from TPODs is used to compare relative frequencies of occurrence/echolocation activity between sites or through time. TPODs were originally deployed at the entrance to the Narrows (outer Lough), in the main body of the Narrows close to SeaGen, and also in the inner Lough. Owing to initial loss of equipment at some sites and low encounter rates at the outer Lough sites, monitoring at the outer Lough sites was sacrificed to prioritize monitoring in the main body of the Narrows and the inner Lough.

General Additive Models (GAMs) within a General Estimating Equation framework were used to explore the factors driving porpoise detection. Any relationships described here imply that the covariates were significant predictors of detection rates in the various models fitted. The biggest turbine-related effect was observed during the short installation phase in 2008, with a large and rapid decline in acoustic activity. This effect was only apparent in the Narrows; activity in the inner Lough remained unaffected. Harbour porpoises are generally considered to be shy of boats, so it may have been the increased levels of boat and human activity that caused the decline in porpoise activity in the Narrows then. There is also the possibility that porpoises may have been avoiding the noise produced by construction activities, although work commissioned by COWRIE suggested that harbour porpoises would be unlikely to hear the generated drilling noise at ranges beyond a few metres because of the high levels of background noise in the Narrows (Nedwell and Brooker 2008). Levels of activity in the Narrows recovered immediately after the installation phase, but remained slightly lower than the baseline level. Levels of activity in the inner Lough increased slightly above the baseline level post-installation and may represent a small change in distribution of porpoises post-installation. Harbour porpoises have continued to frequent the Narrows and the inner Lough throughout the operational phase, indicating that neither the presence nor the operation of the turbine has created a barrier effect to them.

The magnitudes of turbine-related effects were notably small relative to the variation in detections explained by the other covariates such as tide, time of day, month and location.

Monitoring Seal Haul-outs

Aerial surveys of seal haul-out sites along the Northern Ireland coast between Carlingford Lough and Belfast Lough, including Strangford Lough, have been carried out annually by SMRU as part of the EMP since 2006. Comparable information is also available from a survey carried out in August 2002 (Duck 2003) that covered the whole of the Northern Irish coast. The post-2006 surveys were carried out during either the breeding season (July) or the moult (August). July surveys provide information on the number and location of harbour seals breeding within the survey area and on the relative numbers of pups born in different areas, and August surveys provide a minimum population estimate for harbour seals at moult in line with standard SMRU survey procedures. Although not the primary focus, grey seals were also counted during those surveys. The surveys were carried out from a helicopter with a thermal imaging camera. Their aim was to determine the overall numbers of harbour seals and pups and the locations of their haul-out sites between Carlingford Lough and Belfast Lough.

Unpublished historical data from surveys carried out in July over several years by the Northern Ireland Environment and Heritage Service (EHS), the predecessor to the NIEA, and the National Trust show that in the late 1970s there were just under 300 harbour seals in Strangford Lough. Numbers then increased, peaking at just over 600 in the mid-1980s, before declining to ~200 by the mid-1990s (Montgomery-Watson 1999).

There has been a decreasing trend in the numbers of harbour seals in both the breeding and moult seasons across the whole of the Northern Ireland survey region since 2002 (Duck and Morris 2012), but no major changes have been observed in the distributions of haul-out and breeding sites during the same period. Numbers in Strangford Lough and the Narrows follow this region-wide trend, with the start of the decline pre-dating the installation and operation of SeaGen. Monthly boat-based counts in the Lough and Narrows by NIEA and the National Trust in 2010 demonstrated a similar pattern of counts declining annually (Lonergan 2009), although more recent counts up to the end of 2012 suggest that numbers in Strangford Lough and the Narrows are now increasing (Lonergan 2013).

Harbour Seal Telemetry

In all, 36 seals were fitted with electronic tags during the environmental monitoring of SeaGen, glued to the animals' fur such that they detached during the annual moult. The instruments collect GPS (Global Positioning System) location data and information on diving and haul-out behaviour and relay the data through mobile telephones incorporated into each instrument. Three deployments took place: in 2006 (April–July, pre-installation), in 2008 (March–July, during installation and commissioning) and in 2010 (April–July, operation). The seals were captured in the Strangford Narrows and the southern islands in Strangford Lough. The three groups of animals tagged contained similar mixes of age and sex.

The major features of the tracks delivered were broadly consistent between years, with great variability between seals, although individual seals were consistent in their behaviour over time. There was, however, some local avoidance of the turbine, with the spatial distribution of the transit locations changing between 2006, 2008 and 2010. A different sample of animals was tagged in each year, so individual responses to the installation of the turbine could not be tracked, and the assumption was that a representative sample of animals was tagged in each year.

In 2010 when the turbine was operating, individual seals, which regularly transited the Narrows, transited on average 20% less frequently when the turbine was operating than when it was not (this was statistically significant). Data from all three years showed that seals transited relatively more frequently during periods of slack tide, which may have implications for the level of collision risk if seals transit preferentially during periods when the turbine is not operating.

Active Sonar: Monitoring and Mitigation

An active sonar monitoring and mitigation system has been in operation on SeaGen since the turbine was commissioned in 2008. Two Super SeaKing mechanical scanning sonar units provide real time subsurface sonar imagery of large objects within 80 m of the turbine while it is operating. The system is remotely monitored by operators in real time during turbine operation and is used to detect potential marine mammal and other large vertebrate targets that may be at risk of rotor strike when the turbine is operational.

The system development has been somewhat iterative and has gone through a number of stages since it was first installed in July 2008. Initially its effectiveness at detecting marine mammals underwater in close proximity to the turbine was trialled alongside concurrent pile-based MMO visual observations in the early stages of SeaGen commissioning and operation (see above). The trial had two objectives: (i) to determine whether the sonar could detect moving marine mammals in a tidally turbulent environment and provide an effective mitigation tool, and (ii) to determine whether the sonar could be used as a monitoring tool to measure the behaviour of marine mammals around the turbine. The first objective was met successfully and the sonar now forms an integral part of ongoing mitigation. The use of active sonar as a tool to monitor fine-scale behaviour of marine mammals around tidal turbines is discussed in more detail in the chapter by Hastie et al. (2014, Chap. 10).

Underwater Noise

The impact of operational noise on marine mammals was considered as part of the EMP. Noise measurements of SeaGen during operation were carried out with high-precision instruments from a drifting boat. Underwater sound propagation models were used to predict how noise levels would vary with distance from the turbine. The potential effects on marine mammals of underwater noise from SeaGen were predicted by SMRU using information on hearing abilities of marine mammals and observed responses from previous studies.

Instantaneous levels of noise from SeaGen, when operating, were below the levels anticipated to cause auditory injury. When considering cumulative noise exposure, the zones predicted for potential auditory injury were small and residence times within these would need to be very high for any marine mammals to be at risk from injury. Data from the land-based observations and the seal telemetry study suggest that neither seals nor porpoises would likely remain close enough to SeaGen for the length of time required to receive a noise-dose sufficiently high to cause damage.

SeaGen noise was above the thresholds predicted to elicit behavioural responses in seals and porpoises up to several hundred metres from the turbine. However, the predictions of behavioural response need to be viewed in the context of the behaviour of marine mammals observed around the turbine. Land-based observations, telemetry-derived data on seal movements and TPOD detections of harbour porpoise echolocation all show that seals and porpoises regularly visit waters adjacent to SeaGen, where they would be predicted to display a behavioural avoidance response.

Conclusions and Lessons Learned

The marine mammal monitoring associated with the SeaGen turbine has underpinned a detailed and comprehensive study of marine mammal behaviour and activity in response to an anthropogenic impact that is likely unrivalled anywhere in the world. As such, it has not only informed the development of the tidal energy industry but has also provided insights into marine mammal behaviour in tidal environments. The only changes detected in any of the metrics monitored have been relatively small and are largely suggestive of small-scale changes in local distribution in relation to SeaGen presence and operation. Some of the comparisons between operational and baseline phases lacked the desirable level of statistical power for there to be absolute certainty of an absence of underlying differences in the metrics measured. This was generally the result of high levels of natural variation in the metrics measured and underscores the need to consider carefully the power of future monitoring programmes to detect biologically significant changes in the metrics under consideration.

The ongoing requirement for precautionary shutdown of the turbine if a marine mammal approaches the rotors within 30 m means that despite almost four years of operation, we do not yet know the extent to which marine mammals can avoid or evade the moving blades. Therefore, the concomitant uncertainty surrounding collision risk remains a barrier to the development of the tidal energy industry.

Several important insights have been gained, however, not only on the effects of SeaGen installation and operation on marine mammals, but also on the limitations of the survey and analytical methodology for monitoring for impacts at tidal sites. Also, it has been shown that installation activities can influence local harbour porpoise activity but that the effects are short-lived and recovery to baseline levels during operation is rapid. There is also evidence that harbour seals exhibit small-scale local avoidance of the turbine and that turbine operation does not result in a barrier effect, with seals regularly transiting through the Narrows past the site. As is often the case with marine mammal abundance, distribution and behaviour, natural variability was high and power analyses of baseline data are useful for determining whether a given methodology/sample size is adequate for detecting change of a given magnitude.

SeaGen EMP: Seabirds [KEP, RI, NC, SB]

Marine renewable energy installations (MREIs) could have a complex array of effects, both positive and negative, on marine birds (Inger et al. 2009; Grecian et al. 2010; Furness et al. 2012). Changes to movements, behaviour, foraging patterns and population dynamics might all result both directly and indirectly from the installation of tidal turbines such as SeaGen. The magnitude of these effects, however, will depend on the ecology and behaviour of the seabird species utilizing the areas in which the devices are deployed; unfortunately, up to now there have been no scientific data with which to assess these predictions.

Strangford Lough is designated as an SPA because of its global importance to wildfowl, marine birds and wading birds, so it was important to investigate the impacts of the SeaGen turbine on the local avian community. The Lough's rich array of habitats and feeding opportunities support in excess of 40,000 wildfowl and 50,000 waders over winter, making it one of the region's most important wintering grounds for migrant species such as light-bellied Brent geese (*Branta bernicla hrota*), knot (*Calidris canutus*) and redshank (*Tringa tetanus*). In summer, internationally important numbers of common, Arctic and sandwich terns nest colonially on the Lough's sheltered network of islands. Further, resident species of gulls such as the black-headed gull, *Chroicocephalus ridibundus*, the common gull, *Larus canus*, and the herring gull, *L. argentatus*, plus auks (black guillemot, *Cepphus grille*, common guillemot, *Uria aalge*, and razorbill, *Alca torda*), cormorants (*Phalacrocorax carbo*) and shags (*Phalacrocorax aristotelis*) use the Narrows as a feeding area, flight route, breeding ground and/or roosting area throughout the year.

The ecological consequences of SeaGen on avian communities in the Strangford Narrows have been investigated with three complementary approaches: (i) collating historical knowledge gained from long-term avian monitoring in the Strangford Narrows; (ii) conducting field-based observations of seabird distributions while the device was under deployment and operational trials; (iii) implementing a spatially and temporally replicated survey of seabird movements and behaviour while SeaGen was in active operation. This thorough approach was expected to highlight any potential repercussions of SeaGen on the Strangford Narrows avifauna, and help to shape future tidal renewable energy development.

Long-term Avian Monitoring in the Strangford Narrows

In order to assess accurately the possible effects of SeaGen on seabirds, it was important to have an understanding of bird species distributions and population size fluctuations within the areas surrounding the SeaGen deployment site. Researchers used a 12-year dataset of bird counts during winter at six sites surrounding and including the SeaGen deployment site (British Trust for Ornithology 2010). From 2000 to 2011, 10,941 observations were recorded in the Strangford Narrows, representing 54 species. Of these, 12 species with sufficient data for analysis were considered likely to interact directly with or to be affected by SeaGen based on knowledge of their foraging and behavioural ecology. These included black-headed gulls, common gulls, cormorants, great black-backed gulls (*Larus marinus*), grey herons (*Ardea cinerea*), herring gulls, light-bellied Brent geese, oystercatchers (*Haematopus ostralegus*), shags, shelduck (*Tadorna tadorna*), teal (*Anas crecca*) and wigeon (*Anas penelope*).

During the 12-year period, population sizes increased for most of these species, with only teal showing a slight decline and great black-backed gull and oystercatcher populations remaining static. Given that three years of data have been collected since SeaGen's deployment in April 2008, differences in distributions among survey areas before and after deployment could also be interpreted. Nine species showed no difference in population size or distribution in response to SeaGen deployment. The black-headed gull was the most abundant species, showing evidence of increasing population size following SeaGen deployment and relocation to sites farther from the device site, although this was only marginally significant. Numbers of grey herons also increased following the deployment of SeaGen, although that result may simply reflect natural population growth. However, the numbers of both grey herons and cormorants increased notably in the areas surrounding SeaGen. For cormorants, the statistical evidence of this effect was highly significant and showed a definitive shift towards the SeaGen turbine from surrounding areas. One possible interpretation of these results is that SeaGen acts as an artificial reef or fish aggregation device (FAD; Inger et al. 2009), leading to locally enhanced fish abundance and improved foraging opportunities for cormorants and other diving species. Although the

British Trust for Ornithology (BTO) data provide insights into historical population sizes and distributions of a number of bird species, they do not incorporate information on some species for which the Strangford Narrows are an important habitat, notably auks, gannets (*Morus bassanus*) and terns which, as diving species, may be vulnerable to collision with the moving rotors of devices such as SeaGen.

Shore-based Observations During Deployment and Operational Trials

Previous work into the effects of windfarms has identified avoidance distance as the best metric for determining the impacts of such structures (Devereux et al. 2008; Madsen and Boertmann 2008). Therefore, the differences in the distance of observed seabirds from the position of SeaGen before deployment, during deployment and during early operational trials were investigated for eight species or functional groups most likely to be affected by SeaGen (Inger et al. 2010). These included black guillemots, light-bellied Brent geese, cormorants/shags, auks (common guillemot and razorbill), common eider (*Somateria mollissima*), red-breasted merganser (*Mergus serrator*), gannets and terns (Arctic, common, sandwich). As described above, observations were made systematically from a fixed shore location monthly between January 2006 and September 2009. Subsequent data analysis showed some evidence that SeaGen was altering the distributions of avian communities, but although these were statistically significant, the magnitude of the effects was small so it is deemed unlikely that SeaGen produced any impacts of ecological or conservation significance on the species investigated. More specifically, where the presence of the SeaGen device did have an influence on species distributions of black guillemots, Brent geese, cormorants, auks, gannets and terns during or after deployment, its effect was considerably less than the effects of annual and behavioural variation. Moreover, the average maximum displacement distance for the six species/species groups mentioned above was just 38.0 m ± 14.8 s.e. (range 5–100 m), below the level of precision with which location could be estimated. Therefore, this result is unlikely to represent a true effect or to have any notable biological consequences.

When considering the observation data summarized above, it is necessary to bear in mind the caveats attributable to methodological limitations of the sampling design and that surveys were carried out during the operational trials phase of the SeaGen project. During that phase, SeaGen was not continually operating owing to regular maintenance, so there is a possibility that any potential impacts on avian communities may only become apparent once SeaGen has been operational at full capacity over a longer period and evaluated through the use of more rigorous sampling protocols.

Shore-based Observations During SeaGen Operation

The most telling evidence of possible effects of SeaGen on avian distributions came from an analysis of robust datasets obtained using standardized survey techniques, replicated across both time and space, while SeaGen was in full-time operation (Plummer et al. 2012). Despite more than 10 years of study, evidence of the impacts of terrestrial windfarms on birds remains unclear (Stewart et al. 2007), primarily because survey protocols have often failed to include sufficiently long time-scales and neglected to record baseline controls. Similar concerns were raised with respect to the primary avian dataset obtained for the SeaGen programme. Therefore, a second phase of EIA was instigated by MCT and funded by the Technology Strategy Board (TSB) that included further assessment of the consequences of SeaGen on seabirds. Although the datasets so obtained were not part of the original EMP, the survey methods used and the outcomes of their analysis are commented on briefly here for completeness.

During this second phase, seasonal surveys were conducted at three sites on either side of the Strangford Narrows. The central site was in line with the SeaGen turbine and the other two sites 2 km north and south of that location, providing control data for comparison. At each site, point counts were conducted along three designated transect lines at 30° intervals, and behavioural observations of focal individuals were conducted within the boundaries of the outer transects. Data were collected over a 14-day period and repeated across seasons between July 2011 and April 2012, accounting for seasonal variations in species abundance and activity. The effects of SeaGen on the abundance, distribution and foraging behaviour of six avian functional groups, comprising 34 different species, were investigated. These functional groups included auks, cormorants/shags, gulls, terns, waders and wildfowl (Plummer et al., 2012, provide a complete species list). It was predicted that if SeaGen is affecting avian communities, then this would be evident through a difference in bird activity at the SeaGen site relative to the control sites. It was clear, however, that inherent differences between the three sites made it difficult to distinguish the effect SeaGen might have been having on bird communities. For example, statistical analyses using mixed effects modelling indicated that complex interactions among environmental and behavioural factors were influencing the distributions of the six functional groups. However, although survey site significantly influenced the distributions, it was not easy to distinguish whether the differences between distributions at the SeaGen site and at the (control) sites away from the device were a response to the turbine or merely a reflection of natural fluctuation along the Narrows. To improve interpretation of the results, therefore, differences in effect sizes were evaluated closely, comparing, for example, the effects of site with those of seasonal and tidal variation. Further, by applying randomization tests as an additional step in the data analysis, avian displacement could be investigated without relying on a control site comparison. As such, despite the ambiguity of the control site data, it was still possible to draw conclusions regarding SeaGen's impacts. Based on the results generated, it has been concluded that SeaGen does not impact avian communities in the Strangford Narrows significantly, although it is acknowledged that subtle, small-scale changes may have been undetected and that longer-term effects remain unknown.

Avian data collected at SeaGen have also been used to predict the potential impacts future tidal arrays may have on avian communities. As the tidal renewable energy sector expands, multiple turbine arrays are already planned for future deployment. By extrapolating predicted avian displacement values generated from the SeaGen survey results, it has been possible to estimate how avoidance at one turbine might scale-up over multiple devices. Variation in habitat, environmental patterns and bird community dynamics are all likely to interact, however, causing birds to react differently to turbines at different locations, so the results can be interpreted only as a guide to possible impacts on distribution alone. Nevertheless, based on the findings, it appears that turbines located in close proximity to one another (ca. 50 m) may produce a barrier to movement for diving marine birds such as auks and cormorants. How these predicted effects impact upon avian populations in general, though, is unclear, but as the first evidence-based prediction of tidal array impacts, they do underscore the necessity in future to monitor avian activity at turbine arrays carefully before, during and after deployment.

Lessons Learned

Several insights on field survey methods, statistical analysis and the benefits of long-term data have arisen from SeaGen's avian monitoring programme. For example, identified weaknesses in the initial survey design for avian monitoring, such as the possibility of repeated counting of individual birds within the survey area, were rectified by adopting a more strategic sampling approach when investigating the impacts of SeaGen during full-time operation. Notably, it was necessary to introduce fixed transects and control sites to place events at the SeaGen site in context. However, it quickly became apparent that despite best efforts being made to select similar sites for controls, such selections provided a relatively poor basis for comparison owing to their inherent differences. Using randomization tests proved an effective way to circumvent this problem when it came to analysing the data. Similarly, it proved important to consider the ecological relevance of effect sizes (not only their statistical significance) to establish how SeaGen might be influencing avian communities. Further, examining trends in local bird populations in the years prior to SeaGen's deployment was particularly useful in providing the context for assessing local effects of the device. It is evident from SeaGen's avian monitoring programme that carefully planned sampling methods, appropriate statistical testing and local knowledge will be critical to detecting environmental impacts at tidal energy devices in future.

SeaGen EMP: Benthic Studies [RK]

It has been recognized for some time that epifaunal benthic communities in fast-flowing tidal current zones are richly biodiverse (Connor et al. 2004) so have great ecological importance. Moreover, the relatively restricted occurrence of such areas in UK coastal waters was a major consideration in the designation of Strangford Lough as an SAC. The regulator considered these environmental factors when requesting that a benthic survey programme be designed to assess the possible effects on the ambient benthic communities of variation in hydrodynamic flow in the Narrows caused by SeaGen.

The Strangford Narrows are designated Reefs, including tide-swept boulders and bedrock, in the main channel. These substrata are densely covered in suspension-feeding epifauna, notably the soft coral dead-men's fingers *Alcyonium digitatum*, sponges, especially *Pachymatisma johnstonia* and *Cliona celata*, ascidians, notably *Dendrodoa grossularia* and *Corella parallelogramma*, and sea anemones including *Metridium senile*. Boulders tend to be covered with encrusting sponges such as *Myxilla incrustans* and *Myxilla fimbriata*, along with hydroids, especially *Tubularia indivisa*, and sea anemones, including *Sagartia elegans*, *Corynactis viridis* and *Actinothoe sphyrodeta*. Coarse sand scours rock surfaces at the sides and either end of the Narrows, where the characteristic species is the bryozoan *Flustra foliacea*.

The major challenge for the benthic monitoring programme was to test for significant effects against a background of high natural variability in macrobenthic community structure caused by physical disturbance associated with the strong tidal currents in the Narrows. Here, we summarize benthic distribution data collected at four sites adjacent to SeaGen on an annual basis over a four-year period spanning pre-installation, commissioning and operation of the device. The data are interpreted to establish whether the presence of SeaGen in the Narrows and the associated changes in the flow regime influenced benthic communities in its vicinity.

Observational Programme, Techniques and Data Analysis

The benthic component of the EASMP was based on annual sampling by divers at three sites located 20 m, 150 m and 300 m south-southeast of SeaGen along the axis of the main channel of the Narrows along with a reference site 50 m east of the turbine (Fig. 12.1). The water depth was 25–27 m at all sites. At each site, five immediately adjacent video quadrats were sampled using a quadrat of 0.5×0.5 m. Pre-device installation video data were obtained in March 2008 and follow-up surveys in July 2008, in March and July 2009, and in April of 2010 and 2011. The survey sites were relocated precisely using USBL (Ultra Short Base Line) acoustic marking devices attached to weighted marker frames on the seafloor. Preliminary analysis of images indicated strong seasonality in the data, so only spring samples were used for assessment purposes.

Habitat classification at each site was performed in accordance with the European Nature Information System (EUNIS) habitat classification scheme (Connor et al. 2004). A second stage Analysis of Similarity (ANOSIM; Clarke et al. 2006) was carried out to test the null hypothesis of no difference in average multivariate faunal pattern over time between stations. Marker frames on the seafloor allowed the quadrat to be relocated on each sampling trip, so the samples from each quadrat may be treated as a replicate time-series from within the sampling site. Bray–Curtis similarity (Bray and Curtis 1957) was calculated between all temporal samples from each quadrat, giving a pattern of change over time for each. Spearman's rank correlation coefficient was calculated between each pair of quadrats and the derived correlation coefficients used as the distance metric in a resemblance matrix. One-way ANOSIM (Clarke and Green 1988) was used to test the null hypothesis that quadrats within stations were not more correlated to each other than they were to quadrats at other stations.

Outcomes, Conclusions and Lessons Learned

All sites corresponded closely to EUNIS level 5 biotope CR.HCR.FaT, very tide-swept faunal communities, *sensu* Connor et al. (2004). The sites had characteristics similar to both of the level 6 biotopes CR.HCR.FaT.BalTub, *Balanus crenatus* and *Tubularia indivisa* on extremely tide-swept circalittoral rock, and CR.HCR.FaT.CTub, *Tubularia indivisa* on tide-swept circalittoral rock, but could not be confirmed as either because of natural variability. ANOSIM of the Bray–Curtis similarities between sample sites over time showed that >90% of all possible pairwise comparisons between time or site groups were significantly different. The second stage ANOSIM analysis showed no significant difference in the pattern of temporal change between sites (global $R = -0.002$, $p = 0.484$), although all sites changed significantly over the period monitored.

The epifaunal communities of the tide-swept boulders of the Strangford Narrows are highly diverse and subject to major physical disturbance associated with the fast tidal currents. This tends to maintain the communities of the Narrows boulder field in a constant state of succession and produces localized benthic community structure in the study area. The interannual changes observed in the monitoring programme appear to represent random spatial variation that encompasses disturbance, competition and succession. All sites were different from each other and all changed significantly, but the nature and direction of change in the impact stations was not different from the reference site.

The high levels of natural spatial and temporal variability encountered meant that a second stage ANOSIM was an appropriate design for testing for deleterious impacts from SeaGen. Clarke et al. (2006) demonstrated that second stage ANOSIM can determine higher level interaction effects, useful for "before after control impact" (BACI; Green 1979) studies of anthropogenic disturbance. It differs from traditional ANOSIM studies in that sites that are not significantly different tend to have similar fauna, the main effect of similarity among sites having been removed. Testing the higher order correlation between sites over time offers an effective means to carry out a BACI in highly variable communities when there is little possibility of having one representative control, let alone multiple controls (Underwood 1992).

Developments in ocean energy device deployment present new challenges for benthic ecologists owing to their focus on high energy, physically disturbed and highly variable environments and the consequent removal of energy from them. Much of the literature on the effects resulting from disturbance of the seafloor is associated with the introduction of energy into the environment from anthropogenic sources through dredging, trawling and similar activities (Rhoads et al. 1978; Kaiser et al. 2006). Even in relatively low-energy environments, macrobenthic communities are often distributed as a "mosaic of the relics of former disasters" (Johnson 1971, 1972). In very high energy environments, however, the scale of natural variability is less well understood, so more research is needed to constrain the nature of variability at ocean energy extraction sites, particularly in relation to the effects of reduced energy flow. Many of the biotopes associated with high-energy habitats are defined from qualitative, semi-quantitative or very small numbers of quantitative data (Connor et al. 2004), a statement applying equally to reefs and sedimentary habitats. The compilation of a local substantial quantitative dataset is hence likely to be crucial for every ocean energy development.

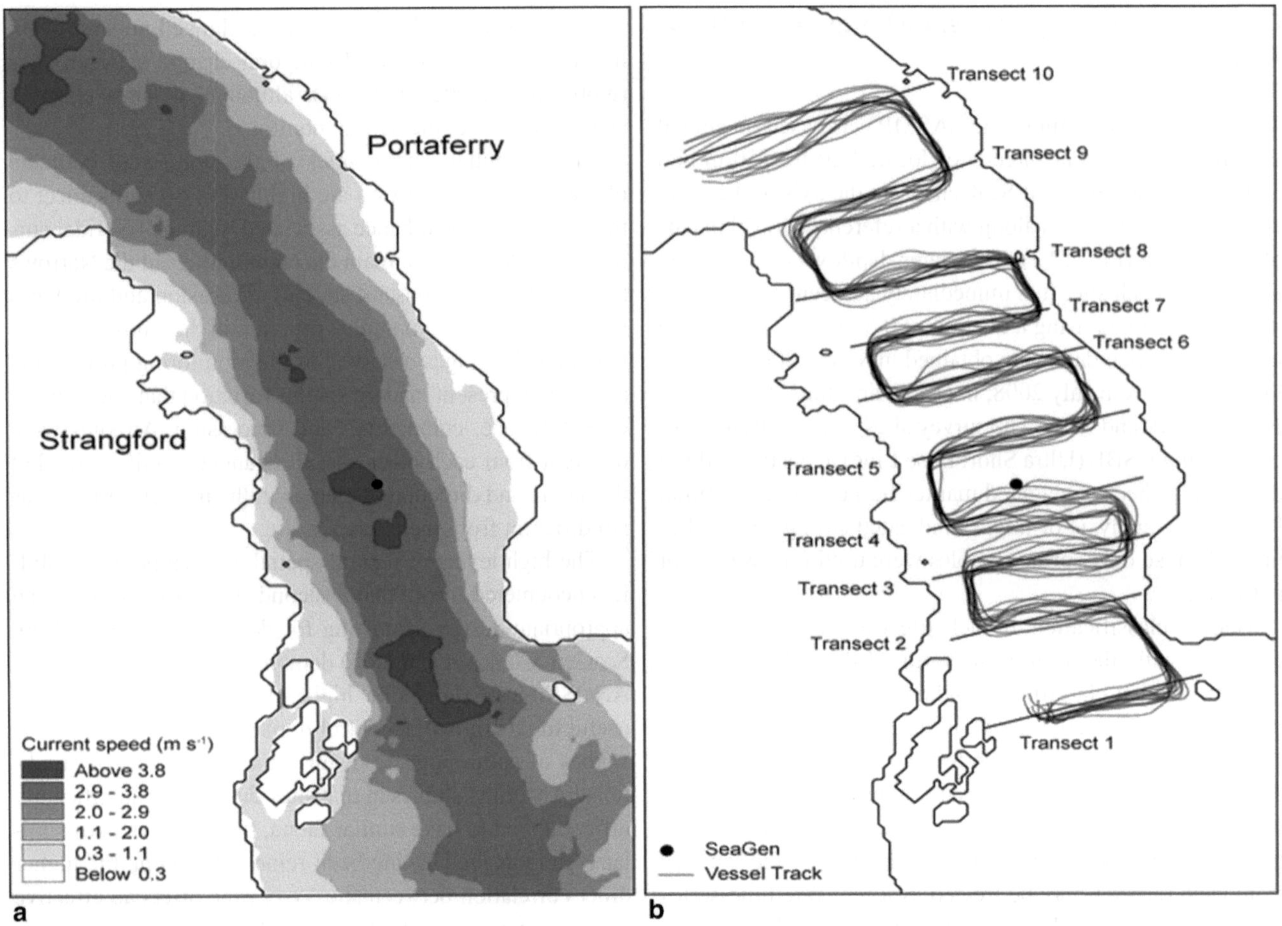

Fig. 12.3 **a** Velocity field of the ebbing spring tide in Strangford Narrows based on numerical predictions (Bell 2005). **b** Proposed survey tracks (*straight lines*) and recorded ADCP vessel tracks (*spaghetti lines*) for 13 passages of the ten transects over one full tidal cycle (Boake et al. 2012). About 40 min was required to cover a full passage of the ten transects, plus another 15 min to return to the starting point, so allowing approximately 1 h between the commencement of each passage of the ten transects

SeaGen and the Flow Regime in the Strangford Narrows [BE, DWP, TJTW]

The wake produced by the pile structure, crossbeam and rotors of SeaGen is widely believed to be one of SeaGen's dominant hydrodynamic impacts, so initial hydrodynamic monitoring focused on capturing the detailed structure of the wake from both the pile and the two rotors. Initial simulations of current flow in the Strangford Narrows were undertaken by Bell (2005) using a finite difference scheme; these revealed typical spring tide velocities peaking around 3.5 m s^{-1} (Fig. 12.3a). The simulation was used to identify a suitable area for turbine installation at the initial development stage of the SeaGen programme and as a basis for designing a vessel- mounted acoustic Doppler current profiler (ADCP) survey programme to investigate the area most likely affected by the wake (Fig. 12.3b). A major objective of the ADCP survey was to obtain detailed current data from an area downstream of the turbine anticipated to cover the most likely detectable spatial extent of the turbine wake.

Comparison of the proposed and actual vessel tracks from the vessel- mounted ADCP survey (Fig. 12.3b) highlights the difficulties encountered during survey vessel operation. It was a survey design requirement that vessel speed be maintained <4 m s^{-1}, to obtain a good representation of tidal flow and signal- to- noise ratios for the ADCP. This requirement, however, resulted in the vessel being only marginally faster than the tidal current in the main flow of the channel, whereas closer inshore, vessel speed exceeded normal navigation speed for small workboats nearshore (typically ≤ 2.5 m s^{-1}). Maintaining the originally planned transects during surveying was, therefore, a challenge.

The ADCP output from a single continuous passage of ten transects across the Strangford Narrows (Fig. 12.4) highlights the marked change in velocity between the channel centre and the shallower margins, with nearshore water velocity often <0.5 m s^{-1} but mid-channel velocities >3 m s^{-1} at the surface. The data also demonstrate the depth relationship of velocity profiles with, in general, decreasing velocities from the surface down to the seabed. Figure 12.4 highlights

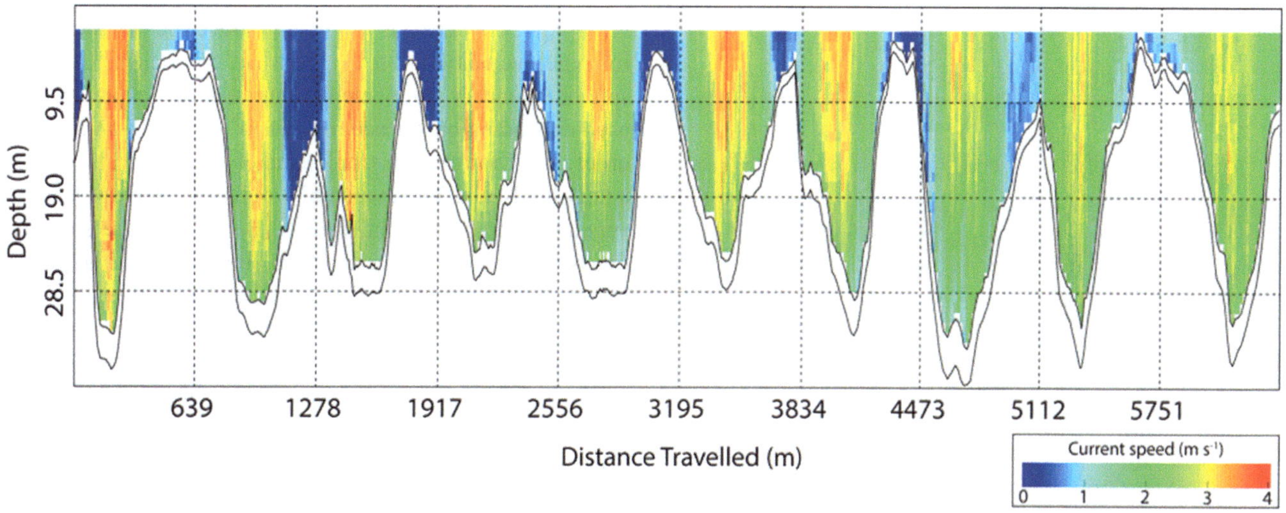

Fig. 12.4 Velocity magnitude in Strangford Narrows covering one passage of the ten continuous transects from south to north during a typical spring flood tide (location of transects is shown in Fig. 12.3b). Data are derived from a vessel-mounted ADCP, and the passage commences at the western end of the most southerly transect. Note that the direction of crossing of the Narrows alternates from the left to the right bank, followed by the right to the left bank, and so on for the other transects of the channel

some limitations of vessel-mounted surveys, notably a large part of the velocity profile not being captured as a consequence of blanking distances and sidelobe interference. In this instance, the first velocity bin is centred 2.55 m from the surface and data are truncated 1.5 m above the seabed because of the echo from the seabed (sidelobe interference).

The patterns in velocity profiles shown in Fig. 12.4 are typical for the velocities observed throughout the measurement campaign. Although flow direction is not shown on the Figure, there was a clear bi-directional flow pattern. In the main channel, flow was dominated by the flood–ebb cycle with flow vectors aligned with the mean longitudinal direction of the channel. Nearer shore, there was considerable local flow in the opposite direction, although the extent varied from one transect to another.

Depth-averaged velocities from the numerical simulations (Fig. 12.3a) and the instantaneous depth-dependent velocities from vessel-mounted ADCP surveys (Fig. 12.4) both show spatially heterogeneous flow in the Strangford Narrows, with notable variations in peak velocity in the centre of the channel. Although depth-dependent velocities generally drop from a maximum near the surface to low values near the seabed, significant fluctuations with depth, in particular in the 3rd, 4th and 6th transects, are clear in Fig. 12.4. As velocity measurements are instantaneous and only averaged over the volume of the bin, the data actually demonstrate some of the turbulent features found in this high-velocity environment. However, the presence of turbulent flow features has limited the ability to compare observed data and numerical model output. Such comparison may only be made using a longer averaging time to filter turbulence to obtain a mean velocity over an interval of some 2 min and requires a vessel to survey at a slower speed; however, in this case it would have proved impossible to navigate the proposed transects.

Velocity differences between predicted values from the numerical model of undisturbed flow (Bell 2005) and those resulting from SeaGen as a drag loss in the ambient flow are shown in Fig. 12.5. Based on estimated forces associated with SeaGen, Bell (2005) obtained a drag coefficient of $c_D = 0.8$, and this value was applied in the model by taking the area of the two rotors and applying the drag loss at their hub centres. It was possible to represent the effect of the turbine on mean flow using a model mesh resolution (15 m) comparable with the width of each SeaGen turbine. In the initial modelling study, the SeaGen support pile was excluded from the model because measurements from the Lynmouth Seaflow single rotor device deployment, the predecessor to SeaGen, indicated a stronger influence of the rotor on the mean flow. However, the vessel-mounted ADCP surveys undertaken for the SeaGen EMP indicated the opposite effect. Here, vessel-mounted surveys identified a velocity reduction up to 300 m downstream of SeaGen in the central streamline of the pile with an across-channel width in the order of magnitude of the pile diameter (Fig. 12.5). In contrast, the wake from the rotors, which should be found either side of the wake and increase its apparent width by up to six times, was difficult to detect and almost unnoticeable in the mean velocity observations. This suggests that the dominant mean-velocity wake effects originate from the pile. Despite a difference in the source of the wake from SeaGen, the extent of the wake and the associated velocity magnitudes are similar to those found by the vessel survey and numerical modelling (Fig. 12.5).

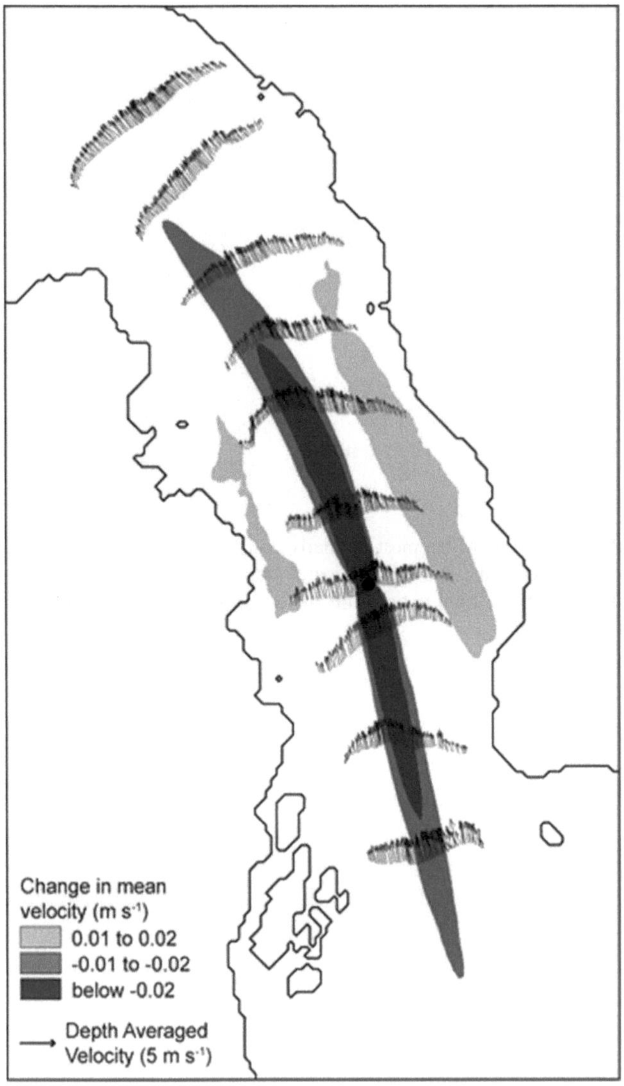

Fig. 12.5 Reduction in current magnitude resulting from the presence of SeaGen in Strangford Narrows based on the original modelling of Bell (2005) and depth-averaged current velocity in the Narrows recorded from ADCP data (Boake et al. 2012)

Quantifying the Characteristics of the Natural Flow Regime

Seabed deployments of ADCPs were also carried out as part of the SeaGen EMP, with two ADCPs deployed on the seabed, 100 m to either side of the eastern rotor in line with the rotor hub and the mean flow direction. The purpose of these deployments was twofold: first, to obtain the incoming flow condition for performance monitoring, certification and design verification and, second, to determine downstream wake effects. Here, we focus on the incoming flow condition and specifically a numerical description of the velocity profile through the water column.

A power law relationship is commonly used in the wind industry to describe the wind profile above a rough surface, where depth-dependent velocity is represented by $u(z)=u_{ref}*z^{\alpha}$, where z is the height above the surface, u_{ref} a reference velocity and α is a constant coefficient. In atmospheric boundary layer calculations, different exponents are used depending on surface roughness type, with an α of the order of 1/7 for smooth arable land or accelerated flow over a hilly landscape, but a value of 1/5 more typical for heavily vegetated shrub land and forests. Because of the similarities between wind power and tidal stream power, the power law relationship with a 1/7 value of α is widely regarded as appropriate and applied in most tidal current profile descriptions, regardless of context. However, with different bed forms and roughness in the coastal environment, a fixed numerical description or set of parameters for profile calculations may not be appropriate.

From the ADCP data it was possible to fit the power law profile to 2-min-averaged velocity data over the depth of the water column. Profiles were fitted to some 9,000 undisturbed (upstream) flow profiles from 30 full flood–ebb tidal cycles but omitting profiles close to slack water. A statistical analysis of the exponential power law parameters was then undertaken and the fit of the observed data to the theoretical curve evaluated. The profile provides a good representation of the vertical velocity distribution through the water column (Fig. 12.6a), and the analysis strongly supports the view that a different velocity profile develops during the ebb relative to the flood tide. The higher exponent of 1/5 during the flood tide suggests greater shear upstream of SeaGen than during the ebb tide's 1/7. Although velocities at the surface are higher during flood flows, the rate of velocity decay with depth is also greater (lower $1/\alpha$; Fig. 12.6b) and bed roughness is higher. This observation is somewhat surprising, but it is assumed to reflect the difference in bed topography and water depth north and south of SeaGen, which previously was not accounted for in predictions of device performance or potential environmental impact.

Lessons Learned

The descriptions above highlight methodologies designed to monitor tidal flow in an attempt to provide essential hydrodynamic background relevant to ecological studies. Much of the underlying hydrodynamic theory related to boundary flows, friction and blockage losses and energy abstraction from free-stream flows is well advanced, but methods to quantify the required parameters in a fast-current environment are less well developed, so considerable experience has been gained from the present study.

The fieldwork demonstrated that mapping spatial variation in current velocity using vessel-mounted equipment and placing monitoring equipment in a fixed location on the seabed are both valuable approaches. The vessel-mounted ADCP survey clearly showed the wake of the turbine in the

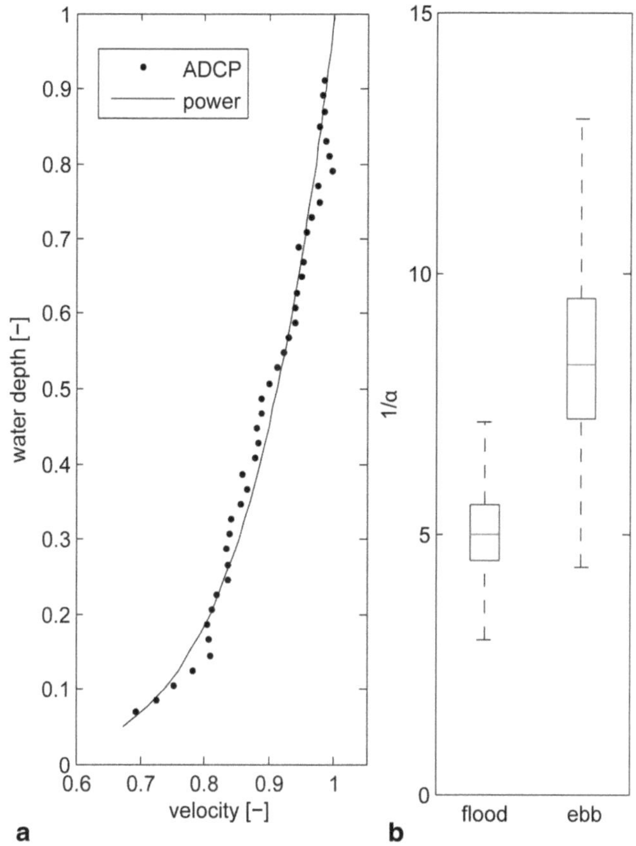

Fig. 12.6 **a** Typical velocity profile from a bottom-mounted ADCP showing observed velocities at the centre of each bin (*points*) and a fitted curve for the power law (*continuous line*). Water depth and velocity have been made non-dimensional (0–1) by scaling to the total water depth and near-surface velocity, respectively. **b** Box-and-whisker plot of fitted power coefficient (1/α) for the power law velocity profile from 2-min-averaged ebb- and flood-tide profiles. The box shows the median and the 25th and 75th percentiles and the upper and lower whiskers encompass 99 % of all values derived. Outliers attributable to erroneous measurements (e.g. driftweed or large-scale eddy features) have been omitted for clarity

nearfield, which extended up to 300 m from the pile (~80 × pile diameter; Boake et al. 2012). In general terms, the extent and the shape of the observed wake matched the numerical prediction. The seabed-mounted ADCP provided information on a longer time-scale and identified previously unrecorded differences in ebb- and flood-flow hydrodynamic conditions.

A significant increase in vertical mixing attributable to SeaGen could not be found. Although lateral shear attributable to the wake would be expected to increase mixing, within the overall context of the natural turbulence in the Narrows, such increased mixing is very limited except in the immediate vicinity of the pile, where extensive Karman vortex shedding may be observed from the surface. More interestingly, a significant difference in the natural turbulent boundary layer between flood and ebb tide was identified. The significant difference in shear obtained from the bed-mounted ADCP indicates that energy losses attributable to bed friction alone are >3 × greater during flood than during ebb in the vicinity of the turbine; this had not been reported before for this location.

Overall, the turbulence induced naturally from the channel is proportionally higher than in lower energy tidal environments because of the roughness of the bed and the many rocky outcrops in the Narrows. This means that any thermal- or density-forced layering from the Irish Sea or the Lough is immediately broken at the entrances to the Narrows, which makes gradients of nutrients and turbidity in the Narrows highly unlikely. The measurement techniques used here failed to yield confident estimates in terms of actual turbulence scaling, because both the spatial and the temporal resolution of standard ADCP equipment is insufficient. Analysis is ongoing to process the vast amount of data collected in the six years of environmental monitoring and to relate the physical data with relevant environmental processes to quantify the possible impacts of larger array deployments in future.

Associated Studies in Strangford Lough [GS]

The deployment of SeaGen in Strangford Lough, the accessibility of the Lough and the presence of support services, together with the variety of current velocity and wind-fetch conditions experienced within a small area, encourage the development of future investigation of the environmental consequences of marine energy extraction in the Lough, including flow perturbation. Given the semi-enclosed nature of the Lough, field trials of marine energy devices other than SeaGen have been restricted so far to trialling scaled tidal and wave energy devices. For this work, advantage has been taken of the interaction between the varied bathymetry of the Lough and current flow to locate appropriately scaled current velocity conditions. Similarly, wind conditions can be scaled by selecting device-deployment locations based on available fetch.

Scaled tidal energy devices that have been trialled in the Lough include Evopod, a single rotor tidal energy turbine tethered on a submerged buoy mid-depth that allows the buoy's alignment in a reversing tidal flow, and also the Minesto seakite. The latter device consists of a turbine rotor and generator pod mounted on a kite-shaped sail tethered to the seabed on a flexible mooring. The pod moves in a figure-of-eight configuration through the tidal flow, resulting in increased velocity of the device through the water relative to the ambient tidal flow. Testing of a SeaGenU device, a development of the full-scale SeaGen device in which 3–5 turbine units are attached in line on a horizontal frame aligned facing the main current flow, is also planned. The mounting frame will be attached by a hinge to a fixed mooring frame on the seabed and will be raised to the surface by flotation, to allow servicing. Permission to deploy these devices has been supported by environmental statements and risk assessments for the deployment sites.

The studies thus far have been complemented by detailed field and modelling investigations to characterize the wake-flow of a midwater tidal turbine. Of particular interest has been the field-testing as part of the PerAWatt project of an array of three scaled tidal turbines investigating the interactive effects of the devices on the flow regime and the implications for the efficiency of energy extraction from the mean flow and also for array design. The field programmes have been complemented by a range of numerical modelling studies carried out by the QUB Marine Energy Research Group (MERG), based on use of the DHI MIKE 21 variable-mesh flow-prediction model and the associated Ecolab model for predicting the ecological consequences of changes in the flowfield. That model has been used extensively in the UK's SuperGen 2 Project to develop predictions of the effects of arrays of wave-energy devices on the incident wavefield in relation to varying wind fetch and direction, in order to allow optimization of array design. This theme is being developed in a further project in which the predictive ability of contrasting hydrodynamic and ecological modelling approaches will be compared.

Several field-based biological investigations have topics related to the environmental impacts of wave- and tidal energy devices. In the case of tidal devices, more detailed attention has been paid to possible effects on the benthos and on the development of active sonar techniques and other approaches to assist in mitigating possible collision effects between marine mammals and tidal turbines. Emphasis on the benthos recognizes that studies on benthic distributions in high-velocity current areas are relatively few compared with studies of more quiescent areas and that such areas exhibit great faunal richness and in general have not been subjected to serious or prolonged human intervention. Two parallel studies on the interaction between current velocities and benthic community structure are in progress: the first is designed to predict the influence of changes in ambient current velocity on benthic communities, and the second is designed to investigate possible effects on the benthos arising from local flow perturbations associated with the rotation of the SeaGen turbine blades, such as tip vortices. Both studies depend on integration of detailed physical modelling of the environments with closely spatially resolved benthic sampling.

The potential effects of inshore wave-energy devices, such as Oyster, on coastal ecology have been investigated in Strangford Lough by focusing on the role of waves in kelp ecology. In mid- to high latitudes, kelp beds input significant energy to local inshore ecosystems as well as providing shelter for populations of juvenile fish and other fauna. Results from a two-year project in which the growth rates of *Laminaria hyperborea* stipes and blades were measured at juxtaposed sheltered and wave-exposed sites revealed that a difference of ~30 % in incident wave energy had no influence on the growth rates of either plant segment. A comparable study was made of populations of *Laminaria digitata*, but over a wider range of wave-exposure conditions predicted from wave-climate and exposure modelling. For that species, the growth rates of the blades were slowest under high current flow or high wave activity. Further investigation based on data obtained from the deployment of three pressure sensors aligned perpendicular to the shore across a *L. hyperborea* bed along with an ADCP sensor was designed to estimate the absorption of incident wave energy by kelp beds. Data from the observations are currently being analysed. All these studies have required close links with high-resolution hydrodynamic modelling.

In addition, in response to ongoing concerns from the environmental regulator, SMRU extended and developed their active sonar programme for sea mammal detection in the vicinity of SeaGen to include trialling improved sonar devices. Discussions are ongoing with regulatory bodies, to establish an acceptable risk strategy in relation to seal–rotor collisions, which would allow use of SeaGen without the need for MMO monitoring of real-time sonar output.

Conclusions and the Future [GS, DA]

When considering the outcomes of the various studies outlined in the earlier sections of this chapter, it is necessary to highlight certain issues. Although several statistically significant effects of SeaGen on the biology of the Narrows have been shown, the magnitude of these effects was generally small. On that basis, it was considered unlikely that any of the effects would have significant ecological consequences for the area. Statistical demonstration of potential effects was, in several instances, challenging because of their small magnitude and the possible influence of secondary controlling factors. In some instances, such as shore observations of harbour porpoises and certain seabirds, insufficient observations were able to be made to allow objective statistical analysis of the data.

It is also necessary to emphasize the fact that there has been considerable evolution of the observational and statistical approaches employed over the five-year period of the EMP, where such evolution was appropriate. The developments reflected not only improvements in equipment available, such as the active sonar system and ADCP instrumentation, and in observational approaches, such as for seals and seabirds, but also an increasing appreciation of the complexity of the physically and ecologically very dynamic system of Strangford Narrows.

It is clearly encouraging for the marine tidal industry that, although there were some statistically significant changes in the ecology of the area adjacent to SeaGen, they were of a minor nature and could not be considered significant in relation to the ecology of the Narrows as a whole. It is recognized that further investigation of several of the topics considered in the EMP is required, for instance in assessing seal-avoidance mechanisms in relation to the turbine blades, diving seabird behaviour patterns and measurements of turbulence, but our opinion is that the overall conclusion of SeaGen having a minimal effect on the local environment is valid. We also recognize that the outcomes of the project may not be of equal relevance to potential future tidal device developments at other sites, where a different complexity of environmental conditions and conservation priorities may arise. However, the SeaGen programme has been notable for its stimulation of detailed hydrodynamic and modelling initiatives relating to the currents in Strangford Lough as a whole and, more especially, in the Narrows. This led to close and valuable integration of physical and biological disciplines in the required environmental research and raised the potential for realistic predictions to be made on the possible ecological effects of marine tidal devices deployed elsewhere. In addition, the variable environmental conditions within the localized area of Strangford Lough, along with excellent logistic facilities and a strong local scientific base, has allowed the active development of the Lough as a site for testing scaled tidal and wave-energy devices.

It is also clear that the fast current speeds in the Strangford Narrows conjoining the Irish Sea and the sheltered nature of the main body of Strangford Lough provided excellent conditions for developing and testing SeaGen, the world's first commercial tidal energy device. The designated conservation status of the location dictated that a particularly broad and detailed EMP be carried out prior to and following installation of the device. Given the breadth and extent of the component studies carried out for the EMP, as required by the regulator, it is likely that the EMP will provide a solid foundation for future developments in the marine tidal energy industry.

Acknowledgements For the Adaptive Management section of this paper, sincere thanks are due to many for the work done by members of the Science Group and colleagues over the eight years of the SeaGen project between 2004 and 2012, particularly by Gemma Keenan. For the marine mammals section, various people were responsible for the design, implementation and analysis of the studies summarized, and detailed methods and results of each study will be published in due course by those concerned. However, thanks are necessary here to Ian Boyd, Bernie McConnell, Callan Duck, Gordon Hastie, Mike Lonergan, Cormac Booth, Beth Mackey, Simon Northridge, Alice Mackay, Monique MacKenzie, Carl Donovan, Andrew Murray, Chris Morris and Daryl Birkett. For the section on SeaGen and the flow regime, thanks are due to Cuan Boake, Jeremy Rodgers and Simon Rodgers for their assistance in carrying out the ADCP measurements and to Cuan Boake for preliminary analysis of the datasets. Finally, David Erwin provided necessary leadership and guidance in seeing through the environmental impact aspects of the whole SeaGen project: his support was invaluable to its success and is gratefully acknowledged.

References

Allan C, Stankey GH (2009) Adaptive environmental management: a practitioner's guide. CSIRO, Collingwood, 368 pp

Bell AK (2005) Tidal energy turbine—Strangford Narrows: hydraulic model studies. Report by RPS Kirk McClure Morton, Belfast

Boake C, Elsaesser B, Whittaker TJT (2012) SeaGen wake survey—ADCP current monitoring campaign. Report of the Marine Energy Research Group, Queen's University, Belfast

Bray JR, Curtis JT (1957) An ordination of the upland forest communities of Southern Wisconsin. Ecol Monogr 27:325–349

British Trust for Ornithology (2010) The wetland bird survey (WeBS). http://www.bto.org/volunteer-surveys/webs

Brown R (1990) Strangford Lough. The wildlife of an Irish Sea Lough. Institute of Irish Studies, Queen's University, Belfast, 288 pp

Clarke KR, Green RH (1988) Statistical design and analysis for a biological effects study. Mar Ecol Prog Ser 46:213–226

Clarke KR, Somerfield PJ, Airoldi L, Warwick RM (2006) Exploring interactions by second-stage community analyses. J Exp Mar Biol Ecol 338:179–192

Connor DW, Allen JH, Golding N, Howell KH, Lieberknecht LM, Northen KO, Reker JB (2004) The marine habitat classification for Britain and Ireland, Version 04.05. JNCC, Peterborough, UK. ISBN 1 861 07561 8 (internet version) www.jncc.gov.uk/MarineHabitatClassification

Devereux CL, Denny MJH, Whittingham MJ (2008) Minimal effects of wind turbines on the distribution of wintering farmland birds. J Appl Ecol 45:1689–1694

Duck CD (2003) Results of the thermal image survey of seals around the coast of Northern Ireland, August (2002). Sea Mammal Research Unit Unpublished Report to Environment and Heritage Service, Northern Ireland

Duck CD, Morris C (2012) Seals in Northern Ireland: helicopter survey of harbour and grey seals, August 2011. Sea Mammal Research Unit Report to the Northern Ireland Environment Agency

EASMP (2011) Marine Current Turbines project, Strangford Lough. Environment Action and Safety Management Plan (EASMP) working document version 3, January 2011

Furness RW, Wade HM, Robbins AMC, Masden EA (2012) Assessing the sensitivity of seabird populations to adverse effects from tidal stream turbines and wave energy devices. ICES J Mar Sci 69:1466–1479

Gotto RV (1951) Some plankton records from Strangford Lough, County Down. Irish Nat J 10:162

Grecian WJ, Inger R, Attrill MJ, Bearhop S, Godley BJ, Witt MJ, Votier SC (2010) Potential impacts of wave-powered marine renewable energy installations on marine birds. Ibis 152:683–697

Green RH (1979) Sampling design and statistical methods for environmental biologists. Wiley, Chichester, 257 pp

Hastie GD, Gillespie DM, Gordon JCD, Macaulay JDJ, McConnell BJ, Sparling CE (2014) Tracking technologies for quantifying marine mammal interactions with tidal turbines: pitfalls and possibilities (Chap. 10 of this volume)

Inger R, Attrill MJ, Bearhop S, Broderick AC, Grecian WJ, Hodgson DJ, Mills C et al. (2009) Marine renewable energy: potential benefits to biodiversity? An urgent call for research. J Appl Ecol 46:1145–1153

Inger R, Harrison XA, Bearhop S (2010) The impacts of the SeaGen tidal turbine on the avian community of the Strangford Lough Narrows. Statistical report prepared for Queen's University Belfast Marine Laboratory

Johnson RG (1971) Animal-sediment relations in shallow water benthic communities. Mar Geol 11:93–104

Johnson RG (1972) Conceptual models of benthic marine communities. In: Schopf TJM (ed) Models in palaeobiology. Freeman and Cooper, San Francisco, pp 149–159

Kaiser MJ, Clarke KR, Hinz H, Austen MCV, Somerfield PJ, Karakassis I (2006) Global analysis and prediction of the response of benthic biota and habitats to fishing. Mar Ecol Prog Ser 311:1–14

Kregting L, Elsaesser B (in review) A hydrodynamic modelling framework for Strangford Lough, Part 1: Tidal model. Coast Eng

Lonergan M (2009) Modelling monthly boat counts of seals in Strangford Lough. Unpublished report to Northern Ireland Environment Agency (Natural Heritage Directorate)

Lonergan M (2013) Patterns in the number of seals counted during boat-based surveys of Strangford Lough and Strangford Narrows (1993–2012). SMRU Ltd report to Marine Current Turbines

Madsen J, Boertmann D (2008) Animal behavioral adaptation to changing landscapes: spring-staging geese habituate to wind farms. Landsc Ecol 23:1007–1011

McErlean T, Crothers N (2007) Harnessing the tides: the early medieval tide mills at Nendrum Monastery, Strangford Lough. The Stationery Office, Norwich, 468 pp

Montgomery-Watson J (1999) Common seal surveys in Strangford Lough and County Down coast (trends in population). Unpublished report to Common Seal Research and Management Workshop Proceedings, Strangford Lough Management Committee

Nedwell JR, Brooker AG (2008) Measurement and assessment of background underwater noise and its comparison with noise from pin pile drilling operations during installation of the SeaGen tidal turbine device, Strangford Lough. Subacoustech Report 724R0120 to COWRIE Ltd. ISBN 978-0-9557501-9-9

Newbould PJ, Carter RWG (1984) Environmental impact assessment of the Strangford Lough tidal power barrage scheme in Northern Ireland. Water Sci Tech 16:455–462

Plummer KE, Inger R, Christen N, Metcalfe K, Bearhop S (2012) Potential impacts of SeaGen on avian communities in the Strangford Lough Narrows (2011–2012). Report prepared for Marine Current Turbines Ltd

Rhoads DC, McCall PL, Yingst JL (1978) Disturbance and production on the estuarine seafloor. Am Sci 66:577–586

Royal Haskoning (2004) SeaGen environmental scoping study. Ref. 9P5161/R/RS/PBor

Royal Haskoning (2012) SeaGen Environmental Monitoring Programme Final Report. Ref. 9S8562/R/303719/Edin. http://seageneration.co.uk/files/SeaGen-Environmental-Monitoring-Programme-Final-Report.pdf

Stewart GB, Pullin AS, Coles CF (2007) Poor evidence-base for assessment of windfarm impacts on birds. Env Conserv 34:1–11

Underwood AJ (1992) Beyond BACI: the detection of environmental impact on populations in the real, but variable, world. J Exp Mar Biol Ecol 161:145–178

Williams BK, Szaro RC, Shapiro CD (2009) Adaptive management: the US Department of the Interior Technical Guide. Adaptive Management Working Group, US Department of the Interior, Washington, DC

Williams G (1954) Fauna of Strangford Lough and neighbouring coasts. Proc Roy Irish Acad 56B:29–133

Index

M. A. Shields, A. I. L. Payne (eds), *Marine Renewable Energy Technology and Environmental Interactions,* Humanity and the Sea,
DOI 10.1007/978-94-017-8002-5, © Springer Science+Business Media Dordrecht 2014

MIX
Papier aus verantwortungsvollen Quellen
Paper from responsible sources
FSC® C105338

If you have any concerns about our products,
you can contact us on
ProductSafety@springernature.com

In case Publisher is established outside the EU,
the EU authorized representative is:
Springer Nature Customer Service Center GmbH
Europaplatz 3, 69115 Heidelberg, Germany

Printed by Libri Plureos GmbH
in Hamburg, Germany